AF581692

Food, Health and Safety in Cross Cultural Consumer Contexts

Food, Health and Safety in Cross Cultural Consumer Contexts

Editor

Derek V. Byrne

MDPI • Basel • Beijing • Wuhan • Barcelona • Belgrade • Manchester • Tokyo • Cluj • Tianjin

Editor
Derek V. Byrne
Department of Food Science
Aarhus University
Aarhus
Denmark

Editorial Office
MDPI
St. Alban-Anlage 66
4052 Basel, Switzerland

This is a reprint of articles from the Special Issue published online in the open access journal *Foods* (ISSN 2304-8158) (available at: www.mdpi.com/journal/foods/special_issues/Food_Health_Safety).

For citation purposes, cite each article independently as indicated on the article page online and as indicated below:

LastName, A.A.; LastName, B.B.; LastName, C.C. Article Title. *Journal Name* **Year**, *Volume Number*, Page Range.

ISBN 978-3-0365-1340-9 (Hbk)
ISBN 978-3-0365-1339-3 (PDF)

Contents

About the Editor

Derek V. Byrne

Derek V. Byrne is professor of sensory and consumer science and a science leader of the Food Quality Perception & Society (FQS) Team at The Department of Food Science, Aarhus University in Denmark. Derek has a PhD in sensory science from the University of Copenhagen and an MSc in food and nutritional science from University College Cork in Ireland. Derek has over 25 years of experience in sensory and consumer science research, from the senses, through to product quality and design, and on to the importance of sensory science in food industry applicability. Derek's team at Aarhus University focuses on understanding food quality and perception via a cross-disciplinary synergy of multisensory human food analysis, experimental psychology, physiological responses, and cognitive neuroscience in the design and development of high quality, better-tasting, more stimulating, more memorable, and healthier food and drink experiences.

Preface to "Food, Health and Safety in Cross Cultural Consumer Contexts"

The global food and food technology markets are rapidly growing, and food investment is central in many governments'growth plans. A core focus is on the food industry's ability to maintain and strengthen its position and to exploit the unique opportunities for export-driven growth, especially in the exportation of high-quality food with relevant sensory, health, and safety properties as key elements. Thus, the concept of cross cultural perspectives in research in food is critically important particularly in relation to human perception in food and health. Food concepts are very different across different jurisdictions. Different markets and cultures have varying perspectives on what is considered a palatable, acceptable, or useful food or food product; in simple terms, one size does not fit all in the majority of cases. Specific markets thus need targeted food design from a myriad of perspectives. In this Special Issue anthology, "Food, Health and Safety in Cross-Cultural Consumer Contexts", we bring together articles that show the wide range of studies from fundamental to market applicability currently in focus in sensory and consumer science in food, health, and safety contexts. The studies included highlight the importance of considering the human senses, consumer preference, and perception across the food stakeholder chain in several cross-cultural contexts. Overall, it is clear that there is a need for much knowledge in specific spaces and that this leads to better positioned food and food products when looking at world markets. In conclusion, the human senses, consumer acceptance, and preferences are core to future food design via integrating human perception and this will continue to be critical to the success of food transfer in modern cross-cultural contexts.

Derek V. Byrne
Editor

MDPI

Editorial

Current Trends in Food Health and Safety in Cross-Cultural Sensory and Consumer Science

Derek Victor Byrne [1,2]

[1] Food Quality Perception and Society Science Team, iSENSE Lab, Department of Food Science, Faculty of Technical Sciences, Aarhus University, Agro Food Park 48, DK-8200 Aarhus N, Denmark; derekv.byrne@food.au.dk

[2] Food & Health Research, Sino-Danish Center (SDC), Niels Jensens Vej 2, DK-8000 Aarhus C, Denmark

Citation: Byrne, D.V. Current Trends in Food Health and Safety in Cross-Cultural Sensory and Consumer Science. *Foods* **2021**, *10*, 965. https://doi.org/10.3390/foods10050965

Received: 19 April 2021
Accepted: 25 April 2021
Published: 28 April 2021

Publisher's Note: MDPI stays neutral with regard to jurisdictional claims in published maps and institutional affiliations.

1. Introduction

The global food and food technology market is in rapid growth, and food investment is central in many governments' growth plans. A core focus can be said to be on the food industry's ability to maintain and strengthen its position and exploit the unique opportunities for export-driven growth, especially in the export of high-quality food with relevant sensory health, safety, and sustainability properties as key elements. Thus, the concept of cross-cultural perspectives in research in food per se is critically important, none more so than in relation to the human perception area in food and health. Food concepts are very different, of course, in different jurisdictions, with respect to different markets and cultures having very different perspectives on what is considered a palatable, acceptable, or useful food or food product. In simple terms, one size does not fit all in the majority of cases.

Cross-cultural studies have been in focus for some time in the food space and in particular in relation to food design via the senses and from a consumer-driven perspective. From one of the earliest overviews by Prescott and Bell (1995) [1], which reviewed the literature on basic cross-cultural determinants of food acceptability focusing on chemosensory perceptions and preferences from the point of view of their ability to explain differences in food selection in different cultures [1]. On to one of the latest perspectives by Rodrigues et al. (2019) on consumers' food decisions and eating habits as well as cross-cultural eating focusing on the application of virtual reality, mobile applications, and social media, amongst other areas [2]. It is, of course, the case that the space over the last 25 years up to the present anthology collection in "Food, Health and Safety in Cross-Cultural Consumer Contexts" is peppered with a collection of works addressing perspectives linked to the senses, and cross-cultural applicability of note would be, e.g., linking the senses to psychology [3], cross-cultural differences in cross-modal correspondences [4], and measurement of food preference and reward in cross-cultural contexts [5], to consumers' associations with wellbeing in a cross-cultural studies [6].

Specific areas of research presented as relevant for the scope of this Special Issue indicate, clearly, the ever-widening area of cross-cultural research with respect to sensory and consumer science regarding food and health. Areas that were in focus for the special issue and cross-cultural research were as follows. (1) Food quality, processing, and production: focusing on understanding food processing, quality, and perception via a synergy of multisensory human food analysis, combined with novel and sustainable production techniques. (2) Microbial food safety and hygiene: dealing with microbial food safety behaviors and focusing on the knowledge to detect foodborne pathogens, sources of outbreaks of foodborne diseases, and novel strategies to ensure food safety for the consumer. (3) Food business, marketing, and the consumer: focusing on research on the development, marketing, and distribution of foods to generate insight into consumer behavior for the benefit of food industries and public policy. (4) Food economics and the supply chain:

dealing with research in logistics and supply chain management regarding the concepts around economic thinking in food production, trade, and the management of food quality and safety across the supply chain. (5) Food sociology and eating: the sociological elements of food safety and quality, including research around the social and cultural aspects of eating, production, and new technologies, as well as the role of legal frameworks and regulations. (6) Nutrition and health: focusing on the effects on health of specific food, food components, and supplements in health and disease prevention, the rationale for nutritional recommendations, and food and nutrition security.

The present Special Issue's focus overall was food, health, and safety in cross-cultural consumer contexts" for innovative food solutions to meet global food challenges, which can be best addressed by research-based synergies linking, e.g., different jurisdictions, countries, and continents in the food area.

Ultimately, in this special issue we have included contributions that encompass key current research on food, health and safety in cross-cultural consumer contexts with respect to food science synergies for sustainable, healthy, and high-quality food supply, security, and consumption scenarios across the entire food chain from "farm to fork" in cross-cultural contexts. Specifically, we have brought together articles that encompass the wide scope of cross-cultural multidisciplinary research as alluded to above with perspectives in the space related to the determination of the key factors involved. The articles included can be considered to cover stakeholders in cross-cultural perception, from the senses, with respect to differences in sensitivity [7–9], on to consumer preferences [10–12], food pleasure and appetite [13,14], perception of food quality and safety [15–17], and finally key factors in relation to consumer adoption and label information in the market itself [18,19]. This collection of articles is in essence a snapshot of the wide focus and general relevance of sensory and consumer science in cross-cultural studies in food health and safety and we hope it inspires researchers to consider this very interesting and ever-growing space in their future work.

2. A Synopsis of Special Issue Research

2.1. *Sensory Differences*

Thus, with respect to sensory differences, Junge et al. (2020) performed research in the area of sweetness and sour interaction [7]. The authors indicated that tastes interact in almost every consumed food or beverage, yet many aspects of interactions, such as sweet–sour interactions, were not well understood. The study investigated the interaction between sweetness from sucrose and sourness from citric and tartaric acids, respectively, in a cross-cultural consumer study conducted in China and Denmark. Overall, it was determined that culture did not impact the suppression of sweetness intensity ratings of citric or tartaric acids, whereas it did influence sourness intensity ratings. While the Danish consumers showed similar suppression of sourness by both acids, the Chinese consumers were more susceptible toward the sourness suppression caused by sucrose in the tartaric acid–sucrose mixture compared to the citric acid–sucrose mixture. These results indicated that individual differences in taste perception might affect perception of sweet–sour taste interactions, at least in aqueous solutions.

Moreover, in relation to sensory differences, authors Nóbrega et al. (2020) looked at two segments within the rather large Brazilian food service industry with respect to best-selling coffees and serving temperatures with respect to health and safety [8]. The serving temperatures of best-selling coffee beverages in 50 low-cost food service establishments (LCFS) and 50 coffee shops (CS) were studied. The bestsellers in the LCFS were dominated by 50 mL shots of sweetened black coffee served in disposable polystyrene (PS) cups from thermos flasks. In the CS, 50 mL shots of freshly brewed espresso served in porcelain cups were the dominant beverage. The serving temperatures of all beverages were on average 90% and 68% above 65 °C in the LCFS and CS, respectively. Furthermore, the cooling periods of hot water systems were investigated. When median temperatures of the best-selling coffees are considered, consumers should allow a minimum cooling time before drinking of about 2 min at both LCFS and CS. Nóbrega et al. (2020) concluded that further studies to complete a nationwide picture of coffee consumption habits and the temperature at which consumption commonly occurs in Brazil could also present an excellent opportunity for an esophageal cancer risk assessment for hot coffee beverage consumption.

Lastly, in a study of texture preference of Chinese, Korean, and US consumers by Wong et al. (2020) [9], the authors aimed to understand the drivers of liking dried apple and pear chips with various textures. The possibility of hedonic transfer from snack texture preferences to fruit-chip texture preferences was also investigated among Chinese and Koreans. Consumers rated their level of liking for each sample and then they performed hedonic-based projective mapping with the same samples. In the hedonic texture transfer investigation, consumers rated their acceptance of nine snacks with various textures but possessing similar textures to those of dried fruit samples. Most consumers disliked samples with a soft or jelly-like texture and liked samples with a crispy texture. Cross-cultural differences were observed in the liking of puffy samples, with both Chinese and Koreans liking puffy samples as much as crispy ones for their melting characteristics in the mouth, while US consumers perceived the puffy samples as being Styrofoam-like and disliked them. Hedonic transfer was observed from snack texture preferences in fruit chips. Individual texture preferences for snacks seemed to significantly affect the texture preferences for fruit chips. Wong et al. (2020) concluded that the overall impact of the study was the potential to predict the potential market in the chosen countries using hedonic-based projective mapping.

2.2. Consumer Preferences

In relation to choice, per se, in cross-cultural contexts, Profeta et al. (2021) looked at consumer preferences for meat hybrids (referred to as meathybrids) in Germany and Belgium [10]. The authors' basis for the study was high levels of meat consumption are increasingly being criticized for ethical, environmental, and social reasons. Plant-based meat substitutes have been identified as healthy sources of protein that, in comparison to meat, offer a number of social, environmental, and health benefits and may play a role in reducing meat consumption. In meathybrids, only a fraction of the meat product (e.g., 20% to 50%) is replaced with plant-based proteins. Profeta et al. (2021) demonstrated that in many countries, consumers are highly attached to meat and consider it as an essential and integral element of their daily diet. For consumers that are not interested in vegan or vegetarian alternatives as meat substitutes, meathybrids could be a low-threshold option for a more sustainable food consumption behavior [10]. The authors showed that more than fifty percent of consumers substitute meat at least occasionally. Thus, about half of the respondents reveal an eligible consumption behavior with respect to sustainability and healthiness, at least sometimes. The applied discrete choice experiment demonstrated that the analyzed meat products are the most preferred by consumers. Nonetheless, the tested meathybrid variants with different shares of plant-based proteins took the second position followed by the vegetarian-based alternatives. Therefore, meathybrids could facilitate the diet transition of meat-eaters in the direction toward a more healthy and sustainable consumption. The analyzed consumer segment was more open-minded to the meathybrid concept in comparison to the vegetarian substitutes [10].

Further to consumer preference in the cross-cultural space, Garvey et al. (2020) looked at perception and liking among Irish, German, and US consumers for salted butter produced from different feed systems [11]. Overall, it was presented there was no significant difference in overall liking of the butters among any of the consumers, although cross-cultural preferences were evident. Sensory attribute differences based on animal diet were evident across the three countries, as identified by German and Irish assessors and trained US panelists, which were likely influenced by familiarity. Of volatiles measured, Garvey et al. 2020 indicated that the abundance of specific volatile aromatic compounds, especially some aldehydes and ketones, were significantly impacted by the feed system and may also contribute to some of the perceived sensory attribute differences in these butters.

Additionally, in relation to preference in relation to a protected geographical indication (PGI) product in the European Union, authors Kelly et al. (2020) investigated the PGI product called Waterford Blaa, which is a bread product specific to Ireland's East Cost, traditionally [12]. This study aimed to determine whether cultural background/product familiarity, gender, and/or age impacted consumer liking of three Waterford Blaa products and explored product acceptability between product-familiar and product-unfamiliar consumer cohorts in Ireland and the UK, respectively. Familiarity with Blaa impacted consumer liking, particularly with respect to characteristic flour dusting, which is a unique property of Waterford Blaa. UK consumers felt that all Blaas had too much flour dusting. Flavor was also important for UK consumers. Irish consumer liking was more influenced by the hardness of the Blaas, with harder products being less preferred. Age and gender did not impact liking for Blaas within Irish consumers, but gender differences were observed among UK consumers, males liking the appearance significantly more than females. In cross-cultural contexts, such PGI products which have largely fixed formats and properties, it is critical to determine if they can at all cross borders in terms of sensory and consumer acceptance or will they simply be niche, linked to only diaspora or food-curious individuals.

2.3. Appetite Pleasure and Ingestion Sensations

Another area that has emerged of late as a focus for cross-cultural food design is the area of a food's influence on appetite through pleasure and ingestion sensations. In the present volume, Duerlund et al. (2020) investigated post-ingestive sensations and how they drive perceived food pleasure [13]. The authors aimed to compare Chinese and Danish consumers in their post-ingestive drivers of post-ingestive food pleasure (PIFP). Duerlund et al. (2020) define PIFP as a "subjective conscious sensation of pleasure and joy experienced after eating". Key results revealed perceived satisfaction as well as mental, overall and physical wellbeing to be highly influential on PIFP in both countries. Moreover, Danish consumers perceived appetite-related sensations, such as satiety, hunger, desire-to-eat and in-need-of-food, to be influential on PIFP, which was not the case in China. In China, more vitality-related sensations, such as energized, relaxation and concentration, were found to be drivers of PIFP. These results suggest similarities but also distinct subtleties in the cultural constructs of PIFP in Denmark and in China. Duerlund et al. (2020) overall suggested that focusing on food pleasure as a post-ingestive measure provides valuable output, deeper insights into what drives food pleasure, and, importantly, takes us beyond the processes only active during the actual eating event.

Furthermore, in the present volume, Laaksonen et al. (2020) in Finnish and Chinese contexts looked at oat product concepts [14]. The authors present that oats and oat-based products were increasingly popular among consumers and the food industry, and whilst studies exist on the sensory characteristics of oats as such, previous studies focusing specifically on the pleasantness of oats, and especially investigations of a wide range of oat products eaten by European and Asian consumers, are scarce. A questionnaire revealed that Finnish consumers rated the pleasantness and familiarity of several oat product categories, such as breads and porridges, higher compared to participants from other cultures. Further, Laaksonen et al. 2020 indicated that sensory tests showed both similarities, e.g., porridges were described as "natural," "healthy," and "oat-like," and differences between countries, e.g., sweet biscuits were described as "crispy" and "hard" by Finnish consumers and "strange" and "musty" by Chinese consumers. Sweet products were unanimously preferred. Moreover, authors indicated that the culture had an important role affecting the rating of pleasantness and familiarity of oat product categories, whereas food neophobia and health interest status also had an influence [14].

2.4. Perception of Quality and Safety

In relation to food safety and quality, authors Haas et al. (2021) investigated perception differences in the western Balkans [15]. The authors present that domestic food markets are of significant importance to Kosovar and Albanian companies because access to export markets is underdeveloped, partly as a result of the gaps in food safety and quality standards. Identifying Kosovar and Albanian consumers' use of food safety attributes and an evaluation of the quality of domestic food versus imported food were the research objectives of this study. Haas et al. 2021 concluded that despite the prevalent problems with food safety, consumers in both countries considered domestic food to be safer as well as of higher quality than imported products. Kosovars were more likely than Albanians to perceive domestic food products to be significantly better than imported products. Female and better-educated consumers used information related to food safety more often. Expiry date, domestic and local origin, and brand reputation were the most frequently used safety and quality cues for both samples. International food standards, such as ISO or HACCP, are less frequently used as quality cues by these consumer groups. Haas et al. (2021) concluded that though this is a good result for these nations internally to have these perceptions, it is important to strengthen the institutional framework related to food safety and quality following best practices from EU countries, which could ultimately perhaps enable further development of export markets [15].

Moreover, in this special issue volume, Wang and Yueh (2020) investigated the area of food safety cognition with respect to consumer behavior, comparing students in Taiwan and mainland China [16]. The purpose of the study was to investigate how optimistic bias, consumption cognition, news attention, information credibility, and social trust affect the purchase intention of food consumption. Results showed that Taiwanese college students did not display optimistic bias, but Chinese students did. The models showed that both Taiwanese and Chinese students' consumption cognition significantly influenced their purchase intentions, and news attention significantly influenced only Chinese students' purchase intentions. The results revealed that optimistic bias can be reduced in different social contexts. This study also confirmed that people had optimistic bias on food safety issues based on which recommendations were made to increase public awareness of food safety as well as to improve the government's certification system [16].

In relation to the area of high-risk food handling behaviors and consumer perception of food safety issues in these contexts, Cho et al. (2020) investigated behaviors and risk perceptions across time in South Korea among primary food handlers [17]. The authors gathered data in 2010 and 2019, and present that 2010 was characterized by a consumers' risk perception–behavior disconnect, that is consumers believed they knew very well what the safest methods for food handling were, but responses regarding their behaviors did not support their confidence in the actual safety of their food. Such that consumers did not wash/trim foods before storage, thawed frozen foods at room temperature, and exposed leftovers to danger zone temperatures. Interestingly, these three particular trends were found to be similar when assessed again in 2019. The year 2010 was also characterized by other common high-risk behaviors: 70.0% of consumers divided a large portion of food into smaller pieces for storage, but few consumers (12.5%) labeled divided foods with relevant information, and they excessively reused kitchen utensils. Whereas in 2019, more consumers (25.7%) labeled food and usage periods for kitchen utensils were shortened. Consumers usually conformed to general food safety rules in both 2010 and 2019 in the following ways: separate storage of foods, storage of foods in the proper places for the proper periods, washing fruits/vegetables before eating, washing hands after handling potentially hazardous foods, and cooking foods and reheating leftovers to eat. Cho et al. 2021 concluded that their findings provided resources for understanding consumers' high-risk behaviors/perceptions at home, highlighting the importance of behavioral control.

2.5. Consumer Adoption and Label Information

In this final section, we look at research included on consumer adoption and labeling. Nathan et al. (2021) indicated as a backdrop in the organic labeling area that in order to meet the rising global demand for food and to ensure food security in line with the United Nations Sustainable Development Goal 2, the elimination of hunger, technological advances have been introduced in the food production industry [18]. The organic food industry has benefitted from advances in food technology and innovation. However, there remains skepticism regarding organic foods on the part of consumers, specifically on consumers' acceptance of food innovation technologies used in the production of organic foods, and this can be extrapolated to different cultural contexts. In the present volume, Nathan et al. (2021) measured factors that influence consumers' food innovation adoption and subsequently their intention to purchase organic foods. Organic foods purchase behavior of Malaysian and Hungarian consumers was compared to examine differences between Asian and European consumers. The findings showed food innovation adoption as the most crucial predictor for the intention to purchase organic foods in Hungary, while the social lifestyle factor was the most influential in Malaysia. Other factors, such as environmental concerns and health consciousness, were also examined in relation to food innovation adoption and organic food consumerism. Overall, Nathan et al. (2021) present key differences between European and Asian organic foods consumers and provided recommendations for stakeholders interested in these markets going forward [18].

The remaining article in our special issue by Magalhaes et al. (2021) profiles knowledge, utility, and preference for beef traceability labeling between Spain and Brazil [19]. The consumer environment determined consumers' buying behavior and product preferences, and understanding these factors would allow businesses in the industry to identify market demands. The authors contended that in both countries there were existing differences in the consumption of beef, in the production and the regulatory process concerning beef, and, in particular, in relation to the traceability systems. Having a traceability system is in fact mandatory in Spain and voluntary in Brazil. From these perspectives, Magalhaes et al. (2021) carried out a cross-cultural study through a self-administered questionnaire aimed at comparing and understanding familiarity with bovine traceability systems and traceability information on labels as a food security indicator. The authors concluded that traceability information was well received by consumers as an attribute of credibility, and consumers were interested in ensuring that the item they buy is of known and reliable origin. However, the authors contended that more incentives may help clarify the advantages of purchasing food with certified traceability, making it more effective for consumers to use this knowledge in different jurisdictions [19].

3. Conclusions

Overall, the research included in this volume covers a wide range of studies, from fundamental research to market applicability, with respect to sensory and consumer studies in food health and safety contexts. Groupings of the studies have been made to point out the diverse and core need to consider the human senses, consumer preferences and perception, across the food stakeholder chain in cross-cultural research. Several of the studies noted the need for more knowledge in their specific spaces and contended that this will lead to better-positioned food and food products when looking at world markets. An overall conclusion with respect to this collection would be that the human senses, consumer acceptance, and preferences are core to future food design with respect to understanding human perception of key aspects critical to the success of food transfer in modern cross-cultural contexts.

Funding: This research was funded by Sino Danish Centre Denmark.

Acknowledgments: The author thanks the Food and Health research area at Sino Danish Centre Denmark for its help in bringing together this special issue and further thanks the Food Quality Perception & Society Team at the Department of Food Science at Aarhus University, Denmark for their support.

Conflicts of Interest: The author declares no conflict of interest.

References

1. Prescott, J.; Bell, G. Cross-cultural determinants of food acceptability: Recent research on sensory perceptions and preferences. *Trends Food Sci. Technol.* **1995**, *6*, 201–205. [CrossRef]
2. Rodrigues, H.; Otterbring, T.; Piqueras-Fiszman, B.; Gómez-Corona, C. Introduction to the special issue on Global perspec-tives on food and consumer science: A cross-cultural approach. *Food Res. Int.* **2019**, *116*, 135–136. [CrossRef] [PubMed]
3. Dijksterhuis, G.B.; Byrne, D.V. Does the Mind Reflect the Mouth? Sensory Profiling and the Future. *Crit. Rev. Food Sci. Nutr.* **2005**, *45*, 527–534. [CrossRef] [PubMed]
4. Wan, X.; Woods, A.T.; Bosch, J.J.F.V.D.; McKenzie, K.J.; Velasco, C.; Spence, C. Cross-cultural differences in crossmodal correspondences between basic tastes and visual features. *Front. Psychol.* **2014**, *5*, 1365. [CrossRef] [PubMed]
5. Oustric, P.; Thivel, D.; Dalton, M.; Beaulieu, K.; Gibbons, C.; Hopkins, M.; Blundell, J.; Finlayson, G. Measuring food preference and reward: Application and cross-cultural adaptation of the Leeds Food Preference Questionnaire in human experimental research. *Food Qual. Prefer.* **2020**, *80*, 103824. [CrossRef]
6. Ares, G.; de Saldamando, L.; Giménez, A.; Claret, A.; Cunha, L.M.; Guerrero, L.; de Moura, A.P.; Oliveira, D.C.; Symoneaux, R.; Deliza, R. Consumers' associations with wellbeing in a food-related context: A cross-cultural study. *Food Qual. Prefer.* **2015**, *40*, 304–315. [CrossRef]
7. Junge, J.Y.; Bertelsen, A.S.; Mielby, L.A.; Zeng, Y.; Sun, Y.-X.; Byrne, D.V.; Kidmose, U. Taste Interactions between Sweetness of Sucrose and Sourness of Citric and Tartaric Acid among Chinese and Danish Consumers. *Foods* **2020**, *9*, 1425. [CrossRef] [PubMed]
8. Nóbrega, I.C.C.; Costa, I.H.L.; Macedo, A.C.; Ishihara, Y.M.; Lachenmeier, D.W. Serving Temperatures of Best-Selling Coffees in Two Segments of the Brazilian Food Service Industry Are "Very Hot". *Foods* **2020**, *9*, 1047. [CrossRef] [PubMed]
9. Wong, R.; Kim, S.; Chung, S.-J.; Cho, M.-S. Texture Preferences of Chinese, Korean and US Consumers: A Case Study with Apple and Pear Dried Fruits. *Foods* **2020**, *9*, 377. [CrossRef] [PubMed]
10. Profeta, A.; Baune, M.-C.; Smetana, S.; Broucke, K.; Van Royen, G.; Weiss, J.; Heinz, V.; Terjung, N. Discrete Choice Analysis of Consumer Preferences for Meathybrids—Findings from Germany and Belgium. *Foods* **2020**, *10*, 71. [CrossRef] [PubMed]
11. Garvey, E.C.; Sander, T.; O'Callaghan, T.F.; Drake, M.; Fox, S.; O'Sullivan, M.G.; Kerry, J.P.; Kilcawley, K.N. A Cross-Cultural Evaluation of Liking and Perception of Salted Butter Produced from Different Feed Systems. *Foods* **2020**, *9*, 1767. [CrossRef] [PubMed]
12. Kelly, R.; Hollowood, T.; Hasted, A.; Pagidas, N.; Markey, A.; Scannell, A.G.M. Using Cross-Cultural Consumer Liking Data to Explore Acceptability of PGI Bread—Waterford Blaa. *Foods* **2020**, *9*, 1214. [CrossRef] [PubMed]
13. Laaksonen, O.; Ma, X.; Pasanen, E.; Zhou, P.; Yang, B.; Linderborg, K.M. Sensory Characteristics Contributing to Pleasantness of Oat Product Concepts by Finnish and Chinese Consumers. *Foods* **2020**, *9*, 1234. [CrossRef] [PubMed]
14. Duerlund, M.; Andersen, B.V.; Wang, K.; Chan, R.C.K.; Byrne, D.V. Post-Ingestive Sensations Driving Post-Ingestive Food Pleasure: A Cross-Cultural Consumer Study Comparing Denmark and China. *Foods* **2020**, *9*, 617. [CrossRef] [PubMed]
15. Haas, R.; Imami, D.; Miftari, I.; Ymeri, P.; Grunert, K.; Meixner, O. Consumer Perception of Food Quality and Safety in Western Balkan Countries: Evidence from Albania and Kosovo. *Foods* **2021**, *10*, 160. [CrossRef] [PubMed]
16. Wang, G.-Y.; Yueh, H.-P. Optimistic Bias, Food Safety Cognition, and Consumer Behavior of College Students in Taiwan and Mainland China. *Foods* **2020**, *9*, 1588. [CrossRef] [PubMed]
17. Cho, T.J.; Kim, S.A.; Kim, H.W.; Rhee, A.M.S. A Closer Look at Changes in High-Risk Food-Handling Behaviors and Perceptions of Primary Food Handlers at Home in South Korea across Time. *Foods* **2020**, *9*, 1457. [CrossRef] [PubMed]
18. Nathan, R.J.; Soekmawati; Victor, V.; Popp, J.; Fekete-Farkas, M.; Oláh, J. Food Innovation Adoption and Organic Food Consumerism—A Cross National Study between Malaysia and Hungary. *Foods* **2021**, *10*, 363. [CrossRef] [PubMed]
19. Magalhaes, D.; Campo, M.; Maza, M. Knowledge, Utility, and Preferences for Beef Label Traceability Information: A Cross-Cultural Market Analysis Comparing Spain and Brazil. *Foods* **2021**, *10*, 232. [CrossRef] [PubMed]

Article

Taste Interactions between Sweetness of Sucrose and Sourness of Citric and Tartaric Acid among Chinese and Danish Consumers

Jonas Yde Junge [1,2], Anne Sjoerup Bertelsen [1], Line Ahm Mielby [1], Yan Zeng [2,3], Yuan-Xia Sun [2,3], Derek Victor Byrne [1,2] and Ulla Kidmose [1,2,*]

[1] Food Quality Perception and Society Team, iSense Lab, Department of Food Science, Faculty of Technical Sciences, Aarhus University, 8200 Aarhus N, Denmark; jonas.junge@food.au.dk (J.Y.J.); annesbertelsen@food.au.dk (A.S.B.); lineh.mielby@food.au.dk (L.A.M.); derekv.byrne@food.au.dk (D.V.B.)
[2] Sino-Danish Center for Education and Research (SDC), 8000 Aarhus, Denmark; zeng_y@tib.cas.cn (Y.Z.); sun_yx@tib.cas.cn (Y.-X.S.)
[3] Tianjin Institute of Industrial Biotechnology, Chinese Academy of Sciences, Tianjin 300308, China
* Correspondence: ulla.kidmose@food.au.dk; Tel.: +45-2086-5197

Received: 10 September 2020; Accepted: 4 October 2020; Published: 9 October 2020 

Abstract: Tastes interact in almost every consumed food or beverage, yet many aspects of interactions, such as sweet-sour interactions, are not well understood. This study investigated the interaction between sweetness from sucrose and sourness from citric and tartaric acid, respectively. A cross-cultural consumer study was conducted in China ($n = 120$) and Denmark ($n = 139$), respectively. Participants evaluated six aqueous samples with no addition (control), sucrose, citric acid, tartaric acid, or a mixture of sucrose and citric acid or sucrose and tartaric acid. No significant difference was found between citric acid and tartaric acid in the suppression of sweetness intensity ratings of sucrose. Further, sucrose suppressed sourness intensity ratings of citric acid and tartaric acid similarly. Culture did not impact the suppression of sweetness intensity ratings of citric or tartaric acid, whereas it did influence sourness intensity ratings. While the Danish consumers showed similar suppression of sourness by both acids, the Chinese consumers were more susceptible towards the sourness suppression caused by sucrose in the tartaric acid-sucrose mixture compared to the citric acid-sucrose mixture. Agglomerative hierarchical cluster analysis revealed clusters of consumers with significant differences in sweetness intensity ratings and sourness intensity ratings. These results indicate that individual differences in taste perception might affect perception of sweet-sour taste interactions, at least in aqueous solutions.

Keywords: cross-culture; individual differences; taste mixtures; model matrix; taste primaries; taste-taste interactions; basic tastes; hierarchical clustering

---

1. Introduction

Taste is an important part of our perception of foods and beverages. Taste perception in humans is considered to consist of the five canonical basic taste qualities: sweet, sour, salt, bitter, and umami [1].

These basic taste qualities interact in almost every consumed food or beverage [2]. Taste interactions occur when at least two tastants are presented simultaneously and affect the perception of each other. Taste interactions can either be enhancing or suppressing, depending on both the taste quality, specific tastants in play, and tastant concentrations [3,4]. Taste interactions are complex, and, even though the interactions between tastes have been extensively researched and reviewed [3–5], the mechanisms are still not well understood [6].

This also applies for sweet-sour taste interactions, where both suppression and enhancement have been proposed for both tastes. The general consensus is that sweetness and sourness suppress each other [3,4,7]; however, sweetness enhancement by acids have also been reported [8–10]. Only a few studies have investigated other sour stimuli than citric acid in taste interactions with sweetness [11,12]. Moreover, also, very few studies have looked at tartaric acid, but with inconclusive results. Early reports found a sweetness enhancing effect of tartaric acid [8], a claim not supported by later research that has found tartaric acid consistently suppressing sweetness [11,12]. To our knowledge, Fabian and Blum's [8] is the only study comparing the taste of citric acid and tartaric acid in relation to sweet-sour taste interactions. These authors found a suppression effect of the sourness of both tartaric- and citric acid by sub-threshold levels sucrose. In the same study, they found an unexpected enhancing effect from sub-threshold levels of both tartaric- and citric acid on the sweetness of sucrose.

This inconsistency in findings both related to suppression or enhancement between sourness and sweetness, and to different tastants, could be ascribed to differences in test protocols [3]. Fabian and Blum [8] used the tastant concentrations below taste threshold for the interacting tastant, whereas most others [9–16] use supra-threshold concentrations for both affected and interacting tastant. As Puputti et al. [17] argued, supra-threshold concentrations of both tastants are the most relevant levels in which to investigate taste perception.

A further reason for inconsistencies might be individual differences in perception of taste interactions. Individual differences have not been well studied for taste interactions. McBride and Finlay [18] investigated the effect of sensory training on sweet-sour taste interactions, in a study comparing novices with experienced trained sensory panelists. They found generally high correlations between the two groups for both sweetness suppression by citric acid and sourness suppression by sucrose. Only for the highest concentration of both citric acid and sucrose did they find a difference in sweetness response between the two groups with novices showing higher suppression. The same was not the case for sourness. This indicates low degrees of individual differences in sweet-sour taste interactions. On the other hand, individual differences have been shown to affect taste interactions in other taste combinations. Prescott et al. [7] showed that differences in bitterness sensitivity to 6-*n*-propylthiouracil (PROP) correlated with bitter compounds' effectiveness in suppressing sweetness, indicating that individual differences might affect taste interactions. In contrast to taste interactions, individual differences in basic taste perception is better understood. Puputti et al. [17] found substantial differences in the perception of basic tastes between individuals. Further, they found evidence suggesting a correlation between supra-threshold taste response to different tastes in one individual, but they suggested, as others before [19], that there is a degree of independent variation in supra-threshold taste response to different tastes in individuals. Studies have found a range of factors that explain some part of supra-threshold taste response, such as genetic differences, age, gender, and weight [20–22]. This could potentially also affect taste interactions.

Another group of individual differences that could also influence taste interactions are cultural differences. As for individual differences in general, there are only few studies investigating the effect of culture on taste interactions. Two studies, Prescott et al. [14] and Prescott et al. [15], indicated that even though taste intensities are perceived similarly between Australian and Japanese panelists, some taste interactions might vary with nationality. Again, there is more evidence related to basic tastes than with taste interactions. For example, differences in supra-threshold taste response has been shown in different ethnic groups [23,24]. Williams et al. [23] found both African-American and Hispanic individuals to have a higher supra-threshold taste response than White individuals for all four tastes included in their study. Risso et al. [24] investigated four different vaguely defined ethnic groups, namely Italians, Northern Europeans, Mahgrebis, and Lankans. They investigated the relationship between ethnicity and supra-threshold taste response to mono-sodium glutamate and different bitter compounds. Even though they did find evidence for differences between the ethnic groups in taste sensitivity, no clear pattern in taste sensitivity was seen. Bertino et al. [25] investigated supra-threshold taste response in American and Taiwanese students and found that sweetness of sucrose were influenced by ethnicity both in complex matrices and a simple model matrix, with Taiwanese students ratings sweetness higher than American students.

Conducting cross-cultural comparisons of differences of sweet-sour taste interactions are important as it enables an investigation of whether these established taste interactions, and their magnitudes, are culturally unique or culturally universal [26]. Therefore, the present study had two aims. The first aim was to investigate the effect of different acids, namely citric and tartaric acid, on sweetness- and sourness taste intensity interactions. The second aim was to investigate the perception of taste interactions between sweetness and sourness in a cross-cultural setting. By using sucrose as sweet stimulus and citric and tartaric acid as sour stimuli, all in aqueous solutions, the suppression effect of sweetness on sourness, and the suppression effect of sourness on sweetness were thus investigated using Chinese and Danish consumers. These cultures were chosen as they represent two quite different cultures with very different eating habits [27,28], thus making it more likely that if there indeed are cross cultural differences in taste interactions, they will be present in this context.

When conducting cross-cultural research, a definition of which culture, and which cultures are investigated, is important. Often, cross cultural studies in sensory and consumer science define culture from a geographical point [29–31]. As the cross-cultural difference in food perception is thought to be related to eating habits [27,28], being at a certain place at the time of study might not be sufficient to adhere to that food culture. Thus, to further narrow down the cultural differences, nationality is also included in the cultural definition used in this study. Thus, respondents that were both located in Denmark at the time of the study and described their culture as Danish were in this study defined as having a Danish culture. Similarly, respondents that were both in China at the time of study and described themselves as Chinese were considered to have a Chinese culture.

2. Materials and Methods

Taste interaction effects between sucrose, citric acid and tartaric acid were investigated using a 2x3 full factorial design resulting in six samples in total. Relatively low but readily perceivable concentrations of sucrose and acids were chosen in order to make them acceptable for consumers. The concentrations of sucrose and citric acid were chosen based on an internal pilot-study indicating that these concentrations were similar in magnitude of intensity of sweetness and sourness, respectively. The procedure of the internal pilot-study where five internal employees tasted different solutions of citric acid and a sucrose solution with the chosen level of sucrose. After tasting, they indicated which citric acid solution, if any, had similar magnitude as the sucrose solution. To determine equal levels of sourness between citric acid and tartaric acid samples, an iso-sourness study was conducted using a trained sensory panel with eight assessors (7 female, age 42–60), following a protocol adapted after the Sucrose-Sweetener Combined protocol for sweetness by Reference [32], using citric acid as the stereotypical sourness agent. Five different concentrations of both citric acid (0.10, 0.12, 0.14, 0.16, 0.18% *w*/*v*) and tartaric acid (0.80, 0.11, 0.14, 0.17, 0.20% *w*/*v*) were presented. Panelists were provided with three sourness references (being the lowest, medium, and highest concentrations of the citric acid samples) as anchors on the 15 cm line scales used for sourness intensity rating. The anchors were placed at 1.5, 7.5, and 13.5 [7].

A study was conducted afterwards to validate that the sourness of the chosen concentrations of citric and tartaric acid was indeed equal. A trained panel of 10 assessors (9 female, 42–60) evaluated the samples for sourness intensity. No significant differences in sourness intensity were found. Both the iso-sourness study and the validation study were conducted at the sensory facilities at the Department of Food Science, Aarhus University, Aarslev, Denmark.

The six samples were then evaluated by Danish (21st May 2019) and Chinese (24th September 2019) consumers to investigate the effect of culture on the taste interactions.

2.1. Samples

Samples were aqueous solutions of either sucrose (Merck KGaA, Darmstadt, Germany), citric acid (Merck KGaA, Darmstadt, Germany) or tartaric acid (L-(+)-Tartaric acid) (Sigma-Aldrich, Co., St. Louis, MO, USA), or combinations of these, as shown in Table 1. Water for the aqueous solutions was San Benedetto, still natural mineral water (San Benedetto S.p.A., Scorzè, Italy). The tastants were weighted out and transferred to 2000 mL volumetric flasks. Tastants were dissolved in approximately 1 L of water. When all tastant was dissolved, the volumetric flask was filled.

Table 1. Sample names and concentrations of tastants. Samples are Water (W), Citric acid (C), Tartaric acid (T), Sucrose (S), Sucrose and Citric acid (SC), and Sucrose and Tartaric acid (ST).

Sample	Sucrose(% *w*/*v*)	Citric Acid (% *w*/*v*)	Tartaric Acid (% *w*/*v*)
W	-	-	-
C	-	0.140	-
T	-	-	0.114
S	2.50	-	-
SC	2.50	0.140	-
ST	2.50	-	0.114

After production, 20-mL sample was dispensed into opaque sample tubes with red lids (Fisher Scientific, Roskilde, Denmark), coded with random three-digit numbers. Samples were stored at 5 °C until evaluation, which was at room temperature. The samples were served the day after production.

2.2. Consumer Studies

A consumer study was conducted at Aarhus University, Aarhus, Denmark, and at Tianjin Institute of Industrial Biotechnology, Tianjin, China. At both locations, participants were recruited through convenience sampling; thus, both students and staff were recruited to participate in the study and rewarded with a minor gratuity for participation. The consumer characteristics can be found in Table 2. In both places, the consumers conducted the evaluation in a canteen-like area outside eating hours.

Table 2. Baseline characteristics of Danish and Chinese consumers. Provided p-values of Student's t-test for difference between Danish and Chinese consumer groups. The age range is presented in parenthesis. Sweet Foods Liking and Sour Foods Liking rated in 7-point Likert scale (7 = most liked).

	Denmark (n = 139)	China (n = 120)	p-Value
Mean Age (years)	23.24 (19–30)	25.34 (21–30)	<0.01
Number of Female	88 (62.9%)	79 (65.3%)	0.59
Mean BMI [1]	23.37	21.47	<0.01
Sweet Foods Liking	5.28	5.06	0.23
Sour Foods Liking	5.03	4.25	<0.01

[1] BMI is Body Mass Index.

Since all ingredients were commercially available food grade products, and the study was conducted in an aqueous model system, the study was exempted from the need of formal ethical approval. However, the panelists did give their verbal consent prior to participation. After consumers agreed to participate in the study, they were asked to complete the questionnaire on their smartphone or on a provided iPad. Consumers received the six samples and were provided with a glass of San Benedetto water.

The participants were given a short verbal introduction to the questionnaire and instructed to check and ensure that the three-digit codes on the tubes matched the ones in the questionnaire, as well as to drink a sip of water between the samples. Consumers were asked to evaluate one sample at a time, and samples were served randomly in accordance with a Williams Latin Square design. All data was collected using the software Compusense (Compusense Inc., Guelph, ON, Canada), which also instructed consumers to cleanse their mouth and wait at least 30 s between the samples.

The questionnaire contained five sample-related questions and nine none-sample related questions. The sample related questions were ratings of sourness and sweetness intensity evaluated on 9-point scales with anchors "Not at all" and "Extremely" [27]. Liking and Just About Right (JAR) for both sourness and sweetness were also evaluated. The questions were presented in the following order: sourness intensity, JAR of sourness, sweetness intensity, JAR of sweetness, and liking. The results from liking and JAR questions are not presented as the focus of this paper is on sweet-sour taste interactions.

Not sample-related questions were culture, gender, age, weight, height, and a 4-question battery of questions using a 7 point Likert scale about general sweet and sour food liking ("I like to eat sweet foods", "I like to eat sour foods", "I like to drink fruit juices", "I like to drink soft drinks") ranging from "Strongly disagree" corresponding to 1 and "Strongly agree" corresponding to 7. These were all asked after all samples had been evaluated.

Body Mass Index (BMI) was calculated based on consumer height and weight (kg)/(height (m)^2).

2.3. Statistical Analysis

All data analyses were performed using R [33] and RStudio [34]. Data formatting and plotting was conducted using the Tidyverse package [35]. In accordance with the aim only consumers with Danish or Chinese culture were included in the analysis. Further, the age ranged was narrowed to only include consumers in the age 18–30 years of age to ensure a more homogeneous consumer sample. Student's *t*-tests for comparison of baseline characteristics were conducted using the R package FactoMineR [36].

2.3.1. Overall Sample Evaluation Analysis

To provide an overview of the ratings for sourness intensity and sweetness intensity, means were calculated using R package SensoMineR [37] and Tukey's Honest Significant Differences (HSD) were used to evaluate pairwise significant differences between samples, considering sample effects as fixed effects and consumer effects as random effects. HSD was determined using the R package lmerTest [38]. Furthermore, Analysis of Variance (ANOVA) was conducted as a mixed model ANOVA using R package SensMixed [39], with Mixed Assessor Model (MAM) to adjust for scaling. Similar to the HSD, in this model, sample effects are considered fixed effects, and consumer effects are considered random effects.

2.3.2. Effect of Acid Quality on Sourness and Sweetness Taste Perception

To investigate the effect of different acids on the sweet-sour taste interaction, interaction plots for the interaction between acid and sucrose for sweetness intensity ratings and sourness intensity ratings, respectively, were compiled. These plots were constructed using R packages lsmeans [40] and ggplot2 [41].

2.3.3. Overall Effect of Culture

Differences in sourness intensity and sweetness intensity, between Danish and Chinese consumers are presented visually using the ggplot2 package in R [42]. They are shown as differences in estimated marginal means between the two groups. Significant differences were calculated using pairwise comparison of estimated marginal means and *p*-values shown are adjusted using Bonferroni corrections [43]. These were calculated using R packages rstatix [44] and emmeans [45].

2.3.4. Effect of Culture on Sweetness- and Sourness Intensity Ratings

To investigate the effect of culture on taste interactions, ratings of sweetness- and sourness intensities were analyzed as follows. Dummy variables were constructed for sucrose and acids indicating whether samples contained this or not. The effects of acid, sucrose, culture and the interactions between these on either sweetness or sourness ratings were analyzed using Mixed model ANOVA with MAM. In the model, the effects of acid, sucrose, and culture were considered as fixed effects, and consumer effects were considered random effects. To further investigate the three-way interactions between acid, sucrose and culture for sourness these were presented visually. This plot was constructed similarly as the interaction plot for the effect of different acids.

2.3.5. Cluster Analysis on Sweetness and Sourness Ratings

Two separate cluster analyses were performed on the ratings of sweetness intensity and sourness intensity, respectively. In both cases, we used Agglomerative Hierarchical Clustering with Euclidean distances and Wards method using the R package cluster [46]. The number of clusters were determined using NbClust package [47] using the majority rule.

Characteristics of consumers in each cluster were summarized, and differences in culture and gender were evaluated using χ2-tests, whereas gender, BMI, sweet food liking, and sour food liking were evaluated with ANOVA. Mean ratings of the different samples were calculated for each cluster, and HSD and ANOVA test for significant differences between clusters for each sample were performed. Mean calculation and HSD were performed similarly to the Overall Sample Evaluation Analysis. ANOVA was performed as one-way ANOVA between samples and clusters for each of the two taste intensity ratings. Lastly, consumers were plotted in PCA, showing both their culture and cluster number using the R package ggfortify [48].

3. Results

3.1. *Sample Evaluation*

For an initial overview, all sample means and significant differences in ratings of sweetness and sourness of samples are shown in Table 3. For sweetness ratings, the sample only containing sucrose (S) was highest, followed by the ones containing both sucrose and an acid (SC or ST). This indicates a suppression effect as expected. Interestingly, the sample containing only citric acid (C) was evaluated higher in sweetness ratings than pure water (W), indicating a small sweetness enhancing effect of citric acid. The sample containing only tartaric acid (T) was placed between the two and were not different from neither W nor C.

Regarding the sourness ratings, the expected suppression of sourness by sucrose occurred. The samples containing only acid (C or T) were rated higher in sourness than those containing both sucrose, as well as an acid (SC and ST). As expected, there was no difference between W and S in terms of sourness ratings.

Table 3. Means of sweetness and sourness ratings for all samples, as well as F- and p-values for Analysis of Variance model. Samples are Water (W), Citric acid (C), Tartaric acid (T), Sucrose (S), Sucrose and Citric acid (SC), and Sucrose and Tartaric acid (ST). Letters are to be read row-wise, and they indicate significant differences between samples based on Honest Significant Differences (HSD).

	W	C	T	S	SC	ST	F-Value	p-Value
Sweetness	1.53a	1.90b	1.82ab	6.32d	4.53c	4.49c	505.34	<0.001
Sourness	1.49a	6.42c	6.36c	1.51a	4.79b	4.70b	496.78	<0.001

3.2. *Effect of Acid Quality on Sour and Sweet Taste Interaction*

Two different acids were evaluated in this study, namely citric acid and tartaric acid. This was done in order to investigate if there is a difference in the taste interactions between sucrose and different acids. In Figure 1, the results for sweetness and sourness intensity ratings are presented.

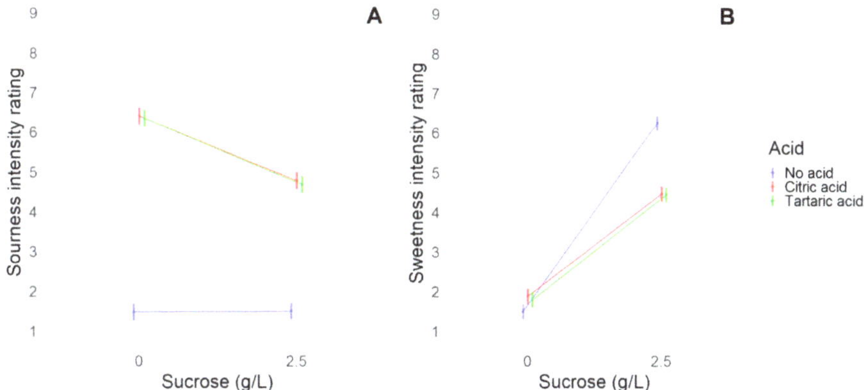

Figure 1. Interaction plots showing the means for the interaction between acids and sucrose. (**A**) Sourness intensity ratings; (**B**) sweetness intensity ratings. Means are full dots. Surrounding faded lines indicate 95% confidence intervals.

Both sourness intensity (Figure 1A) and sweetness intensity (Figure 1B) ratings were not significantly different for the two acids (see Table 3 for significance test). Figure 1A shows that, as expected, citric and tartaric acid differs from the No acid condition, but they are not different from each other. The similarity between the two acids both show that they were rated similarly in sourness intensity but also that citric and tartaric acid had similar sourness suppression by sucrose at the tested concentrations.

In Figure 1B, as expected, the No acid condition differed from the acids when sucrose was present, indicating a sweetness suppression by acids (see Table 3 for significance test). But, also, for sweetness intensity, the acids gave a similar response, indicating a similar suppression effect from both citric and tartaric acid. The citric acid increased sweetness significantly in the No acid condition (Table 3), but Figure 1B shows that the effect of the acids on sweetness intensity ratings are much larger when sucrose is present.

3.3. *Effect of Culture on Sweet and Sour Taste Interaction*

Figure 2 shows differences from the mean in sweetness intensity ratings and sourness intensity ratings (significant differences indicated below) between Danish and Chinese consumers for all samples.

For sweetness intensity ratings, Danish consumers rated the samples consistently higher than the Chinese consumers. This difference is largest for the sucrose containing samples (S, SC, and ST), but still significant for the samples without sucrose (W, C, and T).

There is no clear trend in the differences in the ratings of sourness intensity. Danish consumers rated W and ST highest in sourness intensity, whereas, for T and SC, Chinese consumers rated the highest in sourness intensity. No significant differences were found for C ($p = 0.15$) and S ($p = 0.40$).

Figure 2. Deviations from means of ratings for sweetness intensity and sourness intensity for Chinese (red dots) and Danish (green dots) consumers, respectively. Samples are Water (W), Citric acid (C), Tartaric acid (T), Sucrose (S), Sucrose and Citric acid (SC), and Sucrose and Tartaric acid (ST). Significant differences are pairwise comparison of estimated marginal means corrected using Bonferroni corrections: ns = not significant, * < 0.05, ** < 0.01, *** < 0.001.

3.3.1. Effect of Culture on Sweetness Ratings

The effect of culture on sweet taste ratings was further evaluated by ANOVA. Results can be seen in Table 4.

Table 4. F- and p-values for mixed model Analysis of Variance for Sweetness Intensity Ratings. Significant effects ($p \leq 0.05$) are marked in bold.

	F-Value	p-Value
Acid	53.98	**<0.001**
Sucrose	848.55	**<0.001**
Culture	70.87	**<0.001**
Acid × Sucrose	112.98	**<0.001**
Acid × Culture	0.31	0.74
Sucrose × Culture	9.01	**<0.01**
Acid × Sucrose × Culture	1.55	0.21

The Acid × Sucrose × Culture-interaction was not significant ($p = 0.21$), indicating that there was no effect of culture on the suppression of sweetness intensity ratings by acids. This is further underpinned by the insignificance of the Acid × Culture-interaction (($p = 0.74$), showing that culture of the consumers did not impact the effect of acids on sweetness intensity ratings.

For interactions involving the culture term, only the interaction between culture and sucrose is significant ($p < 0.01$). This means that Chinese and Danish consumers rated the sweetness intensity differently. This was also indicated in Figure 2.

The significant interaction between acid and sucrose for the sweetness intensity ratings ($p < 0.001$) indicates a sweetness suppression effect by acids across the two cultures. This was expected from the results seen in Table 3, where the sweetness intensity of sucrose was suppressed significantly by the acids by a reduction of sweetness intensity ratings.

3.3.2. Effect of Culture on Sourness Ratings

A similar ANOVA was conducted to investigate cultural effects on sour taste ratings (Table 5).

Table 5. F- and *p*-values for mixed model Analysis of Variance for Sourness Intensity. Significant effects ($p \leq 0.05$) are marked in bold.

	F-Value	*P*-Value
Acid	1274.37	**<0.001**
Sucrose	131.60	**<0.001**
Culture	0.02	0.90
Acid × Sucrose	53.72	**<0.001**
Acid × Culture	12.74	**<0.001**
Sucrose × Culture	0.01	0.91
Acid × Sucrose × Culture	8.93	**<0.01**

The table shows that the Acid × Sucrose × Culture-interaction term is significant ($p < 0.01$), indicating that there was an effect of culture on the suppression of sourness intensity ratings by sucrose. To investigate this interaction further, the model is presented graphically in Figure 3.

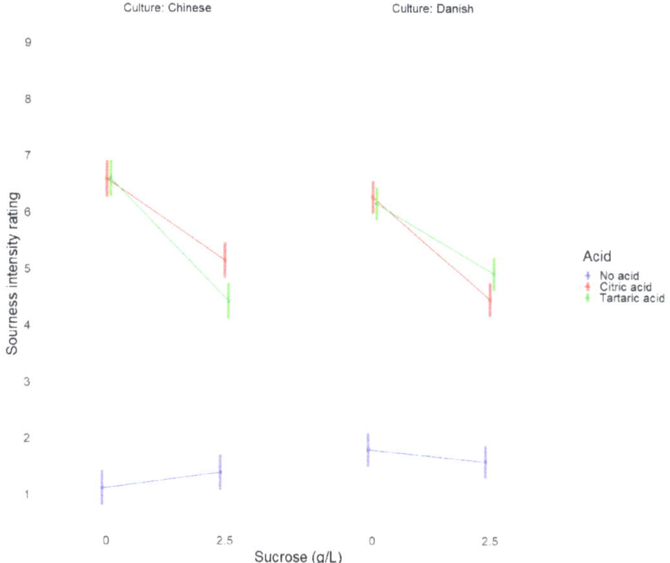

Figure 3. Interaction plot showing the means of the interaction between acids and sucrose for sourness intensity rating for Chinese and Danish consumers, respectively. Means are full dots. Surrounding faded lines indicate 95% confidence intervals.

When investigating Figure 3, two differences between the Danish and Chinese consumer groups should be noted. First, even though the sourness intensity ratings of the two acids appear similar when presented alone, citric and tartaric acid behave differently when combined with sucrose in the two consumer groups. Whereas the Chinese consumers had a significantly higher suppression of the sourness ratings of tartaric acid than of citric acid by sucrose, the Danish consumers do not experience any differences, as the confidence intervals overlap. There might be a trend towards the opposite suppression for the Danish consumers compared to the Chinese consumers, namely that the sourness ratings of citric acid might be more suppressed by sucrose than tartaric acid is.

The other thing that comes to mind is that even though the two consumer groups had similar ratings of sourness when sucrose was present, Chinese consumers found the No acid, No sucrose condition less sour than the Only sucrose condition, whereas Danish consumers seemed to find the No acid, No sucrose condition more sour than the Only sucrose condition. Both seem to be trends and, thus, might not be significant.

3.4. Cluster Analyses across Culture

The cluster analyses are shown below and reveals patterns of individual difference in ratings of the taste intensities, thus enabling an investigation of differences between different groups of individuals.

3.4.1. Cluster Analysis on Sweetness Intensity

A cluster analysis was performed on sweetness ratings of all consumers to further investigate the differences in sweetness ratings. Three clusters were found, and PCA was performed to further visualize these clusters. PC1 accounts for 35.28% of the variance, and PC2 for 18.22% of the variance, respectively. This can be seen in Figure 4, where both the three clusters are identified, and the culture of the consumers are visualized.

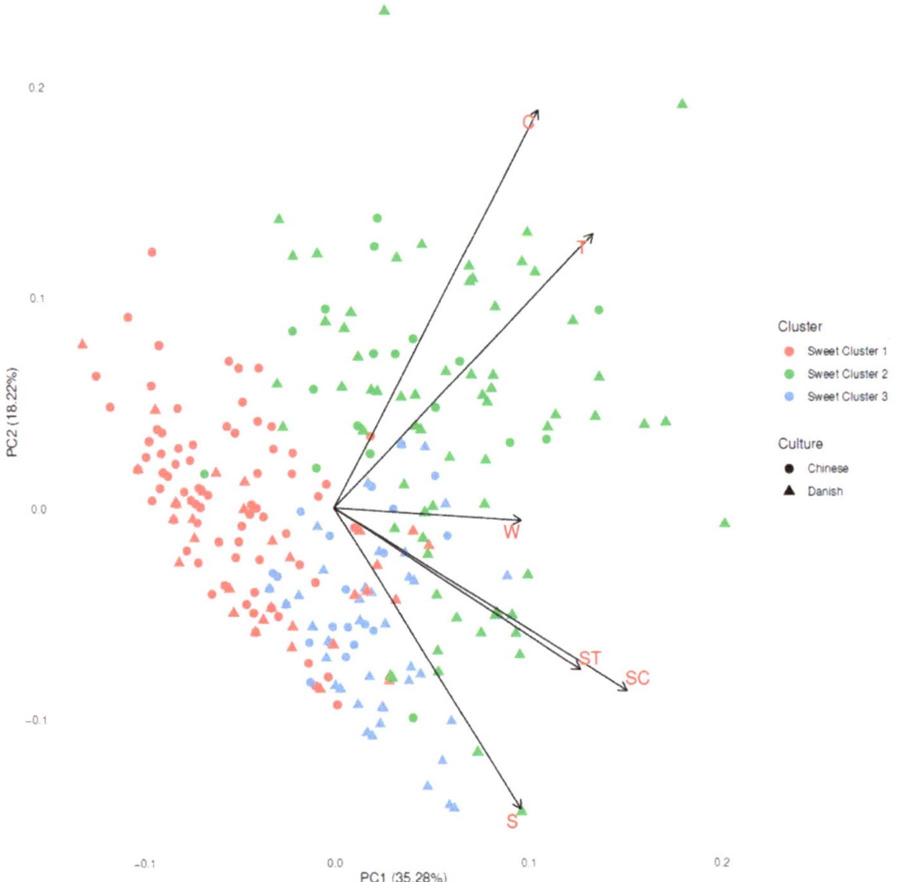

Figure 4. PCA biplot showing consumers sweetness ratings. Both their relation to clusters and culture are visualized. Samples are Water (W), Citric acid (C), Tartaric acid (T), Sucrose (S), Sucrose and Citric acid (SC), and Sucrose and Tartaric acid (ST).

As can be seen in Figure 4, the clusters differentiate rather clearly, even though it should be kept in mind that only slightly more than 50% of the variance is explained by the PCA plot. Further, there might be a visible distinction between Chinese and Danish consumers, where Danish consumers seemed to cluster to the right and Chinese consumers seemed to cluster to the left, though this was not very distinct.

Characteristics of the three clusters are shown in Table 6. The clusters were significantly different for culture ($p < 0.001$) and age ($p < 0.01$). The percentage of female consumers ($p = 0.35$), the BMI ($p = 0.73$), and the liking of sweet ($p = 0.49$) and sour foods ($p = 0.56$) were not significantly different between the different clusters. There was an over-representation of Chinese consumers in Sweet Cluster 1, and an over-representation of Danish consumers in Sweet Cluster 2. It was similar for age, where Sweet Cluster 1 is higher in mean age and Sweet Cluster 2 was lower in mean age. The difference in mean age between those two clusters might be driven by the difference in mean age among the Danish and the Chinese consumer groups.

Table 6. Cluster characteristics and means of ratings for Sweetness in Sweet clusters. Samples are Water (W), Citric acid (C), Tartaric acid (T), Sucrose (S), Sucrose and Citric acid (SC), and Sucrose and Tartaric acid (ST). *P*-values for cluster culture and gender are for $\chi 2$ analysis of difference, whereas *p*-values for Age, BMI, and Sweet and Sour Food Liking are from ANOVA. *P*-values for Sweetness Intensity are derived from mixed model Analysis of Variance. Letters are to be read row wise, and indicate significant differences between samples based on Honest Significant Differences (HSD). Note that, even though the Analysis of Variance find significant differences between clusters for sample S, HSD fail to identify differences. Significant effects ($p \leq 0.05$) are marked in bold.

	Sweet Cluster 1	Sweet Cluster 2	Sweet Cluster 3	*p*-Value
Total Consumers in cluster	110	85	64	-
Mean of Age	24.9	23.6	23.9	**<0.01**
Number of Chinese (%)	78 (70.9)	18 (21.2)	24 (37.5)	**<0.001**
Number of Female (%)	76 (69.1)	52 (61.2)	38 (59.4)	0.35
BMI	22.3	22.8	22.4	0.73
Sweet Food Liking	5.2	5.2	5.3	0.49
Sour Food Liking	4.7	4.8	4.5	0.56
W	1.08a	2.35b	1.20a	**<0.001**
C	1.26a	3.04b	1.50a	**<0.001**
T	1.26a	2.88b	1.36a	**<0.001**
S	5.80a	6.75a	6.62a	**<0.01**
SC	3.25a	5.27b	5.73b	**<0.001**
ST	3.09a	4.95b	6.28c	**<0.001**

Regarding the sweetness intensity ratings, the clusters do differ (Table 6). Sweet Cluster 1 generally had low sweetness ratings and a high degree of sweetness suppression effects by acids (difference between S and SC/ST) compared to the other clusters. Sweet Cluster 1 had the lowest sweetness intensity ratings of all clusters for the samples containing both sucrose and acid (ST and SC).

Sweet Cluster 2 differed from the other clusters by on average having higher sweetness ratings for the non-sucrose containing samples. For the samples containing sucrose, the sweetness ratings are higher than Sweet Cluster 1 for both acid-containing samples. For the sucrose and tartaric acid sample (ST), sweetness ratings for Sweet Cluster 2 lies in between Sweet Cluster 1 and Sweet Cluster 3.

Lastly, Sweet Cluster 3 is similar to Sweet Cluster 1 in the average sweetness ratings of the non-sucrose containing samples, and the sucrose only sample. At the same time, Sweet Cluster 3 has higher sweetness intensity ratings of both sucrose and acid-containing samples than Sweet Cluster 1, and, for the sample containing both sucrose and tartaric acid (ST), it has the highest sweetness intensity rating of all clusters. As Sweet Cluster 3 has low sweetness ratings of sucrose-free samples and similarly high ratings of sucrose-containing samples both with and without acids, this cluster shows low suppression effect of sweetness by acids.

3.4.2. Cluster Analysis on Sourness Intensity

Similar to sweetness, a cluster analysis was performed on sourness ratings of all consumers to further investigate the differences in sourness ratings. The PCA shown in Figure 5 depicts the consumers in relation to their sourness ratings. PC1 account for 32.45% of the variance and PC2 for 18.55% of the variance. The consumers do not separate in culture, though there might be a tendency that Chinese consumers are somewhat more to the upper part and Danish consumers in the lower part. On the other hand, the three clusters are very clearly separated. Sour Cluster 3 is a group of more scattered and, in general, fewer consumers than Sour Cluster 1 and Sour Cluster 2. The loadings indicate that Sour Cluster 3 is explaining variance related to the water-only sample. In Table 7, it is seen that this is due to high sourness intensity ratings of this particular sample by Sour Cluster 3.

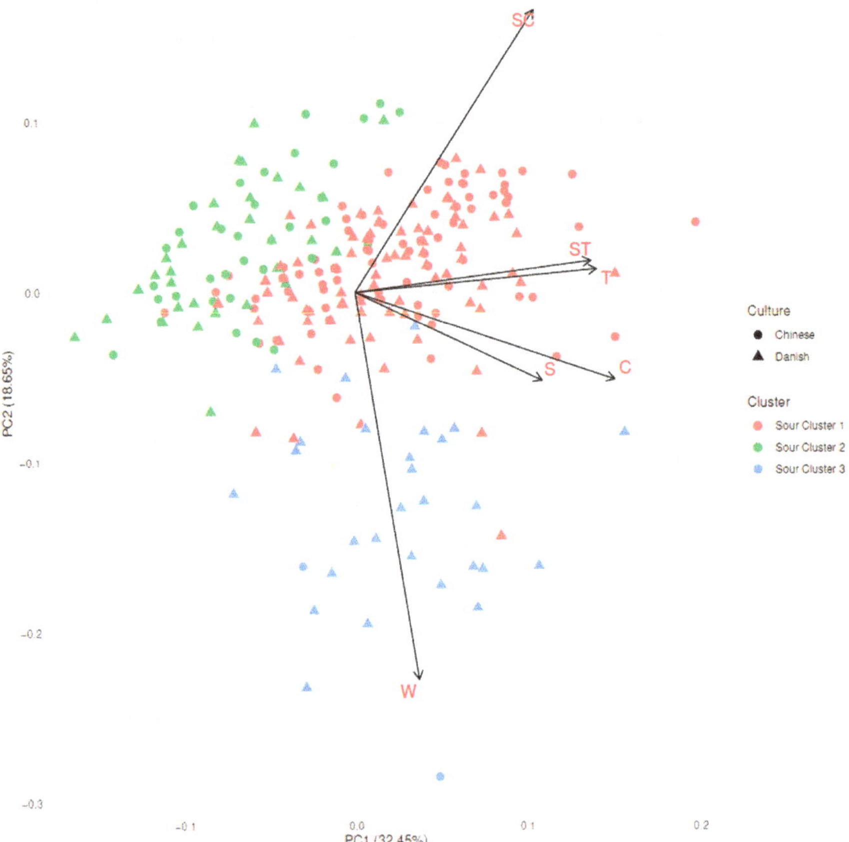

Figure 5. PCA biplot showing consumers sourness ratings. Both their relation to clusters and culture are visualized. Samples are Water (W), Citric acid (C), Tartaric acid (T), Sucrose (S), Sucrose and Citric acid (SC), and Sucrose and Tartaric acid (ST).

In Table 7, the characteristics of the three clusters are shown. In baseline characteristics, the three sour clusters are only significantly different in consumer culture ($p < 0.001$). Only Cluster 3 differs by a very high prevalence of Danish consumers. Thus, the age ($p = 0.70$), BMI ($p = 0.79$), and their liking of sweet ($p = 0.20$) and sour foods ($p = 0.05$) are rather similar for all clusters. Further, it should be noted that Sour Cluster 3 is rather small, with only 30 consumers in this cluster.

Table 7. Cluster characteristics and means of ratings for Sourness Intensity in clusters. Samples are Water (W), Citric acid (C), Tartaric acid (T), Sucrose (S), Sucrose and Citric acid (SC), and Sucrose and Tartaric acid (ST). *P*-values for cluster culture and gender are for $\chi 2$ analysis of difference, whereas *p*-values for Age, BMI, and Sweet and Sour Food Liking are from ANOVA. *P*-values for Sourness Intensity are from mixed model Analysis of Variance. Letters are to be read row wise, and indicate significant differences between samples based on Honest Significant Differences (HSD). Significant effects ($p \leq 0.05$) are marked in bold.

	Sour Cluster 1	Sour Cluster 2	Sour Cluster 3	*P*-Value
Total Consumers in cluster	164	65	30	-
Mean of Age	24.4	24.1	23.6	0.70
Number of Chinese (%)	84 (51.2)	34 (52.3)	2 (6.7)	**<0.001**
Number of Female (%)	108 (64.0)	39 (60.0)	19 (63.3)	0.70
BMI	22.5	22.3	22.8	0.79
Sweet Food Liking	5.3	4.9	5.1	0.20
Sour Food Liking	4.6	4.8	5.2	0.05
W	1.07a	1.03a	4.77b	**<0.001**
C	7.29b	4.06a	6.80b	**<0.001**
T	7.27c	4.18a	6.10b	**<0.001**
S	1.60b	1.11a	1.90b	**<0.01**
SC	4.88a	4.77a	4.33a	0.43
ST	4.87b	3.97a	5.30b	**<0.01**

The sourness intensity ratings certainly differ between the clusters. Sour Cluster 1 and Sour Cluster 2 differ such that Sour Cluster 2 has lower sourness intensity ratings than Sour Cluster 1. This is the case for the two acid-only containing samples (C and T), as well as for the sucrose only containing sample (S) and the sucrose and tartaric acid-containing sample (ST). Interestingly, in Sour Cluster 2 the sourness intensity ratings were not affected by sucrose. Sour Cluster 1 and Sour Cluster 3 only differ in samples W and T. Both Sour Cluster 1 and Sour Cluster 3 have lower sourness intensity ratings of samples containing both sucrose and acids than those just containing acids, indicating a sourness intensity suppression effect from sucrose. Sour Cluster 1 has higher sourness intensity ratings of sample T than Sour Cluster 3. The difference between Sour Cluster 3 and Sour Cluster 1 is mainly due to the oddity of a rather high degree of sourness intensity rating of the water-only sample in Sour Cluster 3.

Interestingly, Sample S is much lower than W in sweetness for Sour Cluster 3, indicating the rated sourness of pure water by Sour Cluster 3 is more or less suppressed by sucrose. It was further investigated whether the high sourness rating of the sample W in Sour Cluster 3 could be due to order effects, but no order pattern were found to be more prevalent in Sour Cluster 3 than in Sour Cluster 1 and 2, indicating that it is not an order effect (data not shown). This was investigated in relation to which number the in W-sample was presented and which sample preceded sample W.

4. Discussion

This study investigated taste interactions between sweetness and sourness in Danish and Chinese consumers. Further, it investigated if, and how, these interactions are dependent on sour tastant, thus determining whether citric and tartaric acid give similar sweet and sour responses. An overall significant suppression effect of sweetness from sucrose by both citric and tartaric acid was found. Likewise, overall suppression effect of sourness from both citric and tartaric acid by sucrose was found. Below, the dynamics of these interactions will be discussed in relation to acid quality, as well as both cultural and individual differences.

4.1. Effect of Acid Quality on Sweet-Sour Taste Interaction

The effect of citric acid and the effect of tartaric acid were found to be the same both related to sweetness intensity and sourness intensity. Both acids were susceptible to sourness intensity suppression by sucrose to a similar extent, which is consistent with earlier findings [8]. Likewise, both acids suppressed sweetness intensity of sucrose to the same extent. This is in disagreement with the findings of Reference [8], who found sweetness enhancing effects of both acids. This discrepancy might be due to differences in protocols on acid concentration. The concentrations of acids used by Reference [8] were below the taste threshold, thus being well below the concentrations used in this study. Further, studies using a similar concentration of either citric acid [2,49,50] or tartaric acid [12,51] as this study have shown suppression effects similar to those found in this study.

Both sub-threshold and supra-threshold protocols provide valuable information for food perception. Concentrations of tastants as those used in the present study are in the ranges found in beverages, such as fruit drinks [52]. Other beverages, such as soft drinks, might contain considerably higher amounts, which, again, could influence the taste interactions occurring.

4.2. Cultural Differences in Taste Interactions

An effect of culture on taste interactions of sweetness on sourness was found but not of sourness on sweetness. Similarities across cultures in taste interactions of sweetness on sourness were also found in an earlier study conducted with Australians and Japanese respondents [15].

Danish consumers, on average, gave higher sweetness intensity ratings than Chinese consumers did. Moreover, an interaction effect was found between sucrose and culture, meaning culture affected the sweetness ratings of sucrose. These effects did not result in effects on taste interactions. This is in agreement with results by Prescott et al. [15], who found a higher sweetness rating by Japanese respondents than Australian respondents but also failed to find taste interactions on sweetness by sourness. Only looking at the taste intensity ratings, and not taste interactions, Bertino et al. [25] similarly found that Taiwanese respondents had a lower sweetness rating than American respondents of aqueous solutions but a higher sweetness response in a complex food matrix.

In contrast to the results found in this study, Prescott et al. [14] did not find cultural differences in taste interactions of sweetness on sourness. The effect found in this study relates to the fact that Chinese consumers displayed a higher sourness suppression by tartaric acid when compared to citric acid, whereas citric and tartaric acid showed the same sourness suppression in Danish consumers. As cultural differences in taste interactions for different acids have not earlier been studied, this is an interesting finding that should be investigated further to see if there are differences for other acids, and if this is similar for other sweeteners.

Besides the difference in susceptibility to sourness suppression of citric and tartaric acid, there was seemingly no consistent effect of culture on sourness ratings.

It is interesting that this study displayed comparable patterns between Danish and Chinese consumers, as Reference [14,15] did between Australians and Japanese respondents. Besides the vague notion of "east versus west", it is difficult to interpret Australian and Danish respondents as similar culture [31]. The same is evident for Japanese and Chinese respondents. This being said, there have been suggestions that taste sensitivity differs between ethnicity, and Yang et al. [53] found an Asian population to have higher overall taste sensitivity than a Caucasian one.

4.3. Individual Differences in Taste Interactions

The cluster analysis for sweetness intensity ratings showed differences between clusters on two parameters, namely sweetness intensity ratings of sucrose-containing samples and sweetness suppression by acids. A possible mechanism could be that these differences could have relation to supra-threshold taste responses. Individual differences in taste response have been reported for all tastes [40,52]. Thus, Sweet Cluster 1 may be said to consist of consumers whom have a lower sweetness sensitivity, as consumers in this cluster respond less to sweetness and experience a higher degree of sweetness suppression from acids. Sweet Cluster 1 includes a larger proportion of Chinese consumers than Danish, indicating that Chinese consumers display a lower supra-threshold taste response. No studies have earlier compared these two groups, but other studies have generally found higher supra-threshold taste response for sweetness in Asian populations [25], and lower in Non-Hispanic White populations [23], indicating the opposite correlation. This could be due to differences in tested concentrations of sucrose, as both Bertino et al. [25] and Williams et al. [23] investigated higher sucrose concentrations than this study.

Further, the low degree of sweetness suppression in Sweet Cluster 3 might be explained by low supra-threshold taste response to sourness, thus affecting lower impact on sweetness. This is more in line with the findings of Williams et al. [23]. Indeed, a relationship between supra-threshold taste response and binary taste interaction have earlier been demonstrated related to PROP sensitivity [7].

Similarly, three factors separate the clusters in the sourness analysis. Here, the main factors are differences in sourness ratings of acids, in suppression of sourness by sweetness, and the surprising sourness of water in Sour Cluster 3. Again, it could be speculated that the differences in sourness is due to differences in taste sensitivity. Sour Cluster 2 show both low sourness response but also low suppression of sourness by sweetness, indicating low sensitivity to both sourness and sweetness. Similarly, Sour Cluster 1 showed high response to sourness and high suppression, which could be an indication of high sensitivity to both sweetness and sourness. This corresponds well to earlier findings that supra-threshold taste response to different tastes for the same individual correlates [54].

The high sourness of water (sample W) in the predominantly Danish Sour Cluster 3 was surprising and difficult to explain. The cluster was somewhat small, so there is a risk that it is just modeling random noise. The effect of order was ruled out by subsequent analysis. Interestingly, the sourness of the water was suppressed by sweetness of sucrose, but Sour Cluster 3 still rated sample S the highest in sourness. As Sour Cluster 3 is only the highest of the clusters in sourness ratings in three of the six samples, it is also unlikely to be an effect of differences in scale use.

This inconsistency in findings both related to suppression or enhancement between sourness and sweetness, as well as to different tastants, could be ascribed to differences in test protocols [3]. Fabian and Blum [8] used the tastant concentrations below taste threshold for the interacting tastant, whereas most others [9–16] use supra-threshold concentrations for both affected and interacting tastant. As Puputti et al. [17] argued, supra-threshold concentrations of both tastants are the most relevant level to investigate taste perception in.

5. Conclusions

This study investigated the effect of different acids on sweet-sour taste interactions. It was shown that citric and tartaric acid both showed a similar overall suppression effect of sweetness from sucrose. Likewise, sucrose showed an overall suppression effect of sourness from both citric and tartaric acid. The effects on sourness from citric and tartaric acid were similar.

Further, this study investigated the perception of taste interactions between sweetness and sourness in a cross-cultural setting with Danish and Chinese consumers. An effect of culture on taste interactions of sweetness on sourness was found, but not of sourness on sweetness. Further analysis revealed that sucrose suppressed sourness of tartaric acid to a larger degree in Chinese than Danish consumers.

Lastly, cluster analysis showed differences in taste interactions for both sweetness and sourness between different clusters, with little cultural selection into the groups.

These findings build to the evidence that differences in the perception of taste interactions in beverages are due to the individual rather than cultural differences. Thus, with a focus in the beverage industry to the diversification of the product range, it might be fruitful to investigate not only different demographic and hedonic consumer segments but also perceptual segments.

Author Contributions: Conceptualization, resources: J.Y.J., L.A.M., Y.Z., Y.-X.S., and U.K.; software, validation, formal analysis, data curation, visualization, project administration, writing–original draft preparation: J.Y.J.; investigation: J.Y.J., A.S.B., and Y.Z.; writing–review and editing: J.Y.J., L.A.M., A.S.B., Y.Z., Y.-X.S., D.V.B., and U.K.; supervision: U.K., L.A.M., A.S.B., Y.Z., Y.-X.S., and D.V.B.; funding acquisition: D.V.B. and U.K. All authors have read and agreed to the published version of the manuscript.

Funding: This work was supported by the Sino Danish Centre (SDC) within the 'Food and Health Research Theme', Aarhus, Denmark, the iFOOD Aarhus University Centre for Innovative Food Research, Aarhus, Denmark, and the Aarhus University Graduate School of Technological Sciences (GSTS), Aarhus, Denmark.

Acknowledgments: The authors would like to thank Chris Kjeldsen for initial conceptual input, Line Elgaard Nielsen, Jens Madsen, Ying Bai, and Lucie Bauderlique for assistance in conducting the study in Denmark, and Yan Men, Peng Chen, Chaoyu Tian, Jiao Li, Junjun Yan, Chenxi Ren, Tong Zhang, Shicheng Mu for assistance in conducting the study in China. Further, we would like to thank Ying Bai and Janice Wang for help with translation between English and Chinese.

Conflicts of Interest: The authors declare no conflict of interest.

Abbreviations

ANOVA	Analysis of variance
C	Citric acid sample
HSD	Tukey's Honest Significant Differences
S	Sucrose sample
SC	Sucrose & Citric acid sample
ST	Sucrose & Tartaric acid sample
T	Tartaric acid sample
W	Water only sample

References

1. Lawless, H.T.; Heymann, H. *Sensory Evaluation of Food: Principles and Practices*, 2nd ed.; Food science texts series; Springer: New York, NY, USA, 2010; ISBN 978-1-4419-6487-8.
2. Green, B.G.; Lim, J.; Osterhoff, F.; Blacher, K.; Nachtigal, D. Taste mixture interactions: Suppression, additivity, and the predominance of sweetness. *Physiol. Behav.* **2010**, *101*, 731–737. [CrossRef] [PubMed]
3. Keast, R.S.J.; Breslin, P.A. An overview of binary taste–taste interactions. *Food Qual. Prefer.* **2003**, *14*, 111–124. [CrossRef]
4. Wilkie, L.M.; Capaldi Phillips, E.D. Heterogeneous binary interactions of taste primaries: Perceptual outcomes, physiology, and future directions. *Neurosci. Biobehav. Rev.* **2014**, *47*, 70–86. [CrossRef] [PubMed]
5. Breslin, P.A. Interactions among salty, sour and bitter compounds. *Trends Food Sci. Technol.* **1996**, *7*, 390–399. [CrossRef]
6. Wang, Q.J.; Mielby, L.A.; Junge, J.Y.; Bertelsen, A.S.; Kidmose, U.; Spence, C.; Byrne, D.V. The Role of Intrinsic and Extrinsic Sensory Factors in Sweetness Perception of Food and Beverages: A Review. *Foods* **2019**, *8*, 211. [CrossRef]
7. Prescott, J.; Ripandelli, N.; Wakeling, I. Binary taste mixture interactions in prop non-tasters, medium-tasters and super-tasters. *Chem. Senses* **2001**, *26*, 993–1003. [CrossRef]
8. Fabian, F.W.; Blum, H.B. Relative Taste Potency of Some Basic Food Constituents and Their Competitive and Compensatory Action1. *J. Food Sci.* **1943**, *8*, 179–193. [CrossRef]
9. Kamen, J.M.; Pilgrim, F.J.; Gutman, N.J.; Kroll, B.J. Interactions of suprathereshold taste stimuli. *J. Exp. Psychol.* **1961**, *62*, 348–356. [CrossRef]
10. Pelletier, C.A.; Lawless, H.T.; Horne, J. Sweet–sour mixture suppression in older and young adults. *Food Qual. Prefer.* **2004**, *15*, 105–116. [CrossRef]
11. Martin, N.; Minard, A.; Brun, O. Sweetness, Sourness, and Total Taste Intensity in Champagne Wine. *Am. J. Enol. Vitic* **2002**, *53*, 6–13.
12. Zamora, M.C.; Goldner, M.C.; Galmarini, M.V. Sourness–Sweetness Interactions in Different Media: White Wine, Ethanol and Water*. *J. Sens. Stud.* **2006**, *21*, 601–611. [CrossRef]
13. Prescott, J. Comparisons of taste perceptions and preferences of Japanese and Australian consumers: Overview and implications for cross-cultural sensory research. *Food Qual. Prefer.* **1998**, *9*, 393–402. [CrossRef]

14. Prescott, J.; Bell, G.A.; Gillmore, R.; Yoshida, M.; O'Sullivan, M.; Korac, S.; Allen, S.; Yamazaki, K. Cross-cultural comparisons of Japanese and Australian responses to manipulations of sweetness in foods. *Food Qual. Prefer.* **1997**, *8*, 45–55. [CrossRef]
15. Prescott, J.; Bell, G.A.; Gillmore, R.; Yoshida, M.; O'Sullivan, M.; Korac, S.; Allen, S.; Yamazaki, K. Cross-cultural comparisons of Japanese and Australian responses to manipulations of sourness, saltiness and bitterness in foods. *Food Qual. Prefer.* **1998**, *9*, 53–66. [CrossRef]
16. Laing, D.G.; Prescott, J.; Bell, G.A.; Gillmore, R.; Allen, S.; Best, D.J.; Yoshida, M.; Yamazaki, K.; Ishii, R. Responses of Japanese and Australians to Sweetness in the Context of Different Foods. *J. Sens. Stud.* **1994**, *9*, 131–155. [CrossRef]
17. Puputti, S.; Aisala, H.; Hoppu, U.; Sandell, M. Multidimensional measurement of individual differences in taste perception. *Food Qual. Prefer.* **2018**, *65*, 10–17. [CrossRef]
18. McBride, R.L.; Finlay, D.C. Perception of Taste Mixtures by Experienced and Novice Assessors. *J. Sens. Stud.* **1989**, *3*, 237–248. [CrossRef]
19. Lim, J.; Urban, L.; Green, B.G. Measures of Individual Differences in Taste and Creaminess Perception. *Chem. Senses* **2008**, *33*, 493–501. [CrossRef]
20. Simchen, U.; Koebnick, C.; Hoyer, S.; Issanchou, S.; Zunft, H.-J. Odour and taste sensitivity is associated with body weight and extent of misreporting of body weight. *Eur. J. Clin. Nutr.* **2006**, *60*, 698–705. [CrossRef]
21. Hansen, J.L.; Reed, D.R.; Wright, M.J.; Martin, N.G.; Breslin, P.A. Heritability and Genetic Covariation of Sensitivity to PROP, SOA, Quinine HCl, and Caffeine. *Chem. Senses* **2006**, *31*, 403–413. [CrossRef]
22. Hyde, R.J.; Feller, R.P. Age and sex effects on taste of sucrose, NaCl, citric acid and caffeine. *Neurobiol. Aging* **1981**, *2*, 315–318. [CrossRef]
23. Williams, J.A.; Bartoshuk, L.M.; Fillingim, R.B.; Dotson, C.D. Exploring Ethnic Differences in Taste Perception. *Chemse* **2016**, *41*, 449–456. [CrossRef] [PubMed]
24. Risso, D.S.; Giuliani, C.; Antinucci, M.; Morini, G.; Garagnani, P.; Tofanelli, S.; Luiselli, D. A bio-cultural approach to the study of food choice: The contribution of taste genetics, population and culture. *Appetite* **2017**, *114*, 240–247. [CrossRef] [PubMed]
25. Bertino, M.; Beauchamp, G.K.; Jen, K.C. Rated taste perception in two cultural groups. *Chem. Senses* **1983**, *8*, 3–15. [CrossRef]
26. Sobal, J. Cultural Comparison Research Designs in Food, Eating, and Nutrition. *Food Qual. Prefer.* **1998**, *9*, 385–392. [CrossRef]
27. Bertelsen, A.S.; Zeng, Y.; Mielby, L.A.; Sun, Y.-X.; Byrne, D.V.; Kidmose, U. Cross-modal Effect of Vanilla Aroma on Sweetness of Different Sweeteners among Chinese and Danish Consumers. *Food Qual. Prefer.* **2020**, *87*, 104036. [CrossRef]
28. Prescott, J. *Taste Matters: Why We Like the Foods We Do*; Reaktion Books: Islington, UK, 2013; ISBN 978-1-86189-951-4.
29. Parr, W.V.; Rodrigues, H. Cross-Cultural Studies in Wine Appreciation. In *Handbook of Eating and Drinking: Interdisciplinary Perspectives*; Meiselman, H.L., Ed.; Springer International Publishing: Cham, Switzerland, 2019; pp. 1–24. ISBN 978-3-319-75388-1.
30. Sáenz-Navajas, M.-P.; Ballester, J.; Pêcher, C.; Peyron, D.; Valentin, D. Sensory drivers of intrinsic quality of red wines: Effect of culture and level of expertise. *Food Res. Int.* **2013**, *54*, 1506–1518. [CrossRef]
31. Peng-Li, D.; Byrne, D.V.; Chan, R.C.K.; Janice Wang, Q. The influence of taste-congruent soundtracks on visual attention and food choice: A cross-cultural eye-tracking study in Chinese and Danish consumers. *Food Qual. Prefer.* **2020**, *85*, 103962. [CrossRef]
32. Choi, J.; Chung, S. Optimal sensory evaluation protocol to model concentration–response curve of sweeteners. *Food Res. Int.* **2014**, *62*, 886–893. [CrossRef]
33. R Core Team. *R: A Language and Environment for Statistical Computing*; R Foundation for Statistical Computing: Vienna, Austria, 2020.
34. RStudio Team. *RStudio: Integrated Development Environment for r*; Rstudio, Inc.: Boston, MA, USA, 2015.
35. Wickham, H.; Averick, M.; Bryan, J.; Chang, W.; McGowan, L.; François, R.; Grolemund, G.; Hayes, A.; Henry, L.; Hester, J.; et al. Welcome to the Tidyverse. *J. Open Source Softw.* **2019**, *4*, 1686. [CrossRef]
36. Husson, F.; Josse, J.; Le, S.; Mazet, J. FactoMineR: Multivariate Exploratory Data Analysis and Data Mining. Available online: https://CRAN.R-project.org/package=FactoMineR (accessed on 3 June 2020).

37. Husson, F.; Le, S.; Cadoret, M. SensoMineR: Sensory Data Analysis. Available online: https://CRAN.R-project.org/package=SensoMineR (accessed on 22 February 2020).
38. Kuznetsova, A.; Brockhoff, P.B.; Christensen, R.H.B.; Jensen, S.P. lmerTest: Tests in Linear Mixed Effects Models. Available online: https://CRAN.R-project.org/package=lmerTest (accessed on 6 May 2020).
39. Kuznetsova, A.; Brockhoff, P.B.; de Sousa Amorim, I.; Mielby, L.; Bech, S.; Ribeiro de Lima, R. SensMixed R package: Easy-to-use application with graphical user interface for analyzing sensory and consumer data within a mixed effects model framework. In Proceedings of the 11th Pangborn Sensory Science Symposium, Gothenburg, Sweden, 23–27 August 2015.
40. Lenth, R.V. Least-Squares Means: The R Package lsmeans. *J. Stat. Softw.* **2016**, *69*, 1–33. [CrossRef]
41. Wickham, H. *ggplot2: Elegant Graphics for Data Analysis; Use R!* 2nd ed.; Springer International Publishing: Cham, Switzerland, 2016; ISBN 978-3-319-24275-0.
42. Wickham, H.; Chang, W.; Henry, L.; Pedersen, T.L.; Takahashi, K.; Wilke, C.; Woo, K.; Yutani, H.; Dunnington, D.; RStudio. ggplot2: Create Elegant Data Visualisations Using the Grammar of Graphics. Available online: https://CRAN.R-project.org/package=ggplot2 (accessed on 3 April 2020).
43. Bower, J.A. *Statistical Methods for Food Science*, 1st ed.; John Wiley & Sons, Ltd.: Hoboken, NJ, USA, 2013; ISBN 978-1-118-54164-7.
44. Kassambara, A. rstatix: Pipe-Friendly Framework for Basic Statistical Tests. Available online: https://CRAN.R-project.org/package=rstatix (accessed on 12 June 2020).
45. Lenth, R.; Singmann, H.; Love, J.; Buerkner, P.; Herve, M. emmeans: Estimated Marginal Means, Aka Least-Squares Means. Available online: https://CRAN.R-project.org/package=emmeans (accessed on 12 June 2020).
46. Maechler, M.; Rousseeuw, P.; Struyf, A.; Hubert, M.; Hornik, K.; Studer, M.; Roudier, P.; Gonzalez, J.; Kozlowski, K.; Schubert, E.; et al. Cluster: "Finding Groups in Data": Cluster Analysis Extended Rousseeuw et al. Available online: https://CRAN.R-project.org/package=cluster (accessed on 3 April 2020).
47. Charrad, M.; Ghazzali, N.; Boiteau, V.; Niknafs, A. NbClust: An R Package for Determining the Relevant Number of Clusters in a Data Set. *J. Stat. Softw.* **2014**, *61*, 1–36. [CrossRef]
48. Tang, Y.; Horikoshi, M.; Li, W. ggfortify: Unified Interface to Visualize Statistical Results of Popular R Packages. *R J.* **2016**, *8*, 474–485. [CrossRef]
49. Schifferstein, H.N.J.; Frijters, J.E.R. Sensory integration in citric acid/sucrose mixtures. *Chem. Senses* **1990**, *15*, 87–109. [CrossRef]
50. McBride, R.L.; Macfie, H.J.H. *Psychological Basis of Sensory Evaluation*, 1st ed.; Elsevier Applied Science: London, UK; New York, NY, USA, 1990; ISBN 978-1-85166-453-5.
51. Pangborn, R.M.; Ough, C.S.; Chrisp, R.B. Taste Interrelationship of Sucrose, Tartaric Acid, and Caffeine in White Table Wine. *Am. J. Enol. Vitic.* **1964**, *15*, 154–161.
52. Belitz, H.-D. Fruits and Fruit Products. In *Food Chemistry*; Belitz, H.-D., Grosch, W., Schieberle, P., Eds.; Springer: Berlin/Heidelberg, Germany, 2009; pp. 807–861, ISBN 978-3-540-69934-7.
53. Yang, Q.; Williamson, A.-M.; Hasted, A.; Hort, J. Exploring the relationships between taste phenotypes, genotypes, ethnicity, gender and taste perception using Chi-square and regression tree analysis. *Food Qual. Prefer.* **2020**, *83*, 103928. [CrossRef]
54. Dinnella, C.; Monteleone, E.; Piochi, M.; Spinelli, S.; Prescott, J.; Pierguidi, L.; Gasperi, F.; Laureati, M.; Pagliarini, E.; Predieri, S.; et al. Individual Variation in PROP Status, Fungiform Papillae Density, and Responsiveness to Taste Stimuli in a Large Population Sample. *Chem. Senses* **2018**, *43*, 697–710. [CrossRef]

Article

Serving Temperatures of Best-Selling Coffees in Two Segments of the Brazilian Food Service Industry Are "Very Hot"

Ian C. C. Nóbrega [1,*], Igor H. L. Costa [1], Axel C. Macedo [1], Yuri M. Ishihara [1] and Dirk W. Lachenmeier [2,*]

1 Department of Food Engineering, Universidade Federal da Paraíba (UFPB), João Pessoa, Paraiba 58051-900, Brazil; igorhenr.98@gmail.com (I.H.L.C.); axeelmacedo@hotmail.com (A.C.M.); yuriufpb@yahoo.com.br (Y.M.I.)

2 Chemisches und Veterinäruntersuchungsamt (CVUA) Karlsruhe, Weissenburger Str. 3, 76187 Karlsruhe, Germany

* Correspondence: ian.nobrega@academico.ufpb.br (I.C.C.N.); lachenmeier@web.de (D.W.L.)

Received: 29 June 2020; Accepted: 31 July 2020; Published: 3 August 2020

Abstract: The International Agency for Research on Cancer has classified the consumption of "very hot" beverages (temperature >65 °C) as "probably carcinogenic to humans", but there is no information regarding the serving temperature of Brazil's most consumed hot beverage—coffee. The serving temperatures of best-selling coffee beverages in 50 low-cost food service establishments (LCFS) and 50 coffee shops (CS) were studied. The bestsellers in the LCFS were dominated by 50 mL shots of sweetened black coffee served in disposable polystyrene (PS) cups from thermos flasks. In the CS, 50 mL shots of freshly brewed espresso served in porcelain cups were the dominant beverage. The serving temperatures of all beverages were on average 90% and 68% above 65 °C in the LCFS and CS, respectively (P95 and median value of measurements: 77 and 70 °C, LCFS; 75 and 69 °C, CS). Furthermore, the cooling periods of hot water systems (50 mL at 75 °C and 69 °C in porcelain cups; 50 mL at 77 °C and 70 °C in PS cups) to 65 °C were investigated. When median temperatures of the best-selling coffees are considered, consumers should allow a minimum cooling time before drinking of about 2 min at both LCFS and CS.

Keywords: coffee; temperature; risk; food service industry; Brazil

---

1. Introduction

It is well established that Brazil is the world's largest producer and exporter of coffee [1,2], but coffee consumption aspects in the country are less known. It has been reported that coffee amounts to 71% of total revenue in the hot drinks market (coffee, tea and cocoa) [3] and is the most commonly consumed food in Brazil after rice [4]. With regard to per capita intakes of the beverage in Brazil, however, data are scattered, limited and not as straight forward to obtain as coffee production and exports. One study, though, based on a 2008–2009 dietary survey, estimated a daily coffee intake of 163 mL per person in Brazil [5] while another, based on data from the Brazilian Coffee Industry Association, estimated the value in 2016 as 220 mL [6] (p. 65). However, these intake values taken alone are out of perspective. Since two key data—namely the annual coffee consumption in kg by regions/selected countries and corresponding populations by age groups—are available [7,8], it is possible to put estimates of Brazilian per capita daily intakes into perspective.

Assuming coffee consumption data from the International Coffee Organization [7] is in green coffee form, we applied a 1.19 conversion factor to roasted coffee (1.19 kg green coffee = 1.0 kg of roasted coffee) [9], then a coffee-to-water ratio of 1 to 12.5 (8 g roasted coffee/100 mL water) [10,11]

(p. 25, p. 3) and distributed the values across populations aged 15 years and older [8] over 365 days. Taking these considerations into account, we estimate Brazilian per capita daily coffee intakes of 223 and 226 mL in 2015 and 2019, respectively. By applying the same approach to other areas, it can be stated with reasonable confidence that coffee intakes in Brazil were above all world regions in 2015 and 2019, for instance, Europe (144 and 153 mL), South America (140 and 140 mL) and Asia-Oceania (17 and 17 mL). Brazilian intakes are also higher than in countries such as the USA (169 and 177 mL), Japan (121 and 117 mL) and neighboring coffee producing Colombia (84 and 77 mL), but lower than Switzerland (251 and 262 mL) and Norway (343 and 348 mL) (see Supplementary Table S1).

In Brazil, retail prices of roasted coffee products reflect their quality certifications (in particular those issued by private coffee associations), such as "traditional" (lower price), "gourmet" and "especial"/specialty coffee (upper prices). Prices of coffee beverages in the Brazilian food service industry also vary with the segment. For instance, a 50 mL shot of espresso—which is typically brewed with "gourmet" or "especial" coffee—costs about BRL 6 in coffee shops; on the other hand, a 50 mL shot of filtered black coffee—commonly brewed with "traditional" coffee or even non-certified coffee—costs around BRL 1 in ordinary (low-cost) food service establishments. In the context of coffee consumption, these considerations are relevant because Brazil, an upper middle-income economy according to the World Bank [12], has a relatively low monthly household income per capita (BRL 1479 or USD 375 in 2019) [13] and 26.5% of its population live below the national poverty line income [14].

In 2016, the WHO's International Agency for Research on Cancer (IARC) classified the consumption of "very hot" beverages (temperature > 65 °C) as "probably carcinogenic to humans" (IARC group 2A), with esophageal cancer particularly associated with the drink habit in human epidemiological studies [6,15]. IARC classifications indicate the weight of the evidence as to whether an agent is capable of causing cancer (technically called "hazard") but it does not measure the likelihood that cancer will occur (technically called "risk"). Category 2A is used for an agent when there is limited evidence of carcinogenicity in humans and sufficient evidence of carcinogenicity in experimental animals [16]. It is important to highlight that 29 human epidemiological studies (on esophageal cancer and drinking very hot beverages) served as scientific evidence for the IARC classification in 2016, most of which related to the consumption of hot tea [6] (p. 451–465). Two prospective cohort studies published after the IARC classification, one by Yu et al. in 2018 [17] and another by Islami et al. in 2020 [18], again with tea, have further strengthened the association between consumption of very hot beverages and increased esophageal cancer risk; however, while the former study reported a positive association between hot tea drinking and esophageal cancer only among those with tobacco or heavy alcohol use, the latter found that the association was independent of these risk factors.

With regard to the risk of very hot beverages to humans, an important cancer risk assessment using the margin of exposure approach (considered the gold standard for assessing agents that are both genotoxic and carcinogenic) was published in 2018. The margin of exposure (MOE) is defined as the ratio between the point on the dose response curve which characterizes low but measurable harmful effects in animal studies (typically, the benchmark dose, or BMD, for a benchmark response of 10%) and the estimated human exposure to the agent, or simply put: MOE = BMD/exposure (clearly, the lower the MOE, the larger the risk for humans and a threshold of 10,000, or higher, is generally considered a low public health concern). The estimated MOE in the 2018 hot beverage study, using the BMD dose, was <1 (meaning that the human exposure had exceeded the dose that may cause a 10% cancer incidence in experimental animals) for individuals that drink as little as 60 mL of very hot beverages five times per week (calculated for a 60 kg person), suggesting very hot beverages pose a significant esophageal cancer risk even for the moderate drinking population [19].

According to Brazil's National Cancer Institute (INCA), a specialized body of the Ministry of Health, the age-standardized incidence rates of esophageal cancer in Brazil (2020 estimates) are approximately seven and three cases in 100,000 men and women, respectively. Furthermore, esophageal cancer is the sixth most common cancer site among men in Brazil (excluding non-melanoma skin cancer) and projections for 2020 indicate 8690 total new cases. The habits listed by INCA as risk factors for the disease are alcohol consumption, smoking and drinking very hot beverages [20] (p. 52; p. 42).

Although drinking a cup of hot coffee is such a common habit in many countries, there are only a very limited number of investigations of the actual serving or dispensing temperatures of the beverage in places such as households, food service industry and workplace, but two publications are worth mentioning. One is a study published in 2007 on the recorded serving temperatures of 164 samples of black coffees in quick service restaurants in the USA, including well-known chains such as McDonald's, Burger King, Wendy's and KFC; the measured temperatures of all samples were on average 9 °C above the IARC threshold for very hot beverages (mean and median of all measurements: 74 °C; standard deviation: 13 °C) [21]. The other is a German study published in 2018 on the recorded serving temperatures of 356 coffee beverages in the food service industry and the dispensing temperatures of 110 coffees in private households; the measured temperatures were on average 10 °C above the IARC threshold temperature both in the household and in the food service industry (mean value of all measurements: 75 °C; standard deviation: 5 °C) [22]. These high serving temperatures are probably related to the commonly recommended temperature of ~93 °C for brewing coffee [23].

Interestingly, it has been reported by several studies that the most preferred or recommended drinking temperatures of black coffee are often well below the mean serving temperatures encountered in the two mentioned studied (~75 °C), for instance (decimals were rounded to the near whole number), 68 °C [24], 72 °C [25], 60 °C [26], 58 °C [27] and 63 °C [28].

Based on the previous considerations, the aim of this paper is to provide an initial set of data on the serving temperatures of best-selling coffee-based hot beverages (e.g., espresso, cappuccino, filtered black coffee, etc.) from two socio-economic segments of the Brazilian food service industry: coffee shops and low-cost food service establishments. The results are mainly discussed in terms of the IARC threshold temperature and in light of practical recommendations for coffee consumers.

2. Materials and Methods

2.1. Laboratory

Experiments described in Sections 2.3 and 2.5 were conducted in the beverage lab of the Department of Food Engineering, UFPB University, Brazil. The laboratory is fully climatized with a controlled air temperature and its GPS coordinates are 7°08′32″ S 34°50′58″ W.

2.2. Field Work (Food Service Industry)

Experiments described in Section 2.4 were carried out between October and December 2019 in 100 food service establishments situated in João Pessoa, capital of the northeastern state of Paraíba, Brazil, which has a population of about 800,000. To obtain a reasonable representation of Brazilian coffee consumers from different socio-economic backgrounds, we sampled equal proportions of two segments of the food service establishments: coffee shops (CS) and low-cost food service establishments (LCFS). The selection process involved research using internet searches and site visits to identify potential establishments in a circular geographic area of approximately 250 km^2 from the laboratory (radius = 9 km).

LCFS were predefined as having these main characteristics: lower-price (up to BRL 2.00 per cup of coffee), a limited offer of coffee-based beverages (typically, no more than 3 types), counter or table service (typically counter service only) and located in lower middle- or lower-income neighborhoods (the typical choice was establishments within or surrounding open air markets or farmers markets). On the other hand, CS were predefined as having a large number of coffee options on the menu (typically more than 10), table service, prices up to BRL 15.00 per cup and located in upper middle- or high-income neighborhoods. All 100 sampled establishments were outside the university campus.

2.3. Calibration Checking of Thermometers

Temperatures were measured using a digital food thermometer device "El Corte Inglés" (EAN:2401662264904) with a stainless steel sensor, heat-resistant cable, range 0 to +300 °C and LCD display (El Corte Inglés, Madrid, Spain).

We assessed the calibration of the digital thermometer using a Braun thermostatic bath model 18 BU coupled with a Thermomix BM-S (B. Braun Biotech International, Melsungen, Germany), internal tank dimensions (length, width, height; mm) 500 × 290 × 130, filled with 14 L deionized water, and regulated at four different temperatures (55, 65, 75 and 85 °C). These temperatures were selected for calibration checking of the digital food thermometer because they cover the range of coffee serving temperatures found in the German food service industry [22].

Given possible inaccuracy in the water bath temperatures, we used, as a standard thermometer, a calibrated mercury filling glass thermometer (−10 °C to +110 °C range; 1 °C division; 303 mm length; Alla France, Chemillé, France) in parallel with the digital thermometer. The calibration of the glass thermometer was checked and confirmed in advance through 5 measurements of boiling ethanol (99.9% purity, boiling point 78 °C, J. T. Baker, Mexico) and chloroform (99.8% purity, boiling point 61 °C, Alphatec, Brazil).

Water bath temperatures from the glass thermometer (calibration standard) and the digital food thermometer were collected in 10 intraday replicate measurements at the same time and proximity in the thermostatic bath, and the coefficient of variation (CV) never exceeded 0.3%. By plotting the data (glass thermometer, x-axis; digital thermometer, y-axis) in a Microsoft Excel 2013 XY scatter graph, a mathematical relationship was obtained and expressed by the linear equation $y = 1.037x - 2.151$, with $R^2 = 0.9998$. The coefficient of determination, or R^2, indicates how well the trend line corresponds to the data (the closer the R^2 to 1, the better the fit). Although the results from the digital thermometer almost perfectly matched the standard glass thermometer, we decided to use the equation to make minor corrections in the temperatures from the digital thermometer (i.e., for a given "y" value, a corrected "x" value was obtained from the equation). Thus, the data from the digital thermometer stated in the results section are already corrected by the equation.

2.4. Temperature Measurements of Best-Selling Coffee-Based Hot Beverages in the Food Service Industry

On arrival at the coffee shop (CS) or low-cost food service establishment (LCFS), the research staff, acting as customers, either sat at a table (CS) or stood at the counter (LCFS), asked the attendant what the most frequently ordered coffee-based beverage (type and volume) was. The "bestselling" beverage was then ordered. The thermometer's sensor was immersed in the beverage as soon as it was handed over. Once the temperature in the thermometer stabilized (i.e., reached a plateau and started to fall; this took approximately 1 min), the upper value was recorded (one measure per coffee per establishment). No sugar, sweetener, milk or water was added to the beverage and therefore it represents the usual serving temperature, or immediate consumption temperature, in a "worst-case scenario". Besides coffee type/volume and temperature, information such as cup material (porcelain, glass, paper, plastic, etc.), beverage price, addition of sugar in the preparation of the beverage (sweetened or unsweetened), type of storage (in cases where the bestselling beverage had been brewed previously and stored ready-to-serve, a procedure typically used in LCFS), type of coffee machine (portafilter machine, capsule machine, etc., typically used in CS for coffee brewing) and ambient temperature were also recorded. Regarding specific construction materials of disposable plastic cups, the information was collected from plastic abbreviations (polypropylene, PP; polystyrene, PS, etc.) which is stamped at the bottom of every cup. The work's main objective ("to investigate the usual serving temperatures of coffee in the food service industry") was given to the attendant or manager afterwards. All ordered coffee beverages were paid for by the research staff.

2.5. Cooling Profile of Hot Water Systems to the IARC Threshold Temperature and Beyond

According to the Brewing Control Chart of the Specialty Coffee Association [29], the solubles concentration in brewed coffee ranges from 0.8% to 1.60%, with a median concentration of 1.20% (1.2 g in 100 mL), which corresponds to the brew strength that should be present in the so called "golden cup standard" [30]. By assuming 1.2% soluble concentration as the reference, a cup of brewed coffee is about 99% water, therefore pure water is a reasonable model to simulate brewed coffee in cooling behavior studies. Moreover, recent experimental results from Langer et al. [31] show that coffee has a similar cooling behavior as hot water.

Cooling behavior studies (temperature as a function of time) to 65 °C, and beyond, were carried out using hot deionized water under different conditions (water temperature, cup construction material and ambient temperature) to simulate hot coffee beverages found in the field work. A water volume of 50 mL was used for all cooling studies. As the working cup volume and materials, a 50 mL disposable white polystyrene (PS) cup (SM-050 model, Copobras S/A, João Pessoa, PB, Brazil; top external diameter = 52 mm; base external diameter = 34 mm; height = 41 mm; wall thickness ~0.3 mm) and a 50 mL white porcelain cup (prisma model, Schmidt Porcelain, Pomerode, SC, Brazil; top external diameter = 58 mm; base external diameter = 34 mm; height = 57 mm; wall thickness ~4.9 mm) were chosen. Initial water temperatures in PS cups of 77 °C and 70 °C, both under an ambient temperature of 30 °C, were chosen to represent higher and median coffee temperatures found in the low-cost food service establishments. Initial water temperatures in porcelain cups of 75 °C and 69 °C, both under ambient temperature of 25 °C, were chosen to simulate higher and median coffee temperatures found in the coffee shops.

We managed to achieve the mentioned initial water temperatures by placing the cup filled with 50 mL water in the thermostatic bath regulated at the specific temperatures. As the water tank depth was higher than the cup's height, we managed to keep it surrounded by water and at the same time stabilized by raising the tank bed with a support. Once the cup was stable, the sensor of the digital thermometer sensor was fixed in it. As the desired temperature in the cup was reached (77, 75, 70 or 69 °C), a chronometer was initiated and, right after, the cup was removed from the tank, wiped with a paper towel (external surface) and placed on the lab bench. Next, the sensor was fixed again in the cup and the cooling temperatures collected in one minute intervals up to 10 min. The procedure for each model system (initial temperature, cup type and ambient temperature) was carried out in three replicates and the CV never exceeded 1.9%.

3. Results

3.1. Coffee Temperatures in Low-Cost Food Service Establishments and Coffee Shops

Table 1 presents an overview of the bestsellers' serving temperatures from 50 low-cost food service establishments (LCFS). Detailed results from each establishment, such as coffee volume, price, type of storage (thermos flask or electric dispenser) and cup material, are presented in Table A1 (Appendix A).

Table 1. Summarized results of bestsellers' serving temperatures (°C) in 50 low-cost food service establishments (LCFS), according to coffee type [1].

Coffee Type [2]	Min. [3]	Max. [4]	Mean ± SD [5]	Median [6]	P95 [7]	% >65°C [8]
Total $n = 50$	62	78	70 ± 4	70	77	90
Sweetened black $n = 34$	62	78	69 ± 4	70	76	88
Unsweetened black $n = 13$	66	77	71 ± 3	71	77	100

Table 1. *Cont.*

Coffee Type [2]	Min. [3]	Max. [4]	Mean ± SD [5]	Median [6]	P95 [7]	% >65°C [8]
Sweetened milk coffee $n = 3$	64	74	69 ± 4	69	74	67

[1] For additional information on data collected from each LCFS (bestseller coffee, volume, price, cup material, etc.), see Table A1 in Appendix A; average ambient temperature in LCFS was 31 °C and all establishments were non-climatized (not equipped with a controlled air temperature). [2] The ordered beverage was always the establishment's bestseller, according to information provided by the attendant previously. [3] Minimum temperature (°C). [4] Maximum temperature (°C). [5] Mean temperature ± standard deviation (°C). [6] Median temperature (°C). [7] 95th percentile (°C). [8] Percentage of samples with serving temperatures above 65°C.

According to Table 1, the temperature range and median value of all best-selling coffees were 62–78 °C and 70 °C, respectively. A total of 90% of the collected values exceeded the IARC threshold of 65 °C. The 95th percentile was 77 °C (95% of temperature data are below this value) and it represents a reasonable estimate of a particular high temperature found in the LCFS. The bestsellers in the LCFS were dominated by sweetened black coffee (68% of total) followed by unsweetened black (26%) and sweetened milk coffee (6%).

It is of interest to highlight that all ordered best-selling coffees in the LCFS had been brewed in advance and stored ready-to-serve in either a thermos flask (82% of samples) or an electric hot coffee dispenser (28%). The mean ± standard deviation of all serving coffee volumes and prices (in Brazilian reais, BRL) was 86 ± 44 mL (median = 50 mL) and BRL 0.84 ± 0.45 (median = BRL 0.50), respectively. Regarding the serving cup material, 66% were made of disposable plastic (40%, polystyrene; 26%, polypropylene), 26% of glass and 8% of porcelain (Table A1).

With respect to the most common coffee in LCFS—sweetened black coffee (34 samples)—62% were ordered in 50 mL shots and 38% in 125–150 mL. The majority of the 50 mL shots of sweetened black coffee were served in 50 mL (approximate working volume) polystyrene (PS) cups (48%), followed by polypropylene (PP) of the same working volume (38%). Thus, a 50 mL shot of sweetened black coffee served at 70 °C in 50 mL PS cups is probably the best representative of the bestseller coffee in LCFS establishments (Table A1).

An overview of the bestsellers' serving temperatures from 50 coffee shops (CS) is shown in Table 2. Detailed results from each CS establishment, such as coffee volume, price, type of machine (used for coffee brewing) and cup material, are presented in Table A2 (Appendix A).

Table 2. Summarized results of bestsellers' serving temperatures (°C) in 50 coffee shops (CS), according to coffee type [1].

Coffee Type [2]	Min. [3]	Max. [4]	Mean ± SD [5]	Median [6]	P95 [7]	% >65°C [8]
Total $n = 50$	52	85	67 ± 7	69	75	68
Espresso $n = 33$	54	77	68 ± 6	69	75	76
Cappuccino $n = 14$	52	85	68 ± 9	71	79	57
Other coffee drinks $n = 3$	57	65	60 ± 3	59	64	33

[1] For additional information on data collected from each CS (bestseller coffee, volume, price, cup material, etc.), see Table A2 in Appendix A; average ambient temperature in CS was 25 °C and all establishments were fully climatized. [2] The ordered beverage was always the bestseller, according to information provided by the attendant previously. [3] Minimum temperature (°C). [4] Maximum temperature (°C). [5] Mean temperature ± standard deviation (°C). [6] Median temperature (°C). [7] 95th percentile (°C). [8] Percentage of coffee samples with serving temperatures above 65°C.

According to Table 2, the temperature range of all best-selling coffees in CS was 52–85 °C. Median (69 °C) and 95th percentile (75 °C) were close to the values found in LCFS. However, in comparison to LCFS, a smaller percentage (68%) of the collected temperatures exceeded the IARC threshold of 65 °C. With respect to cappuccino, particularly high maximum values (85 °C) and P95 (79 °C) were observed; it is noteworthy that hot beverages spills that occur over approximately 82 °C are clearly hazardous and likely to lead to mid-dermal burns which can be serious [32]. The bestsellers in the CS were dominated by espresso (66% of total) followed by cappuccino (28%) and other coffee-based beverages (6%) (Table 2).

All best-selling coffees in CS were freshly brewed on ordering and portafilter machines (espresso machine) were the dominant type (90%) in the establishments. The mean ± standard deviation of all serving coffee volumes and prices (in Brazilian reais, BRL) was 85 ± 59 mL (median = 50 mL) and BRL 5.41 ± 1.95 (median = BRL 4.50), respectively. Regarding the serving vessel material, 88% were made of porcelain and 12% of glass (Table A2).

With respect to the most common coffee in CS (espressos, 33 samples), most of them (97%) were ordered in 50 mL shots while the sizes of cappuccinos (14 samples) were more diverse (150–160 mL, 50%; 225–250 mL, 29%; 50 mL, 21%). A total of 94% of the espressos were extracted in portafilter machines. Regarding cup material, 88% of ordered coffees were served in porcelain, 12% in glass and all 50 mL shots of espresso were served in 50 mL (approximate working volume) porcelain cups. A 50 mL shot of espresso served at 69 °C in 50 mL porcelain cups is therefore the best representative of the bestseller coffee in CS establishments (Table A2).

3.2. Cooling Behavior of Hot Water Systems to the IARC Threshold Temperature and Beyond

Given that drinking beverages with temperatures above 65 °C is a habit classified by the IARC as "probably carcinogenic to humans", this section aimed to obtain practical results for coffee consumers regarding safe drinking. To achieve that, we studied the cooling behavior of hot water system representatives of coffees served at LCFS and CS, and these included variations in cup materials (disposable PS plastic and porcelain), ambient temperatures (30 °C and 25 °C) and initial temperatures (medians and P95). As the cooling process of a hot liquid poured into a drinking vessel is influenced by factors such as the thermal conductivity of the cup material, wall thickness of the cup, liquid temperature, cooling surface areas (of both the cup and open surface of the liquid) and ambient temperature, differences were expected to arise in the results.

Figure 1 shows the cooling behaviors of two hot water system representatives of coffees served at the LCFS. The two LCFS scenarios consisted of 50 mL water in PS cups at initial temperatures of 77 °C (P95) and 70 °C (median), both under a controlled ambient temperature of 30 °C.

Figure 2 presents the cooling behaviors of two hot water system representatives of coffees served at the CS. They consisted of 50 mL water in porcelain cups at initial temperatures of 75 °C (P95) and 69 °C (median), both under a controlled ambient temperature of 25 °C.

Despite the variations in factors affecting heat transfer in our systems, a relatively small difference was observed when the systems were studied at the P95 zone (77 °C/PS and 75 °C/porcelain). In this higher zone of temperature, the cooling time to the safe threshold (65 °C) was 3:12 min (Figure 1) and 2:48 min (Figure 2) in the PS and porcelain cup systems, respectively. Furthermore, when the initial temperatures of the systems were at the median zone (70 °C/PS and 69 °C/porcelain), the cooling behavior of the PS and porcelain cup systems were similar: both systems took 1:36 min to 65 °C (Figures 1 and 2), which may be rounded up to 2 min.

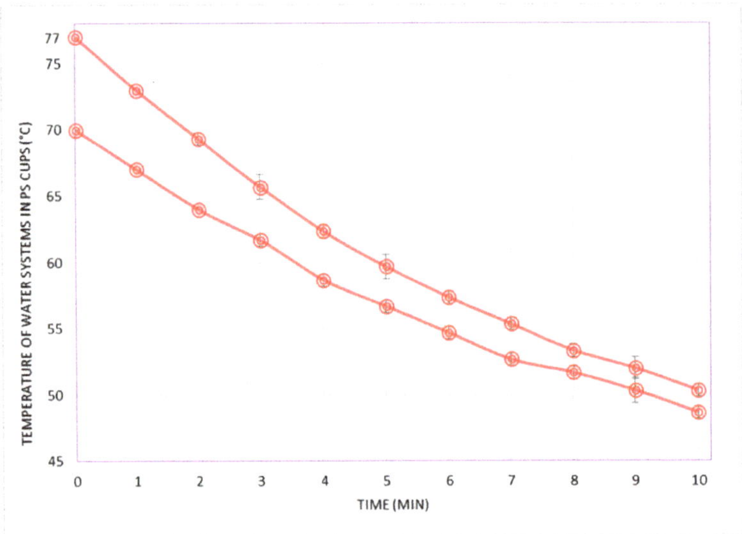

Figure 1. Cooling behaviors (temperature as a function of time) of hot water systems from 77 °C and 70 °C, both in disposable 50 mL polystyrene (PS) cups and at a constant ambient temperature of 30 °C. Temperatures of the water systems were collected in one minute intervals up to 10 min. Point values are shown as the mean of 3 repetitions ± standard deviations of each experiment and the coefficient of variation never exceeded 1.9%.

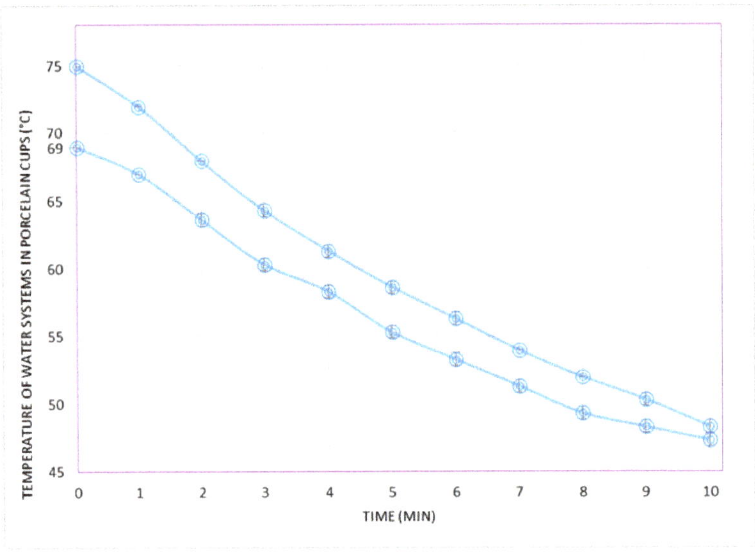

Figure 2. Cooling behaviors (temperature as a function of time) of hot water systems from 75 °C and 69 °C, both in 50 mL porcelain cups and at a constant ambient temperature of 25 °C. Temperatures of the water systems were collected in one minute intervals up to 10 min. Point values are shown as mean of 3 repetitions ± standard deviations of each experiment and the coefficient of variation never exceeded 1.0%.

Although the results suggest similar to close cooling behavior from the porcelain and PS systems, the thermal conductivities of porcelain and PS are 1.5 W m^{-1} K^{-1} [33] and 0.12 W m^{-1} K^{-1} [34], respectively, thus PS insulates about 12 times better than porcelain. On the other hand, the wall thickness of the porcelain and PS vessels in our study are 4.9 and 0.3 mm, respectively, therefore the porcelain cup would have a 16-fold insulation capability advantage in this respect. To complicate comparisons further, the ambient temperatures under which the systems were studied and the shapes of the cups (hence their cooling surface areas to the outer surface) are different. However, in an attempt to clarify the differences in insulation capabilities of the PS and porcelain cups used in our study, we decided to carry out an experiment in which all factors affecting heat transfer remained constant (initial temperature and ambient temperature) except the cups themselves. Figure 3 shows the cooling behaviors of two hot water systems from 77 °C in 50 mL porcelain and polystyrene (disposable) cups, both at an ambient temperature of 25 °C.

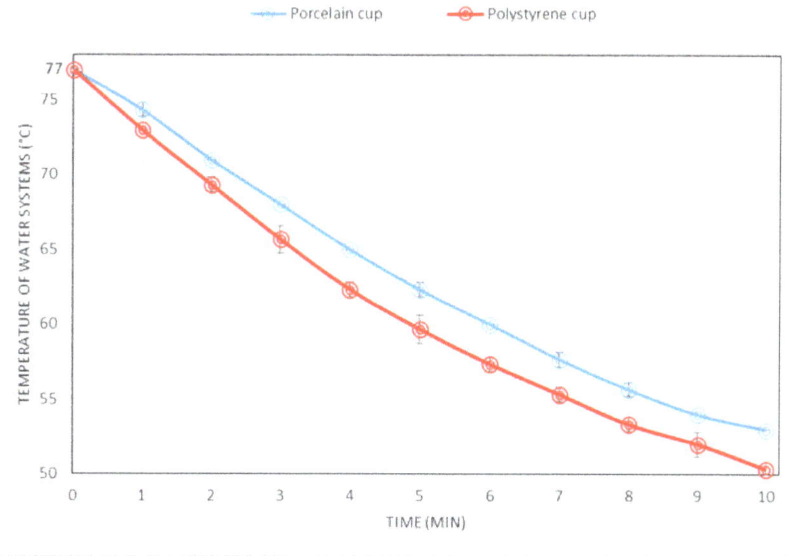

Figure 3. Cooling behavior (temperature as a function of time) of hot water systems from 77 °C in 50 mL porcelain and polystyrene (disposable) cups, both at ambient temperature of 25 °C. Temperatures of the water systems were collected in one minute intervals up to 10 min. Point values are shown as mean of 3 repetitions ± standard deviations of each experiment and the coefficient of variation never exceeded 1.6%.

Under our experimental conditions and once the temperatures of the water and the cups are equalized, Figure 3 shows that the porcelain cup is a better insulator than the PS cup. The cooling time from 77 °C to 65 °C was about 3:12 min and 4:00 min in the PS and porcelain cup systems, respectively (48 s difference). Interestingly, the differences in insulating capabilities of the cups over time (up to 10 min) were more noticeable; for instance, the cooling time to 60 °C was approximately 4:48 and 6:00 min in the PS and porcelain cup systems, respectively (1:12 min difference). In brief, the results in Figure 3 suggest that the larger wall thickness of the porcelain cup probably played a major role in compensating the higher thermal conductivity of porcelain (compared to the PS material). However, for a proper and detailed comparison of the insulation capabilities of the cups, their shape would have to be the same, but this is beyond the scope of the present study.

4. Discussion

This study recorded the serving temperatures of the best-selling coffee-based beverages in 100 food service establishments (50 low-cost food service and 50 coffee shops) in Brazil. The bestsellers in the low-cost food service establishments (LCFS) were dominated by 50 mL shots of sweetened black coffee, which were stored ready-to-drink in thermos flasks and typically served in 50 mL disposable polystyrene cups. In the coffee shops (CS), a 50 mL shot of freshly extracted espresso from portafilter machines, served in 50 mL porcelain cups, was the dominant beverage. The results show that the majority of best-selling coffee beverages are served very hot (>65 °C) in both segments of the food service industry, but a much larger proportion was observed in the LCFS (90%) than in the CS (68%).

This is the first published study on the serving temperature of coffee beverages in the Brazilian food service industry, thus comparisons to previous local studies are not possible at present. In the last 20 years, two research papers on the serving temperature of coffee beverages in food service establishments have been published, one carried out in Germany [22] and the other in the USA [21], but the German establishments seem to be the only ones that bear resemblances to the CS segment studied here (the establishments assessed in the USA were quick service restaurants, such as McDonald's, Burger King and KFC). However, the other segment of the Brazilian food service industry—the LCFS—is probably unique and therefore incomparable to the German and American establishments.

The mean value (± standard deviation) of the measured temperatures in the Brazilian CS (67 ± 7 °C; 50 samples) was below the value found in the German food service industry (75 °C ± 5 °C; 356 samples) and, as a result, the latter achieved a higher percentage of samples above 65°C (98%) [22]. Although the German study measured a much larger number of samples, the types of coffees were not disclosed, therefore specific comparisons cannot be made. However, since 83% of the machines used for coffee brewing in the German food service industry were either portafilter (37%) or fully automated machines (46%), the measured samples were probably Americano, Café crème, espressos or milk-based espresso drinks, such as cappuccinos and lattes. The coffee samples obtained specifically from portafilter machines showed an even higher mean value (77°C ± 5 °C) in the German study; a proposed explanation was the fact that the portafilter machines exhibited pressure tank settings of extremely high temperatures (120 °C), information which is not available in the CS.

Although the majority of the coffee samples in the CS were above the IARC threshold of 65 °C, it can be said that, with differences between the German study and this research kept apart, the Brazilian CS scenario of coffee temperatures is less worrying for consumers than the situation found in the German food service industry. Moreover, most coffee (68%) drunk in the CS were espressos (arithmetic mean, 68 °C; median, 69 °C), therefore it is likely that by simply adding sugar to the beverage (a very common habit in the Brazilian coffee culture), the temperature will drop from 69 °C to below 65 °C. However, in a worst-case scenario (i.e., no addition of sugar and no waiting time once the drink is handed over), the serving temperatures in the CS still pose a hazard to Brazilian consumers.

Regarding the hazard, our findings on the cooling time of water systems representing a typical coffee in the CS (50 mL shot of espresso served at 69 °C in 50 mL porcelain cups and at an ambient temperature of 25 °C) show that consumers should allow a minimum cooling time to 65 °C of 1:36 min, as seen in Figure 2. However, by taking into consideration the recent suggestion showing that consumers perceive coffee most preferable at temperatures of about 63 °C [28], compounded by the fact that the coffee roast beans used in the CS are probably of high sensory quality ("gourmet" or even "specialty" coffee beans), consumers of espresso in CS would enjoy the beverage to the maximum by waiting 2:12 min instead of 1:36. All things considered, including a certain level of uncertainty, a minimum cooling time before drinking of approximately 2 min is recommended.

In comparison to the CS, the distinction of the coffee commonly served in the LCFS (sweetened black coffee stored in thermos flasks) seems to present an increased hazard to Brazilian consumers, as discussed next.

It has been reported by a cross-sectional population-based survey that the most common coffee preparation method in Brazilian households is filter brewing (86%), with a reusable cloth strainer as the dominant type of filter (63%) followed by disposable paper (23%) [35]. The survey also reported that in many circumstances, coffee is prepared in family-size quantities and stored with sugar in thermos flasks in households, a situation we also found in the LCFS establishments.

Although the preparation methods behind the coffees were not specifically targeted in this research, it is reasonable to assume that the sweetened black coffee offered in the LCFS follows the brewing procedure typically carried out in many Brazilian households (i.e., cloth filter brewing, sugar addition and storage in thermos), but details on the preparation were not quite clear. However, a search in a video-sharing platform for "traditional Brazilian cloth-filter brewing coffee recipes", associated with the addition of sugar and storage in thermos flasks, offered some details of the procedure. In brief, a standard recipe for such coffee typically consists of the following: (1) add ~750 mL water into a pot; (2) heat the pot on a stove until the contents simmer; (3) add ~35 g of ordinary ground coffee (fine to medium-fine grind size) and ~120 g granulated sugar to the simmering water and stir; (4) bring the brew to boiling or near boiling (the mixture will froth up at this point); (5) remove the pot from the heat and (6) pour the sweetened hot coffee brew through a reusable cloth strainer (with the handle) directly into a thermos flask of adequate volume. After straining, the yield is about 700 mL of cloth-filtered sweetened black coffee. Instead of adding the sugar together with the coffee into the pot, it may be added between steps 5 and 6. Moreover, a paper filter in a holder may be used instead of a cloth strainer. According to the recipes, the concentration of sugar is about 15 g in 100 mL and the coffee-to-water ratio is 1 to 21 (4.7 g ground coffee/100 mL water). It is of interest to mention that in some LCFS establishments, there were the alternatives of sweetened milk coffee and unsweetened black coffee ready-to-serve in thermos flasks (in parallel to sweetened black coffee), thus we understand that these beverages probably follow the brewing procedure of the sweetened black coffee with minor changes in the recipe.

Due to the nature of the coffee normally used in the traditional recipe of sweetened black coffee (lower price; lower sensory grades of certifications or even non-certified coffee; dark to very dark roasts; etc.), compounded by the brewing procedure itself, the resulting beverage would probably taste quite bitter had no generous quantities of sugar been included in the recipe. Thus, the likely role of sugar in the recipe besides bringing sweetness is to balance the harshness of the brew.

The procedure likely adopted in LCFS for sweetened black coffee brewing poses two main problems for consumers. The first is that the thermos is likely filled with filtered sweetened black coffee in scalding temperatures, thus the first consumers may experience an extremely hot beverage, particularly if the thermos is preheated with boiling water. The second is that the beverage is prepared to be ready to serve and ready to drink, therefore once poured into a cup, there will be no impacting cooling aid into the system (for instance, the addition of sugar) other than the heat loss from the cup system over time. Cold milk was not mentioned as a cooling aid because in Brazil, contrary to many European countries, if milk is to be added to a cup of black coffee, it is often heated beforehand.

Regarding the heat loss, when coffee is poured into a cup, it undergoes a complex transient cooling process, which is determined by the following superimposed sub-processes: heat transfer to the inner surface of the cup, heat conduction and heating and subsequent cooling of the cup material, heat transfer to the outer surface of the cup by convection and radiation and heat transfer to the open surface of the coffee by convection, radiation and evaporation [22]. In turn, the heat transfer processes are influenced by factors such as the type of cup material, its width, the heat transfer area of the cup and the ambient temperature.

Our findings on the cooling time of water systems representing a typical coffee in the LCFS (50 mL shot of sweetened black coffee at 70 °C in 50 mL PS cups) show that consumers should allow a minimum cooling time to 65 °C of approximately 2 min, as seen in Figure 1.

Although the cooling studies of water systems at 77 °C has showed that a standard 50 mL disposable PS cup cools down more easily than a standard 50 mL porcelain cup, it should also be considered that in a real coffee system, the presence of generous quantities of sugar in the beverage (such as the case of sweetened black coffee served in the LCFS) may have an impact on heat loss. This is because the thermal conductivity of an aqueous solution decreases as the concentration of sucrose increases. For instance, the thermal conductivity of pure water at 60 °C is 0.654 W m^{-1} K^{-1} [33] (p. 694) while that of a 20% (m/m) sucrose solution at same temperature is 0.581 W m^{-1} K^{-1} [36] (p. 560). It would be interesting to carry out further cooling investigations on the effect of sugar (already present in a solution or by adding it afterwards) in hot water systems.

Apart from additional investigations of the effect of sugar on cooling, a number of future researches may be carried out from this point. First, and most importantly, data collection on coffee temperature should be expanded to Brazilian households and workplaces, but a second round of data from the food service industry, preferably in a different Brazilian location, would also be interesting to confirm our initial findings. In addition, research should be done on the drinking behaviors of Brazilian coffee consumers (how many cups, how fast it is drunk, what is the frequency, what is added to the coffee and how much, where it is drunk, among others). Finally, wherever possible, future research should also investigate the coffee method preparations on all consumption segments (food service, households and workplace), so that a full picture of coffee consumption can be drawn with corresponding intervention measures proposed.

A more complete picture of coffee consumption habits and the temperature at which consumption commonly occurs in Brazil would also present an excellent opportunity for an esophageal cancer risk assessment of hot beverages consumption in the country, with special consideration to coffee.

Supplementary Materials: The following are available online at http://www.mdpi.com/2304-8158/9/8/1047/s1, Table S1: Estimates of per capita (15+) daily intakes of coffee (in ml of beverage) in Brazil compared to selected countries and world regions, 2015 and 2019.

Author Contributions: I.C.C.N. developed the conception and design of the study, coordinated the research project, supervised I.H.L.C. and A.C.M. and drafted the first version of the manuscript; I.H.L.C. and A.C.M. performed the experiments, organized data acquisition and contributed to the interpretation of the findings; Y.M.I. developed the conception and design of study, co-supervised I.H.L.C. and contributed to the interpretation of the findings; D.W.L. participated in drafting and revising the manuscript and contributed to the interpretation of the findings. All authors have read and agreed to the published version of the manuscript.

Funding: This research received no external funding.

Acknowledgments: I.C.C.N. and Y.M.I. thank the UFPB through Department of Food Engineering (Project No. 23074.086154/2019-31) for covering the basic cost of this research (staff salaries, lab structure and lab maintenance). Marciane Magnani, Stela Mendonça, Anoar El Aouar, Helenice Holanda and Julice Lopes (UFPB) are thanked for their technical assistance and/or helpful advice. Alan Cull (UK) is thanked for a dedicated English language check and corrections of the text.

Conflicts of Interest: The authors declare no conflict of interest.

Appendix A

Table A1. Serving temperatures and other information of best-selling coffee beverages in 50 LCFS.

LCFS	Coffee Type [1]	V (mL) [1]	Price (BRL) [2]	Storage [3]	Cup Material [4]	T (°C) [5]
01	Sweetened milk coffee	125	0.50	Thermos f.	Glass	69
02	Sweetened black	50	0.50	Thermos f.	Glass	68
03	Sweetened milk coffee	50	0.50	Thermos f.	Polystyrene	74
04	Sweetened black	125	0.50	Thermos f.	Porcelain	67
05	Sweetened black	50	0.50	Thermos f.	Polypropylene	75
06	Sweetened black	50	0.50	Electric d.	Porcelain	70
07	Sweetened black	50	0.50	Thermos f.	Polystyrene	77
08	Unsweetened black	50	1.00	Thermos f.	Polystyrene	71
09	Sweetened milk coffee	125	0.50	Thermos f.	Glass	64
10	Sweetened black	50	0.50	Thermos f.	Polystyrene	70
11	Sweetened black	50	0.50	Thermos f.	Polypropylene	78
12	Sweetened black	50	0.50	Thermos f.	Glass	63
13	Unsweetened Black	80	0.50	Electric d.	Glass	70

Table A1. *Cont.*

LCFS	Coffee Type [1]	V (mL) [1]	Price (BRL) [2]	Storage [3]	Cup Material [4]	T (°C) [5]
14	Unsweetened Black	50	0.50	Electric d.	Polystyrene	77
15	Sweetened black	125	1.00	Electric d.	Glass	68
16	Sweetened black	150	1.50	Thermos f.	Glass	72
17	Sweetened black	50	1.50	Thermos f.	Polystyrene	71
18	Sweetened black	150	1.00	Thermos f.	Polypropylene	75
19	Sweetened black	50	0.50	Thermos f.	Polystyrene	67
20	Sweetened black	50	0.50	Thermos f.	Polystyrene	64
21	Sweetened black	150	1.00	Thermos f.	Polystyrene	68
22	Sweetened black	150	1.50	Thermos f.	Glass	71
23	Sweetened black	50	0.50	Thermos f.	Polypropylene	74
24	Sweetened black	50	0.50	Thermos f.	Polypropylene	73
25	Sweetened black	50	0.50	Thermos f.	Polypropylene	70
26	Sweetened black	150	1.00	Thermos f.	Polypropylene	71
27	Sweetened black	50	0.50	Thermos f.	Polystyrene	69
28	Sweetened black	150	1.00	Thermos f.	Glass	68
29	Sweetened black	50	1.00	Thermos f.	Polypropylene	72
30	Sweetened black	50	0.50	Thermos f.	Polystyrene	71
31	Sweetened black	50	0.50	Thermos f.	Polypropylene	72
32	Sweetened black	150	0.50	Thermos f.	Polypropylene	71
33	Sweetened black	50	0.50	Thermos f.	Polypropylene	68
34	Sweetened black	50	0.50	Thermos f.	Polystyrene	69
35	Sweetened black	50	0.50	Thermos f.	Polystyrene	70
36	Sweetened black	150	1.00	Thermos f.	Polystyrene	67
37	Unsweetened black	150	0,50	Thermos f.	Polystyrene	66
38	Sweetened black	125	1.00	Thermos f.	Glass	62
39	Unsweetened black	125	1.00	Thermos f.	Glass	67
40	Unsweetened black	50	0.50	Thermos f.	Polystyrene	72
41	Unsweetened black	150	1.00	Thermos f.	Polystyrene	71
42	Unsweetened black	50	1.50	Electric d.	Polystyrene	73
43	Unsweetened black	50	0.50	Electric d.	Polystyrene	76
44	Unsweetened black	50	1.00	Electric d.	Polypropylene	69
45	Unsweetened black	125	2.00	Electric d.	Porcelain	70
46	Sweetened black	150	1.00	Thermos f.	Polypropylene	68
47	Unsweetened black	125	2.00	Thermos f.	Glass	71
48	Sweetened black	50	1.50	Thermos f.	Polystyrene	68
49	Sweetened black	125	1.50	Electric d.	Glass	63
50	Unsweetened black	50	2.00	Thermos f.	Porcelain	73

[1] The ordered beverage (type and volume) was always the bestseller, according to information provided by the attendant previously. [2] Unit price in Brazilian reais (BRL or R$). [3] All best-selling beverages were brewed previously and stored ready-to-serve in a thermos flask (thermos f.) or electric hot coffee dispenser (electric d.) [4] Plastic cups were disposable. [5] Corrected temperatures in °C, according to a previous calibration curve.

Table A2. Serving temperatures and other information of best-selling coffee beverages in 50 CS.

CS	Coffee Type [1]	V (mL) [1]	Price (BRL) [2]	Machine [3]	Cup Material	T (°C) [4]
51	Brigadeiro coffee [5]	170	10.50	Portafilter m.	Porcelain	59
52	Filtered coffee	120	4.50	Filter m.	Porcelain	65
53	Filtered coffee	50	4.50	Capsule m.	Porcelain	57
54	Cappuccino	50	5.50	Portafilter m.	Porcelain	74
55	Cappuccino	250	5.50	Portafilter m.	Glass	70
56	Cappuccino	150	8.00	Portafilter m.	Glass	85
57	Cappuccino	150	8.00	Portafilter m.	Porcelain	75
58	Cappuccino	225	8.00	Portafilter m.	Glass	63
59	Cappuccino	50	6.50	Portafilter m.	Porcelain	63
60	Cappuccino	50	5.50	Portafilter m.	Glass	75
61	Cappuccino	160	8.00	Portafilter m.	Porcelain	75
62	Cappuccino	160	11.00	Portafilter m.	Porcelain	71

Table A2. *Cont.*

CS	Coffee Type [1]	V (mL) [1]	Price (BRL) [2]	Machine [3]	Cup Material	T (°C) [4]
63	Cappuccino	150	8.00	Portafilter m.	Porcelain	61
64	Cappuccino	150	8.00	Portafilter m.	Porcelain	57
65	Cappuccino	150	8.50	Portafilter m.	Porcelain	57
66	Cappuccino	225	10.00	Portafilter m.	Glass	52
67	Cappuccino	225	8.00	Capsule m.	Glass	71
68	Espresso	150	4.50	Portafilter m.	Porcelain	70
69	Espresso	50	4.00	Portafilter m.	Porcelain	71
70	Espresso	50	4.00	Portafilter m.	Porcelain	61
71	Espresso	50	4.50	Portafilter m.	Porcelain	62
72	Espresso	50	6.00	Portafilter m.	Porcelain	72
73	Espresso	50	4.00	Portafilter m.	Porcelain	57
74	Espresso	50	5.50	Portafilter m.	Porcelain	73
75	Espresso	50	5.80	Portafilter m.	Porcelain	70
76	Espresso	50	4.00	Portafilter m.	Porcelain	60
77	Espresso	50	4.50	Portafilter m.	Porcelain	72
78	Espresso	50	4.00	Capsule m.	Porcelain	66
79	Espresso	50	4.00	Portafilter m.	Porcelain	71
80	Espresso	50	3.20	Portafilter m.	Porcelain	71
81	Espresso	50	4.50	Portafilter m.	Porcelain	71
82	Espresso	50	4.00	Portafilter m.	Porcelain	77
83	Espresso	50	4.00	Portafilter m.	Porcelain	69
84	Espresso	50	4.00	Portafilter m.	Porcelain	75
85	Espresso	50	4.00	Portafilter m.	Porcelain	68
86	Espresso	50	4.00	Portafilter m.	Porcelain	69
87	Espresso	50	5.00	Portafilter m.	Porcelain	66
88	Espresso	50	4.50	Portafilter m.	Porcelain	69
89	Espresso	50	4.00	Portafilter m.	Porcelain	65
90	Espresso	50	4.00	Portafilter m.	Porcelain	70
91	Espresso	50	4,00	Portafilter m.	Porcelain	67
92	Espresso	50	4.00	Portafilter m.	Porcelain	66
93	Espresso	50	5.50	Portafilter m.	Porcelain	74
94	Espresso	50	4.00	Portafilter m.	Porcelain	73
95	Espresso	50	5.00	Portafilter m.	Porcelain	55
96	Espresso	50	4.00	Portafilter m.	Porcelain	69
97	Espresso	50	4.00	Portafilter m.	Porcelain	74
98	Espresso	50	4.00	Capsule m.	Porcelain	68
99	Espresso	50	4.00	Portafilter m.	Porcelain	56
100	Espresso	50	4.00	Portafilter m.	Porcelain	54

[1] The ordered beverage (type and volume) was always the bestseller, according to information provided by the attendant previously. [2] Unit price in Brazilian reais (BRL). [3] Machine type used in coffee brewing. [4] Corrected temperatures, according to a previous calibration curve. [5] According to menu, "brigadeiro coffee" is a beverage consisting of espresso, milk and a typical Brazilian sweet made from cooked condensed milk and cocoa powder.

References

1. World Coffee Consumption. International Coffee Organization. Available online: http://www.ico.org/prices/po-production.pdf (accessed on 12 May 2020).
2. Exports of All Forms of Coffee by All Exporting Countries. International Coffee Organization. Available online: http://www.ico.org/historical/1990%20onwards/PDF/2a-exports.pdf (accessed on 12 May 2020).
3. Hot Drinks. Brazil. Statista Market Forecast. Available online: https://www.statista.com/outlook/30000000/115/hot-drinks/brazil (accessed on 16 May 2020).
4. Souza, A.M.; Pereira, R.A.; Yokoo, E.M.; Levy, R.B.; Sichieri, R. Most consumed foods in Brazil: National Dietary Survey 2008–2009. *Rev. Saúde Pública* **2013**, *47* (Suppl. 1), 190s–199s. [CrossRef] [PubMed]
5. Sousa, A.G.; Costa, T.H.M. Usual coffee intake in Brazil: Results from the National Dietary Survey 2008–9. *Br. J. Nutr.* **2015**, *113*, 1615–1620. [CrossRef] [PubMed]
6. IARC Working Group on the Evaluation of Carcinogenic Risks to Humans. Coffee, mate, and very hot beverages. *IARC Monogr. Eval. Carcinog. Risks Hum.* **2018**, *116*, 1–501.

7. World Coffee Consumption in Thousand 60-kg Bags. International Coffee Organization. Available online: http://www.ico.org/prices/new-consumption-table.pdf (accessed on 10 May 2020).
8. Population by Broad Age Groups—Both Sexes. De Facto Population as of 1 July of the Year Indicated. Department of Economic and Social Affairs. United Nations. Available online: https://population.un.org/wpp/Download/Standard/Population/ (accessed on 11 May 2020).
9. ICC-102-10. Rules on Statistics Statistical Reports. International Coffee Organization. Available online: http://www.ico.org/documents/icc-102-10e-rules-statistical-reports-final.pdf (accessed on 12 May 2020).
10. Farah, A. Coffee constituents. In *Coffee: Emerging Health Effects and Disease Prevention*, 1st ed.; Chu, Y.-F., Ed.; Wiley-Blackwell: Oxford, UK, 2012; pp. 21–58. [CrossRef]
11. Farah, A.; de Paula Lima, J. Consumption of chlorogenic acids through coffee and health implications. *Beverages* **2019**, *5*, 11. [CrossRef]
12. World Bank. Country and Lending Groups. Available online: https://datahelpdesk.worldbank.org/knowledgebase/articles/906519-world-bank-country-and-lending-groups (accessed on 10 May 2020).
13. IBGE Divulga O Rendimento Domiciliar Per Capita 2019. Brazilian Institute of Geography and Statistics (IBGE). Available online: https://agenciadenoticias.ibge.gov.br/agencia-sala-de-imprensa/2013-agencia-de-noticias/releases/26956-ibge-divulga-o-rendimento-domiciliar-per-capita-2019 (accessed on 15 May 2020).
14. Multidimensional Poverty Index. Human Development Indices: A statistical update 2019. United Nations. Available online: http://data.un.org/DocumentData.aspx?q=brazil+poverty&id=421 (accessed on 8 May 2020).
15. Loomis, D.; Guyton, K.Z.; Grosse, Y.; Lauby-Secretan, B.; El, G.F.; Bouvard, V.; Benbrahim-Tallaa, L.; Guha, N.; Mattock, H.; Straif, K.; et al. Carcinogenicity of drinking coffee, mate, and very hot beverages. *Lancet Oncol.* **2016**, *17*, 877–878. [CrossRef]
16. IARC Monographs Questions and Answers. International Agency for Research on Cancer. Available online: https://www.iarc.fr/wp-content/uploads/2018/07/Monographs-QA.pdf (accessed on 16 May 2020).
17. Yu, C.; Tang, H.; Guo, Y.; Bian, Z.; Yang, L.; Chen, Y.; Tang, A.; Zhou, X.; Yang, X.; Chen, J.; et al. Hot tea consumption and its interactions with alcohol and tobacco use on the risk for esophageal cancer: A population-based cohort study. *Ann. Intern. Med.* **2018**, *168*, 489–497. [CrossRef] [PubMed]
18. Islami, F.; Poustchi, H.; Pourshams, A.; Khoshnia, M.; Gharavi, A.; Kamangar, F.; Dawsey, S.M.; Abnet, C.C.; Brennan, P.; Sheikh, M.; et al. A prospective study of tea drinking temperature and risk of esophageal squamous cell carcinoma. *Int. J. Cancer* **2020**, *146*, 18–25. [CrossRef] [PubMed]
19. Okaru, A.O.; Rullmann, A.; Farah, A.; Gonzalez de Mejia, E.; Stern, M.C.; Lachenmeier, D.W. Comparative oesophageal cancer risk assessment of hot beverage consumption (coffee, mate and tea): The margin of exposure of PAH vs. very hot temperatures. *BMC Cancer* **2018**, *18*, 236. [CrossRef] [PubMed]
20. Incidência De Câncer no Brasil. Estimativa 2020. INCA. Available online: https://www.inca.gov.br/sites/ufu.sti.inca.local/files//media/document//estimativa-2020-incidencia-de-cancer-no-brasil.pdf (accessed on 16 June 2020).
21. Borchgrevink, C.P.; Sciarini, M.P.; Susskind, A.M. Hot beverages at quick service restaurant (QSR) drive-thru windows. *J. Hosp. Manag. Tour.* **2007**, *14*, 37–46. [CrossRef]
22. Verst, L.-M.; Winkler, G.; Lachenmeier, D.W. Dispensing and serving temperatures of coffee-based hot beverages. Exploratory survey as a basis for cancer risk assessment. *Ernahr. Umsch.* **2018**, *65*, 64–70. [CrossRef]
23. Abraham, J.; Diller, K. A review of hot beverage temperatures-satisfying consumer preference and safety. *J. Food Sci.* **2019**, *84*, 2011–2014. [CrossRef] [PubMed]
24. Borchgrevink, C.P.; Susskind, A.M.; Tarras, J.M. Consumer preferred hot beverage temperatures. *Food Qual. Prefer.* **1999**, *10*, 117–121. [CrossRef]
25. Pipatsattayanuwong, S.; Lee, H.S.; Lau, S.; O'Mahony, M. Hedonic R-index measurement of temperature preferences for drinking black coffee. *J. Sens. Stud.* **2001**, *16*, 517–536. [CrossRef]
26. Lee, H.S.; O'Mahony, M. At what temperatures do consumers like to drink coffee? Mixing methods. *J. Food Sci.* **2002**, *67*, 2774–2777. [CrossRef]
27. Brown, F.; Diller, K.R. Calculating the optimum temperature for serving hot beverages. *Burns* **2008**, *34*, 648–654. [CrossRef] [PubMed]
28. Dirler, J.; Winkler, G.; Lachenmeier, D.W. What temperature of coffee exceeds the pain threshold? Pilot study of a sensory analysis method as basis for cancer risk assessment. *Foods* **2018**, *7*, 83. [CrossRef] [PubMed]

29. Specialty Coffee Association. Brewing Control Chart. Available online: https://store.sca.coffee/products/brewing-control-chart-grams-per-liter?variant=14732978758 (accessed on 20 July 2020).
30. Specialty Coffee Association. Brewing standards. Available online: https://sca.coffee/research/coffee-standards (accessed on 19 July 2020).
31. Langer, T.; Winkler, G.; Lachenmeier, D.W. Studies on the cooling behavior of hot drinks against the background of temperature-related cancer risk. *Deut. Lebensm. Rundsch.* **2018**, *114*, 307–314. [CrossRef]
32. Abraham, J.P.; Nelson-Cheeseman, B.B.; Sparrow, E.M.; Wentz, J.E.; Gorman, J.M.; Wolf, S.E. Comprehensive method to predict and quantify scald burns from beverage spills. *Int. J. Hyperth.* **2016**, *32*, 900–910. [CrossRef] [PubMed]
33. Spiess, W.E.L.; Walz, E.; Nesvadba, P.; Morley, M.; van Haneghem, I.A.; Salmon, D.R. Thermal conductivity of food materials at elevated temperatures. *High Temp. High Press.* **2001**, *33*, 693–697. [CrossRef]
34. Thermal Properties of Plastic Materials. Available online: http://edge.rit.edu/edge/P15611/public/Detailed%20Design%20Documents/Heat%20Transfer/ThermalPropertiesofPlasticMaterials.pdf (accessed on 4 June 2020).
35. Sousa, A.G.; Machado, L.M.M.; Silva, E.F.; Costa, T.H.M. Personal characteristics of coffee consumers and non-consumers, reasons and preferences for foods eaten with coffee among adults from the Federal District, Brazil. *Food Sci. Technol. (Campinas)* **2016**, *36*, 432–438. [CrossRef]
36. Mohos, F.Á. Data on engineering properties of materials used and made by the confectionery industry (Appendix 1). In *Confectionery and Chocolate Engineering*; Mohos, F.Á., Ed.; Wiley-Blackwell: Chichester, UK, 2010. [CrossRef]

Article

Texture Preferences of Chinese, Korean and US Consumers: A Case Study with Apple and Pear Dried Fruits

Runrou Wong, Seulgi Kim, Seo-Jin Chung * and Mi-Sook Cho

Department of Nutritional Science & Food Management, Ewha Womans University, 52, Ewhayeodae-gil, Seodaemun-Gu, Seoul 03760, Korea; nikowong0222@gmail.com (R.W.); seulgikim@ewhain.net (S.K.); misocho@ewha.ac.kr (M.-S.C.)

* Correspondence: sc79d@ewha.ac.kr; Tel.: +82-2-3277-3454

Received: 7 February 2020; Accepted: 19 March 2020; Published: 24 March 2020

Abstract: The present study aimed to understand the drivers of liking dried apple and pear chips with various textures among Chinese ($n = 58$), Korean ($n = 58$), and US ($n = 56$) consumers. The possibility of hedonic transfer from snack texture preferences to fruit-chip texture preferences was also investigated among Chinese and Koreans. Fourteen fruit-chip samples with four textural properties (crispy, puffy, soft, and jelly-like) were selected. Consumers rated their level of liking for each sample, and then they performed hedonic-based projective mapping with the same samples. In the hedonic texture transfer investigation, consumers rated their acceptance of nine snacks with various textures but possessing similar textures to those of dried fruit samples. The data were analyzed by ANOVA and multiple factor analysis. Most consumers disliked samples with a soft or jelly-like texture, while liked samples with a crispy texture. Cross-cultural differences were observed in the liking of puffy samples, with both Chinese and Koreans liking puffy samples as much as crispy ones for their melting characteristics in the mouth, while US consumers perceived the puffy samples as being Styrofoam-like and disliked them. Hedonic transfer was observed from snack texture preferences to fruit-chip. Individual texture preferences for snacks seem to significantly affect the texture preferences for fruit chips.

Keywords: fruit chips; hedonic based projective mapping; hedonic transfer; cross-culture; consumer liking

1. Introduction

One of the problems facing Korean agricultural businesses is the steady decrease in the consumption of fresh domestic fruits, including apples and pears, which is partly due to increases in the amount of fruit being imported [1]. Specifically, imported bananas and oranges are replacing the consumption of domestic apples and pears [1,2], and so the fruit industry in Korea is actively seeking strategies to increase the consumption of these traditionally grown domestic fruits. Value-added fruit products are a very attractive marketing concept since they meet increasing consumer demands for healthy and natural products [3–5]. Efforts are being made to develop novel fruit products utilizing innovative processing techniques [6–10]. Dried fruit chips have recently become one of the most prominent product categories due to them being marketed as convenient and healthy snack alternatives [11], and the compound annual growth rate of fruit snacks is anticipated to exceed 8% between 2019 and 2025 [12]. Mainland China and US are especially attractive markets for Korea since these two countries are two of the top three countries importing Korean food products [13].

Drying not only extends the shelf life of fresh fruit but also concentrates health functional substances such as antioxidants [14]. Moreover, the sensory characteristics of dried fruit products can be altered markedly even within the same fruit type by using different drying methods [15]. This could create numerous opportunities for fruit industries to develop a broad spectrum of products with various flavor and texture properties. The quality parameters for fresh fruits such as pear and apple are relatively straight forward. Optimal sugar to acid ratio, juiciness, and firmness are some of the key attributes influencing consumer acceptance positively [16,17]. On the contrary, the drivers of liking for dried fruit chips can be less predictable since various textures as well as flavors can be created through different processing methods mentioned above.

Texture is an extremely important sensory modality influencing the liking of a solid food product. Attributes such as crispiness, crunchiness, and viscosity have been reported as key factors affecting the liking of snacks, fruits, and dairy products [18]. Zou et al. [19] reported that a pleasant crispy mouthfeel was one of the reasons for the increased popularity of fruit chips among consumers. Although many researchers adopted cross-cultural designs to understand consumer's liking for various foods, most previous studies have focused on flavor rather than texture aspects due to the native characteristics of target food products [10,20–23].

Various hypotheses have been proposed for the reasons underlying consumer preferences for certain foods. For example, the relationships between consumer genetic sensitivities to specific tastants and food liking have been extensively studied [24,25]. Familiarity with a certain sensory quality is frequently reported to be a prominent factor for explaining food preferences [26–28]. Familiarity (or previous experience) for a food product, flavor, or texture has been a useful component for delineating the cross-cultural discrepancies in food acceptances. Cross-cultural differences in liking are often observed when the familiarity for the target food items differs among different countries. In contrast, cross-cultural agreement in liking are reported when the target food (i.e., fresh apple and apple juice) item is consumed commonly in different countries [16,29].

Additionally, it has been demonstrated that liking a specific taste can serve as an effective predictor for the acceptance of a foodstuff that exhibits the corresponding taste at a high level [30]. For example, likers of sweetness were shown to give higher acceptance scores for sweet foodstuffs with higher sucrose concentrations. In the present study we defined this phenomenon as hedonic transfer, with consumer preferences for certain modalities (e.g., sweetness or bitterness) not being restricted to a specific food category, but instead transferring to other product categories. It may be worth investigating whether such hedonic transfer can also be observed in the context of texture, where the preference for a specific texture in snacks can serve as a predictor for fruit-chip texture preferences.

General projective mapping (also called Napping) [31,32] and variants of this method have been used widely to understand the perceptual configuration of consumers toward target products based on the similarities and dissimilarities [33] when there is a relatively large number of products to investigate. Projective mapping combined with ultraflash profiling (UFP) has become a widely accepted method for profiling the sensory characteristics of target products from the perspective of consumers. Projective mapping using a hedonic framework was recently introduced by several researchers and shown to be effective in identifying the drivers of liking for certain product categories [34,35].

The present study investigated the acceptance as well as the perceptual configuration of dried apple and pear chips with various texture properties among Chinese, Korean, and US consumers, with the aim of understanding the key attributes that drive the liking of fruit chips in a cross-cultural context. Additionally, the possibility of hedonic transfer from snack texture preferences to fruit-chip texture preferences was studied among Chinese and Korean consumers. We hypothesized that the consumers with different cultural background will differ in the acceptance of and the drivers of liking for fruit chips. Additionally, the occurrence of hedonic transfer from snack texture to fruit chip texture was hypothesized.

2. Materials and Methods

2.1. Experimental Overview

This study performed two subexperiments. The first part of the study (Experiment 1) investigated the cross-cultural liking of dried fruits among consumers from China, South Korea, and the US. Consumers were asked to evaluate two types of fruit (apple and pear) produced as fruit chips with four textural properties (crispy, jelly-like, puffy, and soft) in terms of liking and the reasons for liking and disliking by means of hedonic projective mapping. Fourteen samples were evaluated by the Chinese and Korean consumers while the US consumers evaluated 13 samples (1 sample was not available). Thus, 13 products commonly tasted by all consumers were cross-culturally compared for their liking and perceptual configuration.

The second part of the study (Experiment 2) investigated whether the preference for a certain texture in one product category (snack texture preference) would transfer to the texture preference in another product category (dried fruit). Nine types of snack samples with various texture characteristics but sharing texture properties similar to those of the dried fruit samples (i.e., crispy, jelly-like, and puffy) were selected. The same Chinese and Korean consumers who participated in Experiment 1 were asked to evaluate their acceptance of the nine snack samples. Based on the texture preferences for snacks, consumers were grouped into likers of a crispy, jelly-like, or puffy texture. The significance of the relationships between their snack texture preferences and their liking of fruit-chip textures were then analyzed.

This study was approved by the Institutional Review Board (IRB) at Ewha Womans University, Seoul, South Korea (IRB No.: 164-29). The hypotheses of the experiments and analytic plans were specified prior to the actual data collection and statistical analysis.

2.2. Consumers

This study used online community bulletin boards to recruit 172 consumers from Korea, China, and the US who were interested in participating in consumer taste tests of dried fruits. Fifty-eight Korean consumers (8 males and 50 females) and 58 Chinese consumers (7 males and 51 females) ranging in age from 21 and 37 years who resided in Seoul, South Korea were recruited. The taste testing experiments were performed in Seoul. Fifty-five US consumers (12 males and 43 females) aged between 18 and 37 years were recruited in the community of Corvallis, OR, USA, where the taste testing experiments were conducted. All participants from Korea and US were citizens of South Korea and USA, respectively. Chinese participants consisted of subjects from Mainland China (81.1%), Taiwan (15.5%), and Hong Kong (3.4%). The mean ages of the Chinese, Korean, and US consumers were 24.7 (STD ± 2.7), 25.1 (±3.1), and 24.2 (±3.7) years, respectively. After completing the taste testing, all consumers received a small token of appreciation for their participation.

2.3. Samples

2.3.1. Experiment 1: Cross-Cultural Drivers of Liking for Dried Fruit Chips

Fourteen types of fruit chips (seven pear and seven apple chips) with four textural properties (crispy, jelly-like, puffy, and soft) were selected as the products of interests in the hedonic mapping experiments. All of the dried fruit chips from Korea, China, and the US were purchased online. The textural properties of the dried fruit products were affected by the drying method applied in their production: the crispy dried fruits were either baked or fried after air drying, jelly-like dried fruits were dried using wind or forced air, puffy dried fruits were made using freeze-drying, and soft dried fruits were made by combining air drying and microwaving. Chinese and Korean consumers evaluated all 14 samples, while the US consumers evaluated 13 samples since sample Crisp_P_SNC was not available for the experiments performed in the US (Table 1). Detailed information about the samples is provided in Table 1.

Table 1. Product information for the dried fruit chips.

Fruit Type	Product Name	Origin	Manufacturer
Pear	Northwest pear chips (Crisp_P_SF)	USA	Sisters Fruit Company, Cornelius, OR, USA.
	Seneca pear chips (Crisp_P_SNC)	USA	Seneca Foods Corp., New York, NY, USA.
	Ichibiya freeze-dried pear chips (Puff_P_ICBY)	South Korea	Dami Dry Food Industry. Mungyeong, South Korea.
	Heswim pear chips (Puff_P_HS)	South Korea	Good Food Co., Ltd. Naju, South Korea.
	Starbucks real fruits pear (Soft_P_STBS)	South Korea	Midm Agricultural Union Co., Pyeongtaek, South Korea.
	GugGu miao li gan tiao(Jelly_P_GGM)	China	Ji ning shi gu gu miao Food Company, Jining, China.
	Yi cun nong fu (Jelly_P_YCNF)	China	Qing zhou shi ting you fu Food Company, Qingzhou, China.
Apple	Northwest apple chips (Crisp_A_SF)	USA	Sisters Fruit Company, Cornelius, Oregon, USA.
	Seneca apple chips (Crisp_A_SF)	USA	Seneca Foods Co. New York, NY, USA.
	Ichibiya freeze-dried apple Chips (Puff_A_ICBY)	South Korea	Dami Dry Food Industry. Mungyeong, South Korea.
	Apple ring (Puff_A_AR)	South Korea	Shingi Farm, Mungyeong, South Korea.
	Ibisak apple chips (Soft_A_IBS)	South Korea	Hephzibah Food and Beverage, Naju, South Korea.
	Sol to sa rang apple chips (Soft_A_STSR)	South Korea	Daon Food, Seoul, South Korea.
	Kyeong chang nong won apple chips (Crisp_A_KCNW)	South Korea	Safe Food, Busan, South Korea.

In order to maintain the freshness of the produced fruit chips, the samples were opened 30 min prior to the testing sessions. All samples were cut into bite size (3 × 3 × 0.1~0.3 cm), and 2~5 g of each sample was served. The cut dried fruit chips were put in white disposable plastic cups (diameter of 7.0 cm and height of 4 cm; Samboopack Corporation, Incheon, Korea) and lidded. The samples were coded with 3-digit random numbers and were served in room temperature. The serving orders of the samples were determined by a Williams Latin square design [36]. Bottled spring water and unsalted crackers (Carr's Original Table Water, United Biscuits, Carlisle, UK) were provided to the consumers for cleansing the palate between sample evaluations.

2.3.2. Experiment 2: Hedonic Transfer of Texture Preference

All 14 types of dried fruit-chip samples in Experiment 1 and 9 types of snacks were selected as the products of interest. Nine types of snacks with various textural properties (i.e., crispy, jelly-like, and puffy) corresponding to those of the dried fruit-chip samples were selected and purchased online (Table 2). Additionally, sweet/sour-flavored rather than savory snacks were chosen for the experiment in order to minimize the influence of the flavor factor when investigating the relationships between the general preferences of the consumers for snack and fruit-chip textures. The sample preparation method, serving size, serving order, serving methods, and evaluation protocol for the snacks were identical to those for the fruit-chip samples.

Table 2. Product information for the snacks.

Product Name	Origin	Manufacturer
Honey butter maple syrup chips (Crisp_MS)	South Korea	Haitai Confectionery & Foods Co., Seoul, South Korea
Terra sweet potato chips (Crisp_SP2)	USA	The Hain Celestial Group, Inc, New York, NY, USA
Coconut chips (Crisp_C)	Thailand	Thitinan Bee Food Co., Bangkok, Thailand
Korea sweet potato chips (Crisp_SP1)	South Korea	Lotte Confectionery Co. Seoul, South Korea ood Food Co., South Korea
Apple and yogurt cubes (Puff_A)	South Korea	GoodFood Co. Naju, South Korea.
Handmade meringue-white, non-artificial pigment (Puff_M2)	South Korea	Miacake, Busan, South Korea
Aramonde meringue (Puff_M1)	South Korea	Aramonde, Seoul, South Korea
Fruit by the foot jelly -grape and green grape flavor (Jelly_G)	USA	General Mills, Minneapolis, MN, USA
Juicy mango soft jelly (Jelly_MG)	China	Fuijia Foods Co., Quanzhou, China

2.4. Procedure

2.4.1. Experiment 1: Cross-Cultural Drivers of Liking for Dried Fruit Chips

The consumers from the three countries evaluated the fruit-chip samples using projective mapping with a hedonic framework [31] in order to compare the perceptual configurations of sample liking and reasons for liking/disliking. The projective mapping evaluation consisted of three sessions: (1) a projective mapping learning session, (2) sample evaluation concerning the liking level and reasons for liking/disliking, and (3) the main projective mapping task with fruit chips using a hedonic framework.

As recommended in previous studies [37], a learning session for the overall projective mapping procedure was held to help consumers understand the evaluation method prior to participating in the main experiment. Consumers practiced performing projective mapping with a hedonic framework using photographs of 10 different flowers during the learning session. In the second session, consumers tasted and rated the liking (overall, appearance, flavor, and texture) of 14 dried fruit-chip samples (13 samples for the US consumers) on a 9-point hedonic scale whose anchor words were 'dislike extremely' to 'like extremely.' Consumers were also asked to freely write their reasons for liking and disliking each sample. During the third session, consumers performed projective mapping with a hedonic framework on a sheet of paper (40 × 60 cm). That is, they positioned the 14 samples (13 samples for the US consumers) based on the similarities of overall liking levels and reasons for liking or disliking as evaluated during the previous session. Consumers then tasted the samples again to remember and confirm their evaluation from the second session. This procedure resulted in samples being located closer to each other if their liking levels and reasons for liking or disliking were similar, and vice versa. After placing all of the samples on the sheet of paper, consumers were asked to replace the samples with the corresponding three-digit coded stickers and write down the reasons for liking and disliking directly beside each sample on the sheet.

2.4.2. Experiment 2: Hedonic Transfer of Texture Preference

The consumers from Korea and China who participated in Experiment 1 were also asked to taste and rate the nine types of snack samples on the same 9-point hedonic scale on another day in terms of their overall liking, appearance liking, taste/flavor liking, texture liking, and familiarity.

2.5. Demographic Information

The consumers filled in a questionnaire on demographic information after completing all of the tests. The questionnaire queried their age, gender, nationality, and general liking and frequency of consumption of fresh fruit, dried fruit chips, and snacks. In the case of measuring the general liking of these products, consumers were asked to list top 3 favorite products in rank order for each of fresh fruit, fruit chip, and snack category.

2.6. Statistical Analysis

2.6.1. Experiment 1: Cross-Cultural Drivers of Liking for Dried Fruit Chips

Statistical analysis was carried out with data obtained from 13 samples commonly evaluated by all consumers in the three countries. Analysis of variance (ANOVA) using a general linear model (GLM) was used to determine the effects of sample, consumer nationality, and the interaction between nationality and sample on the scores for overall liking, appearance liking, taste/flavor liking, and texture liking. The following GLM was applied: fruit chip acceptance = dried fruit texture type + nationality + dried fruit texture type × nationality + nationality × panelist. Duncan's multiple-range test was applied as a post-hoc test when a sample was significant for a specific sensory attribute. The data were analyzed using IBM SPSS Statistics software (version 21, SPSS, Chicago, IL, USA).

Multiple factor analysis (MFA) was used to visually summarize the perceptual sample configuration of the consumers from the three countries. The descriptive terms generated by UFP to express the reasons for liking and disliking each sample were analyzed as supplementary variables in the MFA. For the hedonic projective mapping data, the position of stickers on the sheet of paper were measured as X and Y coordinates relative to the bottom-left corner of the sheet, which were entered as X and Y values in a contingency table. Text mining was used to analyze UFP data by calculating the number of occurrences of each term mentioned by the consumers to describe the reasons for liking and disliking each sample. Comments with long sentences were first shortened to words (e.g., "it is sweet" was simplified to "sweet") and terms with similar meanings were combined (e.g., "skin or peel" was shortened to "skin"). During this preliminary process, two native speakers of each of Chinese, Korean, and English reviewed, discussed and transcoded the descriptors. After refining the terms for statistical analysis, terms that were mentioned by more than 10% of the consumers in each country were chosen for further analysis. A cut off value of 10% was used to disclose the drivers of (dis) liking since many attributes commented for (dis)liking reasons showed frequencies between 10%~20%. Korean and Chinese descriptors were translated into English, and the finalized descriptors were used as the supplementary variables in MFA. This analysis was performed with the FactoMineR package in R Studio software [38].

2.6.2. Experiment 2: Hedonic Transfer of Texture Preference

As in Experiment 1, ANOVA using a GLM was applied to the data obtained from the consumer acceptance tests of snacks and dried fruits. In the case of dried fruits, all 14 samples were subjected for analysis since Chinese and Koreans evaluated these samples. To find the correlation between snack and dried fruit texture preferences, the snack texture preference of each consumer was identified (i.e., crispy, jelly-like, or puffy texture) based on the highest mean texture liking scores among crispy, jelly-like, and puffy snacks. The data from consumers with no particular snack texture preference were removed prior to the analysis. For example, the data were removed if the consumer's mean texture liking scores of crispy and puffy snacks were identical and scored the highest. Based on this criterion, data from five Chinese and two Korean consumers were removed. ANOVA using a GLM was then conducted with the following model: fruit chip texture acceptance = snack texture preference + dried fruit texture type + nationality + dried fruit texture type × nationality + snack texture preference × nationality + snack texture preference × dried fruit texture type + snack texture preference × dried fruit texture type × nationality + snack texture preference × nationality × panelist. IBM SPSS Statistics software (version 21, SPSS) was again used to analyze these data.

3. Results

3.1. Consumption Frequencies of Fresh Fruits, Dried Fruits and Snacks

The favorite dried fruit (Figure 1) of the Chinese consumers was mango (29.3%), followed by apple and banana (10.3%) and durian (8.6%). The Korean consumers mostly preferred mango (31%), followed by banana (12.1%) and apple/coconut/persimmons/strawberry (6.9%). Similar to the Chinese consumers, US consumers mostly preferred dried mango (23.6%) and apple (21.8%), followed by apricot/banana/raisin (9.1%). The consumption frequencies of dried fruit were slightly higher for US consumers than for Chinese and Korean consumers. Dried fruits were eaten at least monthly by approximately 70%, 55%, and 35% of the US, Chinese, and Korean consumers, respectively.

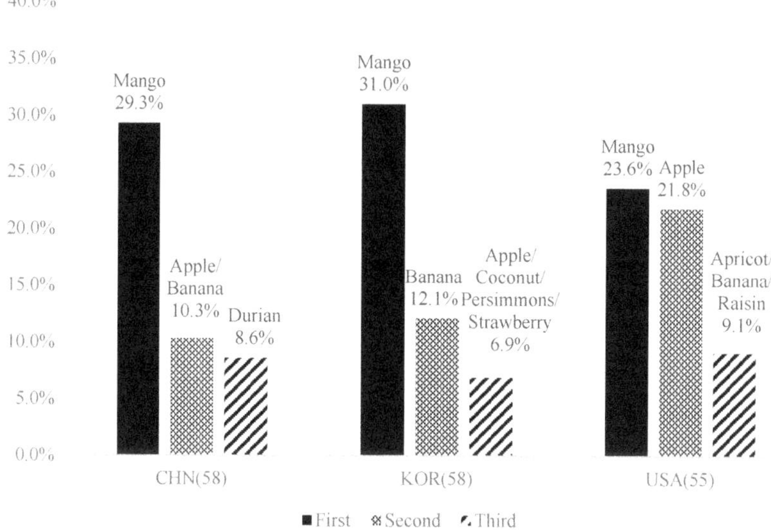

Figure 1. Top ranked dried fruits preferred by Chinese, Korean, and US consumers.

The top-three snacks among Chinese consumers were chips (39.7%), extruded snacks (12.1%), and chocolate (8.6%), while those were extruded snacks (39.7%), chips (36.2%), and chocolate snacks (6.9%) among Korean consumers. Unlike the Chinese and Korean consumers, the US consumers liked crackers (27.3%) the most, followed by chips (20%) and a dried fruit snack mixture.

3.2. Cross-Cultural Drivers of Liking for Dried Fruit Chips

3.2.1. Cross-Cultural Comparisons of Consumer Acceptance of Dried Fruit Chips

A GLM was used to analyze the effects of consumer nationality, sample, and the interaction between nationality and sample on the acceptance scores (i.e., overall, appearance, taste/flavor, and texture) (Table 3). US consumers generally gave significantly higher scores for the attributes compared to Chinese and Koreans (mean overall liking scores [OLs] of 5.2^a, 5.4^a, and 5.7^b (values sharing a same alphabet are not significantly different among countries) for Chinese, Korean, and US consumers, respectively); the main exception was for texture liking (OL = 5.1^a, 5.4^b, and 5.4^b, respectively). The nationality and sample interaction effect differed significantly in all of the attributes. Overall, crispy type fruit chips made from apple were commonly scored high in overall liking among consumers in all three countries (Table 4). Nevertheless, cross-cultural differences were observed for the specific acceptance levels of several samples; for example, sample Crisp_A_SNC was liked significantly more by Korean and US consumers than by Chinese consumers, while sample Soft_A_STSR was liked by US consumers but disliked by Chinese and Korean consumers.

Table 3. F- and p- values associated with statistical significance for 13 types of dried fruit.

Factor	Attributes	F-Value	p-Value
Sample	Overall liking	33.092	0.000
	Appearance liking	51.220	0.000
	Taste and flavor liking	18.295	0.000
	Texture liking	40.588	0.000
Nationality	Overall liking	5.131	0.007
	Appearance liking	3.755	0.025
	Taste and flavor liking	7.817	0.001
	Texture liking	1.358	0.260
Nationality × Sample	Overall liking	3.596	0.000
	Appearance liking	8.340	0.000
	Taste and flavor liking	2.881	0.000
	Texture liking	3.318	0.000
Nationality × Panelist	Overall liking	3.193	0.000
	Appearance liking	3.189	0.000
	Taste and flavor liking	3.141	0.000
	Texture liking	3.634	0.000

× refers to interaction effect of the independent factors

Specifically, in terms of overall liking, the Chinese consumers liked samples Crisp_A_SF and Puff_A_AR significantly more than the other samples, but disliked all of the soft dried fruit samples regardless of their fruit type (OL = 4.3, 4.1, and 4.5 for samples Soft_A_IBS, Soft_A_STSR, and Soft_P_STBS, respectively) and also disliked sample Jelly_P_YCNF (OL = 4.4). Korean consumers liked crispy apple samples (OL = 6.3 and 6.2 for samples Crisp_A_SNC and Crisp_A_SF, respectively), all of the puffy apple and pear samples (OL = 6.0, 6.1, 6.1, and 6.2 for samples Puff_A_AR, Puff_A_ICBY, Puff_P_HS, and Puff_P_ICBY, respectively), and sample Jelly_P_GGM (OL = 5.8). All of the other samples were rated below 5.0, with scores clustered between 4.0 and 4.8. US consumers noticeably liked sample Crispy_A_SF (OL = 7.6) among all of the samples. The scores were relatively low for sample Puff_A_ICBY (OL = 5.4), soft dried fruit samples (OL = 4.9 and 4.8 for samples Soft_A_IBS and Soft_P_STBS, respectively), and sample Jell_P_YCNF (OL = 4.7).

A GLM was separately applied to each of the pear and apple chip data sets to investigate whether the consumer liking of chips was affected by their texture type. The results (Appendix A) showed that texture type significantly affected ($p < 0.05$) all attributes for both apple and pear chips in all three countries. Overall, US consumers gave clearly higher mean liking scores for crispy than other texture types for apple and pear chips for most of the attributes (Table 4). In contrast, the Chinese and Korean consumers commonly gave higher liking scores for puffy than for other types of dried fruits. In addition to puffy, the Chinese consumers also liked crispy samples.

Table 4. Mean ± standard deviation of sample liking scores evaluated by Chinese, Korean, and US consumers.

Attributes	Overall Liking			Appearance Liking			Taste and Flavor Liking			Texture Liking		
Sample \ Nationality	CHN	KOR	USA	CHN	KOR	USA	CHN	KOR	USA	CHN	KOR	USA
Crisp_A_KCNW	$4.2 \pm 1.4^{a*}$	4.6 ± 2.0^{ab}	4.8 ± 1.9^{ab}	3.8 ± 1.3^{a}	3.7 ± 1.3^{a}	4.7 ± 1.8^{ab}	4.6 ± 1.5^{a}	5.2 ± 2.0^{cd}	5.7 ± 2.1^{abc}	4.4 ± 1.8^{ab}	4.7 ± 2.2^{abc}	4.4 ± 2.2^{ab}
Crisp_A_SF	6.3 ± 1.8^{f}	6.2 ± 1.8^{c}	7.6 ± 1.4^{f}	6.4 ± 1.6^{e}	6.5 ± 1.5^{g}	7.5 ± 1.4^{g}	6.1 ± 1.9^{de}	5.7 ± 2.0^{defg}	7.5 ± 1.7^{e}	6.6 ± 1.8^{f}	6.9 ± 1.4^{g}	7.4 ± 1.5^{f}
Crisp_A_SNC	5.6 ± 1.7^{de}	6.3 ± 1.7^{c}	6.5 ± 1.8^{e}	5.7 ± 1.6^{d}	5.4 ± 1.6^{de}	6.8 ± 1.9^{f}	5.6 ± 1.7^{bcd}	6.3 ± 1.6^{g}	6.4 ± 1.9^{cd}	5.9 ± 1.5^{e}	6.2 ± 1.8^{fg}	6.8 ± 1.8^{ef}
Puff_A_AR	6.2 ± 1.5^{ef}	6.0 ± 2.0^{c}	5.9 ± 1.9^{cde}	6.5 ± 1.5^{e}	6.6 ± 1.4^{g}	6.6 ± 1.6^{ef}	6.4 ± 1.6^{e}	6.4 ± 1.6^{g}	6.2 ± 1.8^{cd}	5.6 ± 2.2^{de}	5.4 ± 2.3^{de}	5.2 ± 2.3^{bc}
Puff_A_ICBY	5.6 ± 1.7^{de}	6.1 ± 2.0^{c}	5.7 ± 2.0^{cd}	6 ± 1.4^{de}	6.3 ± 1.7^{fg}	5.3 ± 2.1^{bc}	5.8 ± 1.5^{cde}	6.0 ± 1.9^{efg}	6.3 ± 1.9^{cd}	5.5 ± 1.7^{de}	6.0 ± 2.1^{ef}	5.3 ± 2.2^{cd}
Soft_A_IBS	4.3 ± 1.6^{ab}	4.4 ± 1.8^{ab}	4.9 ± 1.7^{ab}	4.8 ± 1.3^{c}	4.8 ± 1.4^{bc}	5.9 ± 1.7^{cd}	4.8 ± 1.7^{a}	4.7 ± 1.9^{bc}	5.1 ± 2.1^{ab}	3.8 ± 1.9^{a}	4.1 ± 1.8^{a}	4.1 ± 1.7^{a}
Soft_A_STSR	4.1 ± 1.5^{a}	4.5 ± 1.9^{ab}	5.7 ± 1.8^{cd}	4.1 ± 1.3^{a}	4.6 ± 1.6^{bc}	6.0 ± 1.7^{de}	4.8 ± 1.7^{ab}	5.4 ± 1.7^{de}	6.1 ± 1.9^{cd}	3.9 ± 1.6^{a}	4.2 ± 1.9^{a}	5.1 ± 2.4^{bc}
Crisp_P_SF	5.7 ± 2.1^{de}	4.8 ± 2.0^{b}	6.1 ± 2.2^{cde}	6.0 ± 1.6^{de}	5.6 ± 1.7^{de}	7.0 ± 1.4^{fg}	5.8 ± 2.3^{cde}	5.2 ± 2.1^{cd}	5.9 ± 2.5^{c}	5.7 ± 2.0^{de}	5.1 ± 2.1^{cd}	6.0 ± 2.1^{de}
Jelly_P_GGM	5.1 ± 1.9^{cd}	5.8 ± 2.0^{c}	6.4 ± 2.0^{de}	5.0 ± 1.9^{c}	5.4 ± 1.8^{de}	5.6 ± 2.1^{cd}	5.2 ± 1.9^{abc}	5.8 ± 1.9^{defg}	6.8 ± 1.9^{d}	5.1 ± 1.9^{bcd}	5.9 ± 1.8^{ef}	6.4 ± 2.1^{e}
Jelly_P_YCNF	4.4 ± 2.1^{ab}	4.0 ± 1.7^{a}	4.7 ± 2.3^{a}	4.6 ± 1.8^{bc}	4.4 ± 1.5^{b}	4.9 ± 1.9^{ab}	4.9 ± 1.9^{a}	4.1 ± 1.9^{a}	4.9 ± 2.2^{a}	4.1 ± 1.9^{a}	4.9 ± 1.9^{bcd}	4.9 ± 2.4^{bc}
Puff_P_HS	5.7 ± 1.5^{de}	6.1 ± 2.0^{c}	5.5 ± 2.0^{bc}	5.7 ± 1.6^{d}	5.9 ± 2.0^{ef}	4.4 ± 2.0^{a}	5.9 ± 1.3^{de}	5.9 ± 1.9^{efg}	6.1 ± 2.0^{cd}	5.8 ± 1.6^{de}	6.3 ± 2.0^{fg}	5.1 ± 2.4^{bc}
Puff_P_ICBY	5.7 ± 1.6^{de}	6.2 ± 2.0^{c}	5.4 ± 2.2^{abc}	5.7 ± 1.4^{d}	5.9 ± 1.6^{ef}	4.6 ± 2.1^{ab}	5.8 ± 1.5^{cde}	6.1 ± 1.6^{fg}	5.8 ± 2.2^{bc}	5.7 ± 1.6^{de}	5.9 ± 2.1^{ef}	5.1 ± 2.4^{bc}
Soft_P_STBS	4.5 ± 1.6^{abc}	4.5 ± 1.7^{ab}	4.8 ± 1.9^{a}	4.2 ± 1.4^{ab}	4.3 ± 1.4^{b}	4.3 ± 1.9^{a}	5.1 ± 1.7^{a}	5.4 ± 1.6^{def}	5.7 ± 1.8^{abc}	4.5 ± 1.8^{ab}	$4.4 \pm 1.9n^{ab}$	3.9 ± 2.1^{a}

* Means not sharing a common lowercase letter within the same column are significantly different among samples at $p < 0.05$.

3.2.2. Multiple Factor Analysis of the Consumer Perceptual Configuration of Dried Fruit Chips

Liking and Disliking Terms Generated for Profiling Dried Fruit Samples

The Chinese, Korean, and US consumers used 293, 240, and 241 terms, respectively, to describe their reasons for liking the samples. These terms were merged into, 85, 98, and 80 terms, respectively, for the initial analysis. In the case of disliking terms, Chinese, Korean, and US consumers initially used 345, 352, and 380 terms, respectively, which were reduced to 84, 81, and 89 terms. As mentioned above, the combined terms commonly chosen by more than 10% of the consumers were further selected as supplementary variables in the MFA. Thus, 37 terms (19 liking and 18 disliking) for Chinese consumers, 41 terms (18 liking and 23 disliking) for Korean consumers, and 31 terms (13 liking and 18 disliking) for US consumers were used in the final MFA (Table 5). Concerning the terminology used, consumers from all three countries frequently answered "none" as the reasons for liking when the sample was disliked, and also "none" as the reasons for disliking when the sample was liked. The usage frequencies of this term were strongly correlated with the levels of liking and disliking the samples.

Cross-Cultural Comparison of Consumer's Perceptual Configuration of Dried Fruit Chips

Overall, the MFA map of all three countries defined factor 1 (F1) axis by the degree of liking the samples (higher rated samples on the positive and lower rated samples on the negative axis, respectively) and factor 2 (F2) axis by the texture type of the samples. MFA roughly clustered the samples into three groups: crispy, puffy, and soft/jelly.

A total of 47.9% variance was explained by Factor 1 (30.4%) and Factor 2 (17.5%) for Chinese consumers (Figure 2). F1 was mainly defined by the degree of liking of the samples. Thus, most of the crispy and puffy samples with OL > 5.5 were located on the positive F1 axis, while most of the soft and jelly-like samples as well as sample Crisp_A_KCNW were positioned on the negative F1. The crispy and puffy samples on the positive F1 axis showed strong positive correlations with liking the attributes of appearance and texture and no disliking. The jelly-like and soft samples showed strong positive correlations with off taste, looks unfresh, no liking, unappetizing, and generally disliking the appearance. F2 mainly indicated the contrast between the crispy samples on the positive axis and the puffy samples on the negative axis. Crispy samples with relatively high acceptance scores were characterized by liking reasons of crispiness, sourness, and sweetness/sourness, and disliking reasons of too sour, artificial taste, and sticking to the teeth. Puffy samples showed strong positive correlations with the liking reasons of soft, fresh fruit texture, clean appearance, natural fruity flavor, and melting in the mouth.

The MFA for Korean consumers showed that 45.2% of the variance was explained by F1 (29.5%) and F2 (15.7%) (Figure 3). Similar to the results for Chinese consumers, F1 was mainly defined by the degree of liking of the samples. The samples located on the positive F1 axis were characterized by liking attributes of appearance and appropriate sweetness and no disliking reasons, while those on the negative F1 axis were characterized by strange flavor, sticky, disliking the appearance/taste, and no liking reasons. The F2 axis was mainly defined by the level of crispiness present in the sample. Most of the crispy samples were positioned on the positive F2 axis, and the jelly-like and puffy samples were mainly on the negative F2 axis. The crispy samples were liked for their crispiness, thin size, and fresh taste, but disliked for their sour taste. Puffy samples located on the positive F1 and positive F2 axes were specifically described as melting in the mouth and soft as reasons for liking, while their sponginess was both liked and disliked. The soft and jelly-like samples located on the negative F1 and negative F2 axes were delineated by their stickiness, looking unfresh, toughness, not crisp, soggy, and browning as reasons for disliking.

Table 5. Liking and disliking terms selected by more than 10% of the Chinese, Korean, and US consumers.

CHN		KOR		USA	
Liking	**Disliking**	**Liking**	**Disliking**	**Liking**	**Disliking**
appearance	appearance	appearance	appearance	appearance	appearance
apple flavor/aroma	artificial taste/flavor	apple flavor/aroma	artificial	chewy	artificial
appropriate sweetness	color	appropriate sweetness	browning	color	bland taste/flavor
chewy	feels old	chewy	cosmetic taste/flavor/aroma	crispy	brown color
color	looks unfresh	color	feels like sponge	crunchy texture	dissolves too fast in mouth
crispy	mushy	crispy	hard	fruity flavor/aroma	feel like Styrofoam
fresh fruit texture	none	looks delicious	looks unfresh	none	nasty
looks clean & neat	off taste/flavor	looks like apple	none	skin	none
melt in mouth	oily	melt in mouth	not crispy	sweet	soft texture
natural fruity flavor	skin	none	off taste/flavor/aroma	taste/flavor	stick to teeth
none	soft texture	refreshing taste/flavor/odor	oily	taste like fresh fruit	taste/flavor
smell good	stick to teeth	soft	skin	tastes like candy	tastes like candy
soft	sticky	spongy	soggy	texture	texture
sour	texture	sweet	sour taste/flavor/aroma		too chewy
sweet and sour	too hard	taste/flavor	stick to teeth		too crunchy
sweet flavor	too sour	taste like snack	sticky		too hard
taste/flavor	too sweet	texture	strong taste/flavor/aroma		too sweet
texture	unappetizing appearance	thin	taste/flavor		too thick
white color			texture		
			too sour		
			too sweet		
			tough		
			unknown taste/flavor		

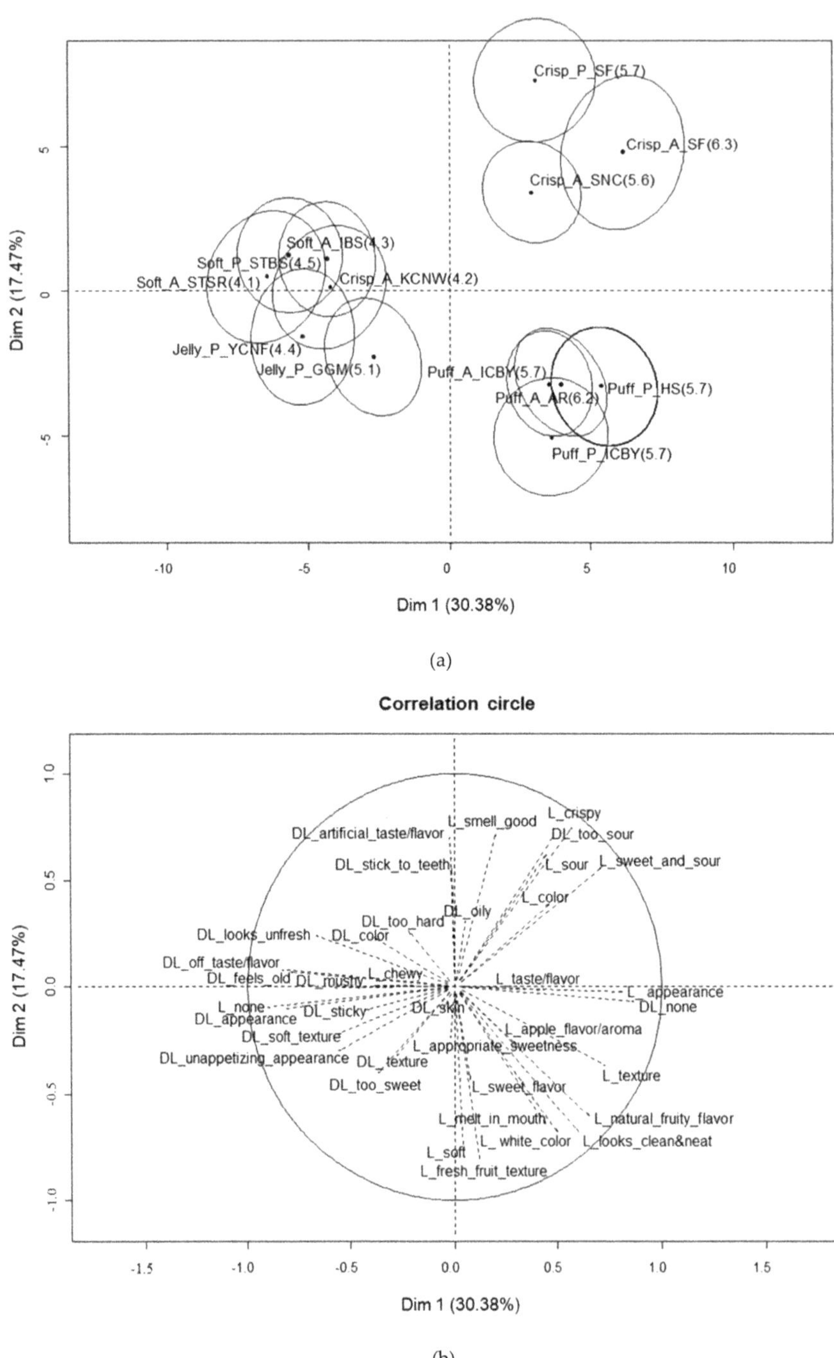

Figure 2. MFA plot of 13 dried fruit chips (**a**) and the reasons for liking and disliking them (**b**) among Chinese consumers.

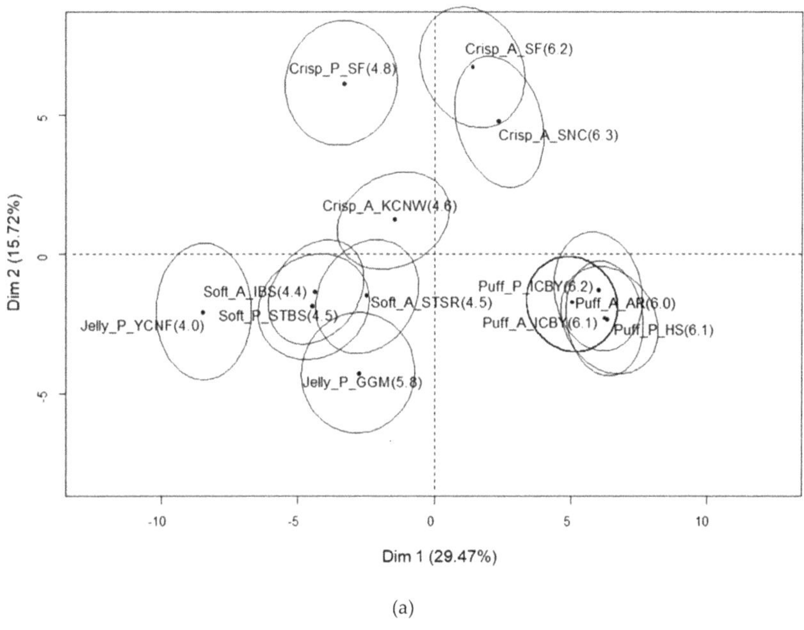

(a)

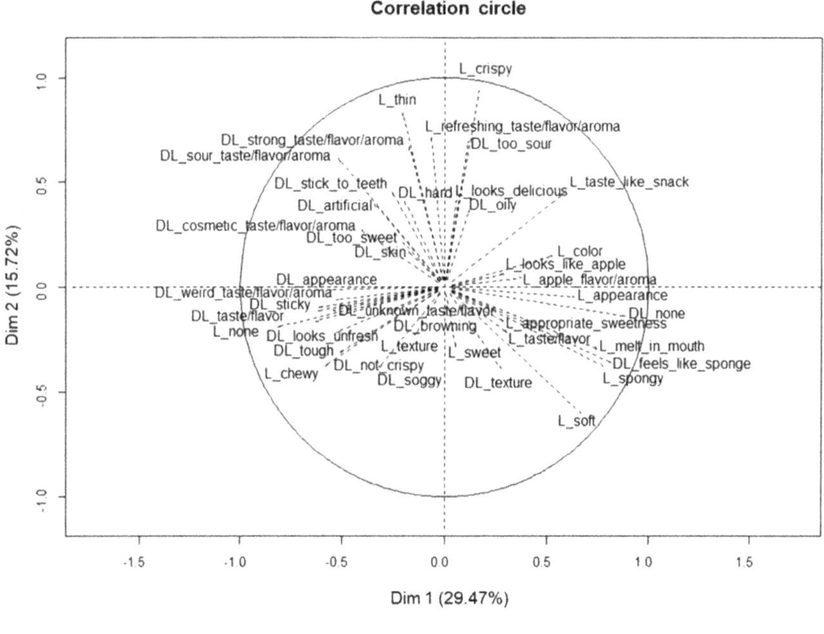

(b)

Figure 3. MFA plot of 13 dried fruit chips (**a**) and the reasons for liking and disliking them (**b**) among Korean consumers.

A total of 40.8% of the variance was explained by F1 (21.8%) and F2 (19.0%) in the MFA of US consumers (Figure 4). Similar to the consumers in the other two countries, the samples that the US consumers scored higher than 6.0 points for overall liking were positioned on the positive F1 axis, while other samples tended to be loaded on the negative F2 axis. These well-liked samples were commonly characterized by liking attributes of sweetness and appearance. The crispy samples were additionally liked for their crispy and crunchiness, whereas sample Jelly_P_GGM was liked for its color but disliked for tasting too sweet. F2 was mainly defined by the presence of a chewy texture. The soft and jelly-like samples were located on the negative F2 axis and showed a strong positive correlation with chewiness, which was both liked and disliked. Sticking to teeth and general taste/flavor were additional reasons for disliking the samples. Crispy and puffy samples were positioned on the positive F2 axis. Puffy samples were specifically characterized by the disliking attributes of dissolving too fast in the mouth and tasting like Styrofoam.

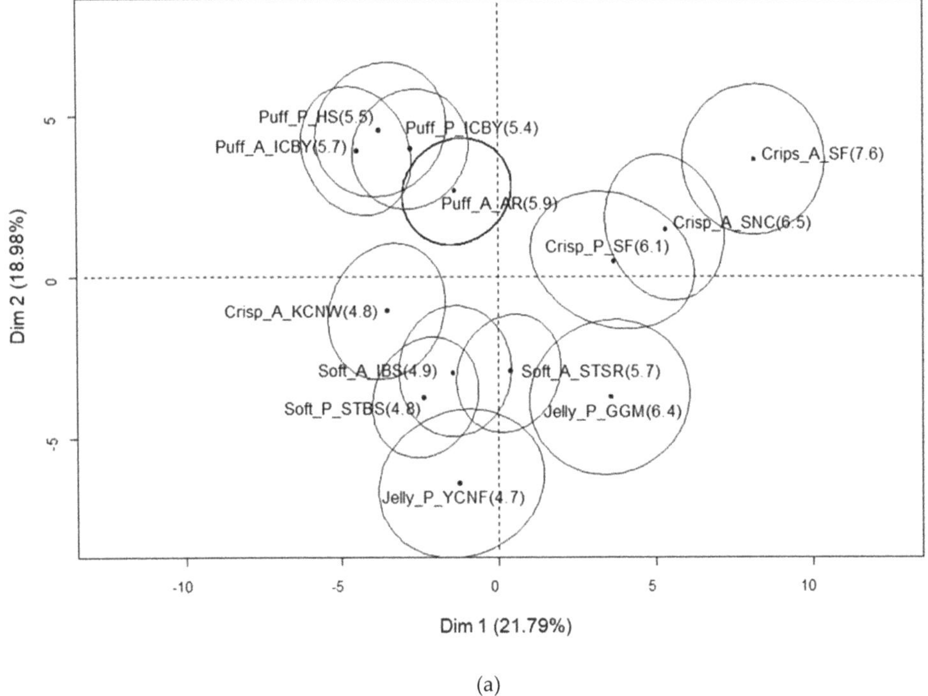

(a)

Figure 4. *Cont.*

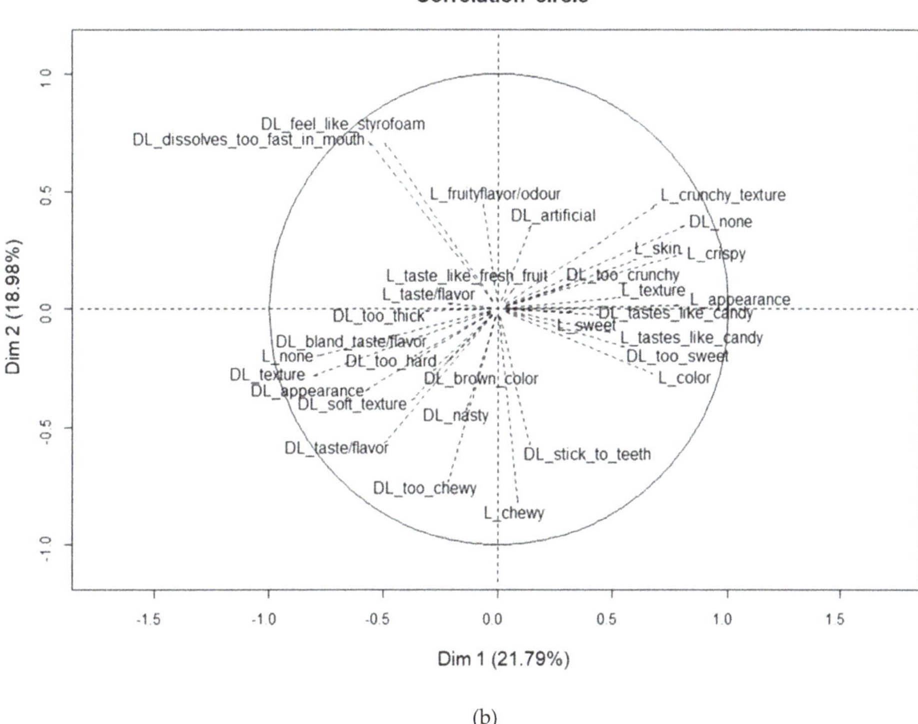

(b)

Figure 4. MFA plot of 13 dried fruit chips (**a**) and the reasons for liking and disliking them (**b**) among US consumers.

3.3. *Hedonic Transfer of Texture Preference*

The second aim of the study was to determine whether the preference for certain textures would transfer across different product categories. The individual texture preferences of consumers within the snack category were identified and their significance on delineating the fruit-chip texture acceptance was investigated using GLM analysis. Based on the snack texture preference, both Chinese and Korean consumers were grouped into likers of crispy, jelly-like, and puffy textures. The numbers of consumers who preferred crispy, puffy, and jelly-like snacks were similar in China and Korea. More than half of the consumers were classified as liking a crispy texture (27 Chinese and 32 Korean consumers), while smaller numbers of consumers were grouped into likers of a puffy texture (14 and 15, respectively) followed by likers of a jelly-like texture (12 and 9, respectively).

The results obtained when the GLM was applied separately to apple and pear chips showed that the OLs and texture liking (TL) scores were significantly affected by the interaction effect of snack texture preference and fruit texture type for both sample types (Table 6). The relationship between snack texture preference and the liking of apple and pear chips of various textures among Chinese and Korean consumers are shown in Figure 5a,b.

Table 6. *F*- and *p*-values associated with statistical significance for seven types of dried apple and pear chips.

		Apple		Pear	
Factor	**Attributes**	***F*-Value**	***p*-Value**	***F*-Value**	***p*-Value**
Texture preference of snack	Overall liking	1.117	0.331	0.970	0.382
	Appearance liking	1.669	0.193	0.342	0.711
	Taste/flavor liking	0.208	0.813	1.848	0.162
	Texture liking	2.761	0.068	1.142	0.323
Dried fruit texture type	Overall liking	35.642	0.000	15.185	0.000
	Appearance liking	58.803	0.000	22.827	0.000
	Taste/flavor liking	17.929	0.000	9.990	0.000
	Texture liking	47.874	0.000	12.850	0.000
Nationality	Overall liking	0.513	0.475	0.188	0.666
	Appearance liking	0.160	0.690	0.341	0.561
	Taste/flavor liking	0.526	0.470	0.000	0.984
	Texture liking	0.732	0.394	0.993	0.321
Dried fruit texture type × Nationality	Overall liking	0.948	0.388	2.541	0.055
	Appearance liking	0.745	0.475	1.694	0.167
	Taste/flavor liking	0.857	0.425	2.009	0.111
	Texture liking	0.577	0.562	1.922	0.125
Texture preference of snack × Nationality	Overall liking	1.435	0.243	0.090	0.914
	Appearance liking	0.928	0.398	0.675	0.511
	Taste/flavor liking	1.528	0.222	0.842	0.434
	Texture liking	1.661	0.195	0.566	0.569
Texture preference of snack × Dried fruit texture type	Overall liking	3.773	0.005	2.393	0.027
	Appearance liking	1.371	0.242	4.107	0.000
	Taste/flavor liking	2.935	0.020	1.012	0.417
	Texture liking	3.425	**0.009**	2.731	**0.013**
Texture preference of snack × Dried fruit texture type × Nationality	Overall liking	0.634	0.638	0.393	0.883
	Appearance liking	1.443	0.218	0.469	0.831
	Taste/flavor liking	0.758	0.553	0.777	0.588
	Texture liking	0.612	0.654	0.857	0.526
Texture preference of snack × Nationality × Panelist	Overall liking	2.767	0.000	2.391	0.000
	Appearance liking	1.955	0.000	2.487	0.000
	Taste/flavor liking	2.606	0.000	2.085	0.000
	Texture liking	2.720	0.000	2.444	0.000

× refers to interaction effect of the independent factors.

In the case of Chinese consumers, likers of a crispy texture scored the texture of crispy dried apple chips higher (TL = 5.7) than did the likers of puffy (TL = 5.2) and jelly-like (TL = 3.5) textures. Likers of a puffy texture scored the puffy dried fruit chips the highest for both apple and pear chips. Likers of a jelly-like texture gave a higher liking score than the other two groups for soft apple and pear dried fruit, although soft chips were still the least-preferred samples. For dried pear chips, Chinese likers of a crispy texture generally gave lower texture liking scores than likers of a jelly-like or puffy texture. In contrast, likers of a jelly-like texture tended to give higher scores for jelly-like fruit or compared to the other consumer groups.

The fruit-chip texture liking tendencies were slightly different for Korean consumers (Figure 5b). These likers of a jelly-like texture gave lower texture liking scores for crispy apple chips, puffy apple and pear chips compared to the other two groups, while likers of a puffy texture gave the highest score to all puffy fruit chips and a higher score to soft dried apple than did the other two groups. The texture likings of crispy or jelly-like dried pear were not strongly affected by their texture preferences, while likers of a jelly-like texture gave the highest score to soft dried pear.

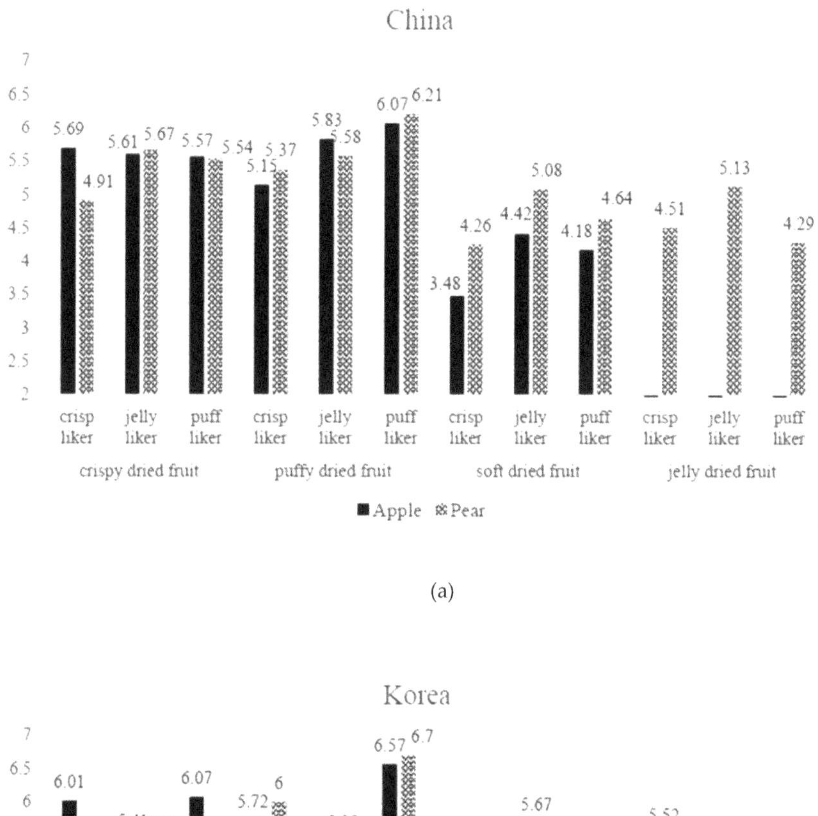

(a)

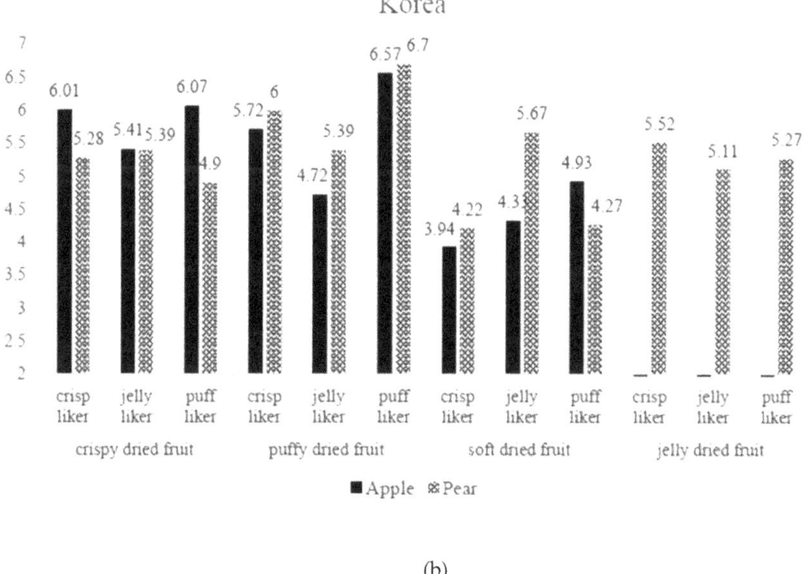

(b)

Figure 5. Effects of interaction between snack texture preferences and dried fruit-chip texture on the texture liking of dried apple and pear among Chinese (**a**) and Korean (**b**).

4. Discussion

Many studies have investigated the drivers of food liking within a cross-cultural context. However, most of these studies have focused on flavor rather than texture aspects of foodstuff when investigating their acceptance, partly due to the nature of the product (e.g., beverages, sauces, and yogurts) but also due to the critical role that flavor quality plays in determining the acceptance level of a foodstuff [26,39–42]. Only a few studies have focused on the cross-cultural acceptance of foods with various texture properties [27,43]. The present study specifically performed experiments designed to understand the effect of texture characteristics on the perceptions of and preferences for fruit chips in a cross-cultural context. Although the flavors of the chips were intentionally varied by testing both apple and pear fruit chips together, flavor characteristics seemed to play a smaller role than texture since the variation in the acceptance score was smaller between apple and pear chips than among the different texture types.

We observed noticeable cross-cultural differences between Asian (Chinese and Korean) and US consumers in terms of fruit-chip liking. US consumers tended to give higher mean liking scores than Chinese and Korean consumers, which might be due to US consumers consuming fruit chips more frequently. Concerning the preferences for specific textures, general agreements were observed in the preference patterns for crispy and soft samples, with most consumers from all three countries giving higher acceptance scores to chips with crispy textures and lower scores to chips with soft textures.

In contrast, cross-cultural differences were found for the liking of puffy samples. The Chinese and Korean consumers liked puffy fruit chips as much as crispy ones, while US consumers gave significantly lower liking scores to puffy than crispy samples. When MFA was applied to the hedonic-based projective mapping with UFP, Chinese, Korean, and US consumers all described that the crispiness was the reason for liking crispy samples. This universal liking of a crispy texture is consistent with previous studies identifying this as one of the most important texture characteristics in snacks and fresh fruits [18,44,45]. Their appearance and the characteristic of melting in the mouth were the main drivers of liking puffy samples among Chinese and Korean consumers. However, US consumers disliked puffy samples because they considered that the samples dissolved too fast in the mouth and tasted like Styrofoam.

While familiarity per se was not frequently mentioned as a driver of liking or disliking in UFP, the consumers tended to find fruit chips more acceptable if the texture was familiar to them in the category of fruit chips or in terms of a broader snack category. That is, not only the familiarity with the fruit-chip texture itself, but also the contextual familiarity of the fruit-chip texture with that of the snacks available in the local markets of China, Korea, and the US seemed to play an important role in determining the cross-cultural differences in the liking of certain fruit-chip textures. Similar results were found in a cross-cultural acceptance study of yackwa, a traditional Korean sweet cookie [27]. Cross-cultural disagreements were shown for the types of texture preferred by Japanese, Korean, and French consumers. Consumers with different cultural backgrounds tend to like samples that have characteristics similar to those of products that they have consumed previously. As mentioned above, the US consumers in the present study generally consumed more fruit chips than did the Asian consumers. Moreover, the snacks with the largest market share in the US are chips (e.g., potato and tortilla), while extruded snacks have the largest share in the Chinese and Korean markets [46–48]. Chinese and Korean consumers are therefore more likely to be frequently exposed to extruded snacks eliciting puffy texture than are US consumers. Thus, consumers may have used the frame of reference constructed for fruit chips if they were frequent consumers of fruit chips, but also used the frame of reference for the broader snack category to evaluate their liking of fruit chips. However, this specific hypothesis remains to be validated in the future.

One of the similarities among the three consumer groups in this study is that they all disliked the soft dried fruits, using similar descriptions of being too chewy, soggy, and soft. These lower scores for soft dried fruits might be influenced by the frame of reference of the consumers for the fruit chips, which were crispy and crunchy. Josh et al. (2011) reported that consumers generally favored crispy-fried apple snack over less crispy one [49]. Similarly, in an investigation of the relationship between oral breakdown patterns and the preferences for biscuits, the 50 panelists demonstrated a consistent dislike of the softened samples, which could be seen from the results of the preference map of the biscuits [50]. Additionally, not all of the crisp samples were liked, and the consumers from all three countries described sample Crisp_A_KCNW being too hard as the reason for disliking it despite them liking crispy samples. The appropriateness of certain texture intensity was also the main criterion used by the consumers in the evaluations. Similarly, Jaworska and Hoffmann [51] found that the level of consumer acceptance of potato chips was only affected by the crispy texture intensity of the chips.

Previous studies have attempted to use the liking status for a certain flavor (e.g., sweetness or bitterness) as a predictor for the liking of food in another category that has similar sensory qualities. For example, Kim et al. [26] classified consumers based on their liking of sweet taste (as measured using sucrose solutions) and investigated the correlation with sweetness liking for orange juice as well as with the liking behavior in other food categories. Harwood et al. [52,53] classified consumers based on the self-declared liking status for milk chocolate versus dark chocolate, and used this to identify the rejection level of bitterness in chocolate-flavored products. These studies successfully demonstrated hedonic transfer from one product category to another; that is, significant differences in the acceptance level for different sweetness or bitterness of samples were observed among consumers with different sweetness or bitterness liking statuses, respectively. The present study also observed hedonic transfer from snack textures to fruit-chip textures. The presence of different texture liking statuses significantly affected the liking of consumers for the fruit-chip texture. For example, consumers who liked a puffy texture in a snack also gave higher liking scores for puffy fruit-chip samples.

While some successful cases of hedonic transfer have been described above, the liking status for specific attributes does not always effectively delineate how consumers will behave toward similar attributes. Several studies have attempted to use the sweet liking status to predict the consumption frequencies of sweet foods among children as well as adults [30,54]. Overall, the link between the sweet liking status and sweet food consumption has varied depending on the food category examined. Clearer hedonic transfer from one product category to another was observed if the product categories shared similar sensory characteristics or similar contextual product usage (e.g., from a water solution to orange juice, from solid chocolate to chocolate ice cream, or from snacks to fruit chips) [26,48]. Further investigations are necessary before drawing reliable conclusions about the food product categories in which hedonic transfer occurs.

Recently, Jeltima et al. (2015) classified consumers into four groups based on their oral behavior (cruchers, chewers, smooshers, and suckers) and proposed that these differences in oral behavior may influence the preference for food texture [55]. Our study has classified consumers into likers of crispy, jelly-like, or puffy texture, and shown that the liking of fruit chips was significantly affected by the consumer's texture liking status of snacks. Investigating the oral behavior of the likers of crispy, jelly-like, or puffy texture may provide another way of understanding the hedonic transfer of textures in different product categories.

The present study carries several limitations to generalize the findings to a wider population due to a relatively small number of consumer participants from each country, a narrow age range of these participants (mostly in their 20s), and an imbalanced female to male ratio. Conducting a hedonic transfer experiment with only Asian consumers is another weakness of the study. Additionally, a lack of physico-chemical measurements (i.e., aroma compounds, free sugar content, total acidity, texture profile analysis, etc.) of the samples confines the interpretation of our findings on the drivers of (dis)liking for fruit chips from the practical perspective of product developers.

5. Conclusions

This study was designed to cross-culturally compare the acceptance and perceptual configuration of dried fruit chips among Chinese, Korean, and US consumers with the aim of predicting the potential market in these countries using hedonic-based projective mapping. The possibility of hedonic transfer concerning texture preferences across different product categories was also investigated. The key drivers of the liking of dried fruit chips was their crispness among the consumers from all three countries, with soft dried fruits being the least preferred dried fruits. Cross-cultural differences were observed in the acceptance of puffy fruit chips, with Chinese and Korean consumers liking puffy chips significantly more than did US consumers. These differences were mainly attributed to different degrees of familiarity with the texture in the context of fruit chips or snacks. Hedonic texture transfer from snacks to fruit chips was observed, with individual snack texture preferences significantly affecting the fruit-chip texture preferences of the consumers.

Author Contributions: Conceptualization, S.-J.C. and M.-S.C.; methodology, R.W.; software, S.K.; validation, S.K.; formal analysis, R.W.; investigation, R.W.; resources, S.-J.C.; data curation, R.W.; writing—original draft preparation, R.W.; writing—review and editing, S.-J.C.; visualization, R.W.; supervision, S.-J.C.; project administration, S.-J.C. & R.W.; funding acquisition, M.-S.C. All authors have read and agreed to the published version of the manuscript.

Funding: This research was carried out with the support of "Cooperative Research Program for Agriculture Science & Technology Development (Project No. PJ012416)" Rural Development Administration, Republic of Korea.

Acknowledgments: We thank the anonymous reviewers for the valuable comments to improve the manuscript.

Conflicts of Interest: The authors declare no conflict of interest.

Appendix A

Table A1. *F*- and *p*- values associated with statistical significance for dried fruit texture types on the liking ratings of apple and pear chip samples.

Fruit Type	Factor	Attributes	*F*-Value	*p*-Value
Apple	Dried fruit texture type	Overall liking	57.759	0.000
		Appearance liking	42.182	0.000
		Taste and flavor liking	34.661	0.000
		Texture liking	87.569	0.000
	Nationality	Overall liking	5.965	0.003
		Appearance liking	12.009	0.000
		Taste and flavor liking	7.824	0.001
		Texture liking	1.451	0.237
	Dried fruit texture type ×[1] Nationality	Overall liking	4.799	0.001
		Appearance liking	13.194	0.000
		Taste and flavor liking	2.623	0.033
		Texture liking	3.053	0.016
	Nationality ×[1] Panelist	Overall liking	2.042	0.000
		Appearance liking	1.766	0.000
		Taste and flavor liking	2.058	0.000
		Texture liking	2.152	0.000
Pear	Dried fruit texture type	Overall liking	17.353	0.000
		Appearance liking	41.936	0.000
		Taste and flavor liking	8.850	0.000
		Texture liking	22.692	0.000
	Nationality	Overall liking	2.070	0.129
		Appearance liking	0.075	0.928
		Taste and flavor liking	3.100	0.047
		Texture liking	0.151	0.860
	Dried fruit texture type ×[1] Nationality	Overall liking	5.156	0.000
		Appearance liking	12.378	0.000
		Taste and flavor liking	2.383	0.027
		Texture liking	7.124	0.000
	Nationality ×[1] Panelist	Overall liking	1.804	0.000
		Appearance liking	1.915	0.000
		Taste and flavor liking	1.895	0.000
		Texture liking	1.922	0.000

×[1] refers to interaction effect of the independent factors.

References

1. Lee, Y.S.; Park, J.W. The Increase of Imported Fruits, What Is the Countermeasure of Consumption and Production? Available online: https://dbpia.co.kr/pdf/pdfView.do?nodeId=NODE06597563. (accessed on 4 October 2019).
2. Kim, S.W.; Park, M.S.; Park, H.W.; Choi, G.L.; Hong, S.P. Agricultural Outlook 2018: Trends and Prospects of Fruit Supply and Demand. Available online: http://library.krei.re.kr/pyxis-api/1/digital-files/66306611-d1b4-4144-875b-885204a3ae2c. (accessed on 4 October 2019).
3. Grunert, K.G. Trends in food choice and nutrition. In *Consumer Attitudes to Food Quality Products: Emphasis on Southern Europe*; Klopčič, M., Kuipers, A., Hocquette, J.-F., Eds.; Wageningen Academic Publishers: Wageningen, The Netherlands, 2013; pp. 23–30.
4. Ares, G.; Saldamando, L.; Giménez, A.; Claret, A.; Cunha, L.M.; Guerrero, L.; Moura, A.; Oliveira, R.; Symoneaux, D.C.R.; Deliza, R. Consumers' associations with wellbeing in a food-related context: A cross-cultural study. *Food Qual. Prefer.* **2015**, *40B*, 304–315. [CrossRef]
5. Román, S.; Sánchez-Siles, L.M.; Siegrist, M. The importance of food naturalness for consumers: Results of a systematic review. *Trends Food Sci. Technol.* **2017**, *67*, 44–57. [CrossRef]
6. Luckow, T.; Sheehan, V.; Fitzgerald, G.; Delahunty, C. Exposure, health information and flavor-masking strategies for improving the sensory quality of probiotic juice. *Appetite* **2006**, *47*, 315–325. [CrossRef] [PubMed]
7. Threlfall, R.; Morris, J.; Meullenet, J.F. Product development and nutraceutical analysis to enhance the value of dried fruit. *J. Food Qual.* **2007**, *30*, 552–566. [CrossRef]
8. Sabbe, S.; Verbeke, W.; Van Damme, P. Confirmation/disconfirmation of consumers' expectations of fresh and processed tropical fruit products. *Int. J. Food Sci. Technol.* **2009**, *44*, 539–551. [CrossRef]
9. Michon, C.; O'Sullivan, M.G.; Sheehan, E.; Delahunty, C.M.; Kerry, J.P. Investigation of the influence of age, gender and consumption habits on the liking of jam-filled cakes. *Food Qual. Prefer.* **2010**, *21*, 553–561. [CrossRef]
10. Koppel, K.; Chambers, E.; Vázquez-Araújo, L.; Timberg, L.; Carbonell-Barrachina, A.; Suwansichon, S. Cross-country comparison of pomegranate juice acceptance in Estonia, Spain, Thailand, and United States. *Food Qual. Prefer.* **2014**, *31*, 16–123. [CrossRef]
11. Morais, R.M.S.C.; Morais, A.M.M.B.; Dammak, I.; Bonilla, J.; Sobral, P.J.A.; Laguerre, J.C.; Afonso, M.J.; Ramalhosa, E.C.D. Functional dehydrated foods for health preservation. *J. Food Qual.* **2018**, 1–29. [CrossRef]
12. Global Market Insight. 2019. Available online: https://www.gminsights.com/industry-analysis/fruit-snacks-market (accessed on 12 October 2019).
13. Food Information Statistics System. Agriculture, Forestry, and Fisheries Import & Export & Statistics. 2018. Available online: https://www.atfis.or.kr/article/M001040000/view.do?articleId=3218&page=&searchKey=&searchString=&searchCategory= (accessed on 2 March 2020).
14. Rababah, T.M.; Ereifej, K.I.; Howard, L. Effect of ascorbic acid and dehydration on concentrations of total phenolics, antioxidant capacity, anthocyanins, and color in fruits. *J. Agric. Food Chem.* **2005**, *53*, 4444–4447. [CrossRef]
15. Asioli, D.; Rocha, C.; Wongprawmas, R.; Popa, M.; Gogus, F.; Almli, V.L. Microwave-Dried or Air-Dried? Consumers' stated preferences and attitudes for organic dried strawberries. A multi-country investigation in Europe. *Food Res. Int.* **2019**, *120*, 763–775. [CrossRef]
16. Jaeger, S.R.; Andani, Z.; Wakeling, I.N.; MacFie, H.J.H. Consumer preferences for fresh and aged apples: A cross-cultural comparison. *Food Qual Prefer.* **1998**, *9*, 355–366. [CrossRef]
17. Pasquariello, M.S.; Rega, P.; Migliozzi, T.; Capuano, L.R.; Scortichini, M.; Petriccione, M. Effect of cold storage and shelf life on physiological and quality traits of early ripening pear cultivars. *Sci. Hortic.* **2013**, *162*, 341–350. [CrossRef]
18. Tunick, M.H.; Onwulata, C.I.; Thomas, A.E.; Phillips, J.G.; Mukhopadhyay, S.; Sheen, S.; Liu, C.; Latona, N.; Pimentel, M.R.; Cooke, P.H. Critical evaluation of crispy and crunchy textures: A review. *Int. J. Food Prop.* **2012**, *16*, 949–963. [CrossRef]
19. Zou, K.; Teng, J.; Huang, L.; Dai, X.; Wei, B. Effect of osmotic pretreatment on quality of mango chips by explosion puffing drying. *Lwt Food Sci. Technol.* **2013**, *51*, 253–259. [CrossRef]

20. Prescott, J. Comparisons of taste perceptions and preferences of Japanese and Australian consumers: Overview and implications for cross-cultural sensory research. *Food Qual. Prefer.* **1998**, *9*, 393–402. [CrossRef]
21. Kohno, K.; Hayakawa, F.; Xichang, W.; Shunsheng, C.; Yokoyama, M.; Kasai, M.; Hatae, K. Comparative study on flavor preference between Japanese and Chinese for dried bonito stock and chicken bouillon. *J. Food Sci.* **2005**, *70*, S193–S198. [CrossRef]
22. Wan, X.; Woods, A.T.; Jacquot, M.; Knoeferle, K.; Kikutani, M.; Spence, C. The effects of receptacle on the expected flavor of a colored beverage: Cross-cultural comparison among French, Japanese, and Norwegian consumers. *J. Sens. Stud.* **2016**, *31*, 233–244. [CrossRef]
23. Lee, S.M.; Kim, S.E.; Guinard, J.X.; Kim, K.O. Exploration of flavor familiarity effect in Korean and US consumers' hot sauces perceptions. *Food Sci. Biotechnol.* **2016**, *25*, 745–756. [CrossRef]
24. Tepper, B.J.; White, E.A.; Koelliker, Y.; Lanzara, C.; d'Adamo, P.; Gasparini, P. Genetic variation in taste sensitivity to 6-n-propylthiouracil and its relationship to taste perception and food selection. International Symposium on Olfaction and Taste. *Ann. NY Acad. Sci.* **2009**, *1170*, 126–139. [CrossRef]
25. Bajec, M.R.; Pickering, G.J. Association of thermal taste and PROP responsiveness with food liking, neophobia, body mass index, and waist circumference. *Food Qual. Prefer.* **2010**, *21*, 589–601. [CrossRef]
26. Chung, L.; Chung, S.J.; Kim, J.Y.; Kim, K.O.; O'Mahony, M.; Vickers, Z.; Cha, S.M.; Ishii, R.; Baures, K.; Kim, H.R. Comparing the liking for Korean style salad dressings and beverages between US and Korean consumers: Effects of sensory and non-sensory factors. *Food Qual. Prefer.* **2012**, *26*, 105–118. [CrossRef]
27. Hong, J.H.; Park, H.S.; Chung, S.J.; Chung, L.; Cha, S.M.; Lê, S.; Kim, K.O. Effect of familiarity on a cross-cultural acceptance of a sweet ethnic food: A case study with Korean traditional cookie (Yackwa). *J. Sens. Stud.* **2014**, *29*, 110–125. [CrossRef]
28. Torrico, D.D.; Fuentes, S.; Gonzalez Viejo, C.; Ashman, H.; Dunshea, F.R. Cross-cultural effects of food product familiarity on sensory acceptability and non-invasive physiological responses of consumers. *Food Res. Int.* **2019**, *115*, 439–450. [CrossRef] [PubMed]
29. Rødbotten, M.; Martinsen, B.K.; Borge, G.I.; Mortvedt, H.S.; Knutsen, S.H.; Lea, P.; Næs, T. A cross-cultural study of preference for apple juice with different sugar and acid contents. *Food Qual. Prefer.* **2009**, *20*, 277–284. [CrossRef]
30. Kim, J.Y.; Prescott, J.; Kim, O. Patterns of sweet liking in sucrose solutions and beverages. *Food Qual. Prefer.* **2014**, *36*, 96–103. [CrossRef]
31. Risvik, E.; McEwan, J.A.; Colwill, J.S.; Rogers, R.; Lyon, D.H. Projective mapping: A tool for sensory analysis and consumer research. *Food Qual. Prefer.* **1994**, *5*, 263–269. [CrossRef]
32. Pagès, J. Collection and analysis of perceived product inter-distances using multiple factor analysis: Application to the study of 10 white wines from the Loire Valley. *Food Qual. Prefer.* **2005**, *16*, 642–649. [CrossRef]
33. Dehlholm, C.; Brockhoff, P.B.; Meinert, L.; Aaslyng, M.D.; Bredie, W.L. Rapid descriptive sensory methods–comparison of free multiple sorting, partial napping, napping, flash profiling and conventional profiling. *Food Qual. Prefer.* **2012**, *26*, 267–277. [CrossRef]
34. Varela, P.; Berget, I.; Hersleth, M.; Carlehög, M.; Asioli, D.; Næs, T. Projective mapping based on choice or preference: An affective approach to projective mapping. *Food Res. Int.* **2017**, *100*, 241–251. [CrossRef]
35. Kim, M.-R.; Kim, K.-P.; Chung, S.-J. Utilizing hedonic frame for projective mapping: A case study with Korean fermented soybean paste soup. *Food Qual. Prefer.* **2019**, *71*, 279–285. [CrossRef]
36. Williams, E.J. Experimental designs balanced for the estimation of residual effects of treatments. *Aust. J. Sci. Res.* **1949**, *2*, 149–168. [CrossRef]
37. Nestrud, M.A.; Lawless, H.T. Perceptual mapping of apples and cheeses using projective mapping and sorting. *J. Sens. Stud.* **2010**, *25*, 390–405. [CrossRef]
38. Husson, F.; Josse, J.; Lê, S.; Mazet, J. FactoMineR: Multivariate Exploratory Data Analysis and Data Mining with R. R Package Version 1.16. 2011. Available online: http://factominer.free.fr, http://www.agrocampus-rennes.fr/math/ (accessed on 4 October 2019).
39. Prescott, J.; Young, O.; Zhang, S.; Cummings, T. Effects of added "flavour principles" on liking and familiarity of a sheepmeat product: A comparison of Singaporean and New Zealand consumers. *Food Qual. Prefer.* **2004**, *15*, 187–194. [CrossRef]
40. Tu, V.P.; Valentin, D.; Husson, F.; Dacremont, C. Cultural differences in food description and preference: Contrasting Vietnamese and French panellists on soy yogurts. *Food Qual. Prefer.* **2010**, *21*, 602–610. [CrossRef]

41. Choi, J.H.; Gwak, M.J.; Chung, S.J.; Kim, K.O.; O'Mahony, M.; Ishii, R.; Bae, Y.W. Identifying the drivers of liking by investigating the reasons for (dis)liking using CATA in cross-cultural context: A case study on barbecue sauce. *J. Sci. Food Agric.* **2015**, *95*, 1613–1625. [CrossRef] [PubMed]
42. Kim, H.-J.; Chung, S.-J.; Kim, K.-O.; Nielsen, B.; Ishii, R.; O'Mahony, M. A cross-cultural study of acceptability and food pairing for hot sauces. *Appetite* **2018**, *123*, 306–316. [CrossRef]
43. Blancher, G.; Chollet, S.; Kesteloot, R.; Nguyen, D.; Cuvelier, G.; Sieffermann, J.-M. French and Vietnamese: How do they describe texture characteristics of the same food? A case study with jellies. *Food Qual. Prefer.* **2007**, *18*, 560–575. [CrossRef]
44. Symoneaux, R.; Galmarini, M.V.; Mehinagic, E. Comment analysis of consumer's likes and dislikes as an alternative tool to preference mapping. A case study on apples. *Food Qual. Prefer.* **2012**, *24*, 59–66. [CrossRef]
45. Seppä, L.; Railio, J.; Vehkalahti, K.; Tahvonen, R.; Tuorila, H. Hedonic responses and individual definitions of an ideal apple as predictors of choice. *J. Sens. Stud.* **2013**, *28*, 346–357. [CrossRef]
46. Food Information Statistics System. POS Retail Store Sales by Items. 2017. Available online: https://www.atfis.or.kr/sales/M002020000/ (accessed on 15 January 2020).
47. Wang, H. *Sector Trend Analysis- Savoury Snacks in the United States*; Agriculture and Agri-Food Canada: Ottawa, ON, Canada, 2018.
48. Wong, B. *China's Packaged Food Market: Consumer Snack Preferences*; Hong Kong Trade Development Council Research: Hong Kong, China, 2018.
49. Joshi, A.; Rupasinghe, H.; Khanizadeh, S. Impact of drying processes on bioactive phenolics, vitamin C and antioxidant capacity of red-fleshed apple slices. *J. Food Process. Preserv.* **2011**, *35*, 453–457. [CrossRef]
50. Brown, W.E.; Braxton, D. Dynamics of food breakdown during eating in relation to perceptions of texture and preference: A study on biscuits. *Food Qual. Prefer.* **2000**, *11*, 259–267. [CrossRef]
51. Jaworska, D.; Hoffmann, M. Relative importance of texture properties in the sensory quality and acceptance of commercial crispy products. *J. Sci. Food Agric.* **2008**, *88*, 1804–1818. [CrossRef]
52. Harwood, M.L.; Ziegler, G.R.; Hayes, J.E. Rejection thresholds in chocolate milk: Evidence for segmentation. *Food Qual. Prefer.* **2012**, *26*, 128–133. [CrossRef]
53. Harwood, M.L.; Loquasto, J.R.; Roberts, R.F.; Ziegler, G.R.; Hayes, J.E. Explaining tolerance for bitterness in chocolate ice cream using solid chocolate preferences. *J. Dairy Sci.* **2013**, *96*, 4938–4944. [CrossRef] [PubMed]
54. Divert, C.; Chabanet, C.; Schoumacker, R.; Martin, C.; Lange, C.; Issanchou, S.; Nicklaus, S. Relation between sweet food consumption and liking for sweet taste in French children. *Food Qual. Prefer.* **2017**, *56*, 18–27. [CrossRef]
55. Jeltema, M.; Beckley, J.; Vahalik, J. Model for understanding consumer textural food choice. *Food Sci. Nutr.* **2015**, *3*, 202–212. [CrossRef]

Article

Discrete Choice Analysis of Consumer Preferences for Meathybrids—Findings from Germany and Belgium

Adriano Profeta [1,*], Marie-Christin Baune [1], Sergiy Smetana [1], Keshia Broucke [2], Geert Van Royen [2], Jochen Weiss [3], Volker Heinz [1] and Nino Terjung [1]

1 DIL e.V.–German Institute of Food Technology, Prof.-von-Klitzing-Straße 7, D-49610 Quakenbrück, Germany; m.baune@dil-ev.de (M.-C.B.); s.smetana@dil-ev.de (S.S.); v.heinz@dil-ev.de (V.H.); n.terjung@dil-ev.de (N.T.)

2 Institute for Agricultural and Fisheries Research (ILVO), Brusselsesteenweg, 370, 9090 Melle, Belgium; keshia.broucke@ilvo.vlaanderen.be (K.B.); geert.vanroyen@ilvo.vlaanderen.be (G.V.R.)

3 Institute of Food Science and Biotechnology, Department of Food Structure and Functionality, University of Hohenheim, Garbenstraße 21/25, 70599 Stuttgart, Germany; j.weiss@uni-hohenheim.de

* Correspondence: a.profeta@dil-ev.de; Tel.: +49-5431-183-326

Abstract: High levels of meat consumption are increasingly being criticised for ethical, environmental and social reasons. Plant-based meat substitutes have been identified as healthy sources of protein that, in comparison to meat, offer a number of social, environmental and health benefits and may play a role in reducing meat consumption. However, there has been a lack of research on the role they can play in the policy agenda and how specific meat substitute attributes can influence consumers to partially replace meat in their diets. This paper is focused on consumers' preferences for so-called meathybrid or plant-meathybrid products. In meathybrids, only a fraction of the meat product (e.g., 20% to 50%) is replaced with plant-based proteins. Research demonstrates that in many countries, consumers are highly attached to meat and consider it as an essential and integral element of their daily diet. For these consumers that are not interested in vegan or vegetarian alternatives as meat substitutes, meathybrids could be a low-threshold option for a more sustainable food consumption behaviour. In this paper, the results of an online survey with 500 German and 501 Belgian consumers are presented. The results show that more than fifty percent of consumers substitute meat at least occasionally. Thus, about half of the respondents reveal an eligible consumption behaviour with respect to sustainability and healthiness, at least sometimes. The applied discrete choice experiment demonstrated that the analysed meat products are the most preferred by consumers. Nonetheless, the tested meathybrid variants with different shares of plant-based proteins took the second position followed by the vegetarian-based alternatives. Therefore, meathybrids could facilitate the diet transition of meat-eaters in the direction toward a more healthy and sustainable consumption. The analysed consumer segment is more open-minded to the meathybrid concept in comparison to the vegetarian substitutes.

Keywords: meat substitute; meathybrid; consumer preference; plant-based proteins

Citation: Profeta, A.; Baune, M.-C.; Smetana, S.; Broucke, K.; Van Royen, G.; Weiss, J.; Heinz, V.; Terjung, N. Discrete Choice Analysis of Consumer Preferences for Meathybrids—Findings from Germany and Belgium. *Foods* **2021**, *10*, 71. https://doi.org/10.3390/foods10010071

Received: 9 December 2020
Accepted: 25 December 2020
Published: 31 December 2020

1. Introduction

There are more than 7.7 billion people on this planet, with forecasts predicting the population to grow to 9.7 billion by 2050 [1]. Securing a sustainable food supply for humankind is therefore becoming a major challenge. Diets with a high share of animal proteins must be adapted in order to ensure that demand is not outstripping production [2,3]. Furthermore, the consumption of meat and meat products in larger portions is associated with higher risk of cardiovascular, coronary and cerebrovascular diseases, stroke, diabetes type 2 and colorectal cancer [4].

In addition to these health issues, meat production chains have a considerable impact on the environment through the use of land, the application of fertilisers, greenhouse emissions and water consumption, resulting in the loss of biodiversity and enhancing climate change [5–8]. It causes more emissions per unit of energy compared with plant-based foods because energy is lost at each trophic level. Meat production is the most important source of methane, which has a relatively high global warming potential, but a lower half-life in the environment compared with that of CO_2 [9]. The carbon footprint of plant-based foods on average is twice as low as the impact of pork [10], while the impact in some other categories can be more than 60 times lower [11]. We also highlight that meat and meat products are associated with severe animal welfare issues, such as pigtail docking, poultry debeaking, calve separation and mistreatment in slaughterhouses [12,13].

Integrating new protein sources into the diet as a solution for the mentioned problems means overcoming barriers such as traditional meat consumption across many cultures [14]. Recent research put forward the idea that consumers have an affective connection with meat (meat attachment) that may play a role in their willingness to change consumption habits [15]. It is argued that the affection towards meat may represent a continuum in which one end refers to disgust (i.e., negative affect and repulsion, related to moral internalization), while the other shows a pattern of attachment (i.e., high positive affect and dependence towards meat, as well as feelings of sadness and deprivation when considering abstaining from meat consumption) that may hinder a change in consumption habits [15]. Likewise, food neophobia, which refers to the reluctance to eat unfamiliar foods [16], may represent a barrier for a transition to a more sustainable diet. According to Apostolidis and McLeay [17], low levels of acceptance of meat substitutes have been associated with high levels of the construct food neophobia.

For increasing the share of plant proteins in the diet, there are several options. An approach could be the usage of textured soy protein, mushrooms, wheat gluten, pulses, etc., as a complete substitute for animal protein. Another opportunity is to replace only a fraction of the meat product (e.g., 20% to 50%) with plant-based proteins [18]. As mentioned in many countries, consumers are highly attached to meat and consider it as an essential and integral element of their daily diet [15]. So-called meathybrids may be an option for the broad consumer segment that is not interested in totally vegan or vegetarian alternatives to meat. Therefore, meathybrids could serve as a low-threshold offer for this group, facilitating the transition in the direction toward a more healthy and sustainable diet. In this context, it has to be mentioned that consumer preferences are in particular affected by the products' sensory characteristics. An inferior or low sensory quality can constitute a critical barrier for the market entry of meat substitutes [19,20]. Therefore, meat substitutes, respectively meathybrids, must catch up with real meat products concerning sensory characteristics.

As with many novel technologies, consumers' lack of understanding of hybrid meat products may lead to scepticism and ultimately to the rejection of these. Through early integration of consumer demand and preferences into the development process, more suitable hybrid products can be designed. Understanding the decision-making process will help to develop tailored communication messages that highlight its benefits as a sustainable and healthy alternative to regular meat products.

The study aims at identifying consumer attitudes and preferences for meat alternatives such as meathybrids. Based on a concise literature overview, a representative online survey was carried out in Germany and Belgium including a Discrete Choice Experiment (DCE) for four product categories (meat balls, chicken nuggets, salami, and mortadella).

2. Data Collection and Methods of Data Analysis

Consumer data were collected using a quantitative online survey approach. The respondents were panellists and were recruited by the market research company Savanta (London, UK). The questionnaire comprised questions about general meat consumption, on the one hand, and specific questions concerning preferences for meat substitutes, on the other.

So-called choice experiments were integrated in the survey for measuring the importance and preference of different levels of plant-based protein shares in mortadella, salami, chicken nuggets, and meat balls. Choice experiments have been shown to reduce social desirability bias [21], as individuals often display socially desirable preferences in surveys [22].

The online survey was carried out in Germany with 500 and in Belgium with 501 respondents. Participants had to be meat eaters, and thus, vegetarians and vegans were sorted out a priori. Furthermore, the participants had to be 50% responsible for food shopping in the household. Concerning the age, respondents had to be in the range of 18 to 69 years. Data collection took place in the time period from 8 November until 19 November 2019 (see Table 1). Both samples are approximately representative in relation to gender and the region of residence (federal states). For the age, we highlight that the age group from 50–59 years was somewhat under-represented, whereas the age group from 60–69 years was over-represented in both countries.

Table 1. Sample.

			Germany			Belgium	
			Sample	pop. *		Sample	pop. *
Attribute	**Characteristics**	*n*	%	%	*n*	%	%
gender	male	245	49.0	49.5	245	48.9	50.0
	female	255	51.0	50.5	256	51.1	50.0
federal state	Baden-Württemberg I Bruxelles	66	13.2	12.9	62	12.4	10.8
	Bayern I Brabant wallon	75	15.0	15.9	27	5.4	3.5
	Berlin I Hainaut	22	4.4	4.4	68	13.6	11.7
	Brandenburg I Liège	15	3.0	3.0	44	8.8	9.7
	Bremen I Luxembourg	5	1.0	0.8	13	2.6	2.5
	Hamburg I Namur	10	2.0	2.2	27	5.4	4.4
	Hessen I Antwerpen	37	7.4	7.4	88	17.6	16.2
	Mecklenburg-Vorp. I Provincie Limb.	10	2.0	1.9	20	4.0	7.8
	Niedersachsen I Oost-Vlaand.	50	10.0	9.7	58	11.6	10.6
	Nordrhein-Westfalen I Vlaams-Brab.	114	22.8	22.0	60	11.9	10.0
	Rheinland-Pfalz I West-Vlaand.	20	4.0	5.1	34	6.8	10.3
	Saarland	5	1.0	1.2			
	Sachsen	26	5.2	4.9			
	Sachsen-Anhalt	15	3.0	2.6			
	Schleswig-Holstein	15	3.0	3.4			
	Thüringen	15	3.0	2.6			
age	18–29 years	95	19.0	20.6	106	21.2	19.3
	30–39 years	88	17.6	18.7	98	19.6	20.3
	40–49 years	84	16.8	18.4	106	21.2	20.6
	50–59 years	88	17.6	23.9	76	15.2	21.8
	60–69 years	145	29.0	18.3	115	23.0	18.1
education	no school qualifications	2	0.4		22	4.4	
	still in school	4	0.8		18	3.6	
	junior high diploma	88	17.6		20	4.0	
	high school diploma	193	38.6		229	45.7	
	university-entrance diploma	105	21.0		78	15.6	
	bachelor's or master's degree	89	17.8		122	24.4	
	other degree	19	3.8		12	2.4	
net income	no income	26	5.2		39	7.8	
	less than 500€	30	6.0		19	3.8	
	500€ up to 1000€	46	9.2		36	7.2	
	1000€ up to 1500€	95	19.0		98	19.6	
	1500€ up to 2000€	92	18.4		115	23.0	
	2000€ up to 2500€	69	13.8		89	17.8	
	2500€ up to 3000€	57	11.4		38	7.6	
	3000€ up to 3500€	27	5.4		27	5.4	
	3500€ up to 4000€	25	5.0		23	4.6	
	4000€ or more	33	6.6		17	3.4	
household size	1	121	24.2		112	22.4	
	2	207	41.4		164	32.7	
	3	92	18.4		96	19.2	
	4	55	11.0		82	16.4	
	5	20	4.0		33	6.6	
	6	4	0.8		9	1.8	
	>6	1	0.2		5	1.0	

* Sources: www.statbel.fgov.be and b4p2019 I—Strukturanalyse (www.gik.media/best-4-planning).

In the Results Section, we report descriptive results. For scale development, Cronbach's alpha was applied [23]. Furthermore, confirmatory factor analyses were run to confirm the validity of the scales by using the R-package psych [24]. For measuring food neophobia, the Food Neophobia Scale (FNS) of Pliner and Hobden [16] was selected. The wording of the German version was chosen from a study by [25]. Participants answered on a five-point response scale that was verbally and numerically anchored (1 = totally disagree, 2 = disagree, 3 = neither disagree nor agree, 4 = agree, 5 = totally agree). The five-point scale was used instead of the originally used seven-point scale for a better display of the questionnaire on tablets and smartphones. The items indicated with (r) in Table 6 were inversely re-coded. Considering that the inclusion of invalid items creates the risk of invalid conclusion [26], a principal components analysis (varimax rotation, eigenvalues greater than one) was carried out to explain the variability of the FNS followed by a confirmatory factor analysis [27]. For measuring consumers' meat attachment, participants answered a five-point Meat Attachment Scale (MEAS) [28] that was verbally and numerically anchored (1 = strongly disagree, 2 = disagree, 3 = neither disagree nor agree, 4 = agree, 5 = strongly agree). The items indicated with (r) in Table 5 were inversely re-coded. In this study, the Health Consciousness Scale (HS) in the style of Visschers et al. [29] was selected for measuring the impact of this psychometric construct on the choice of hybrid products.

Furthermore, a multinomial logistic regression model was applied for measuring the impact of several parameters on the the choice of meat and meathybrid products. Data were collected via a Discrete Choice Experiment (DCE).

2.1. Discrete Choice Experiment and Experimental Design

The DCE method is based on micro-economic theory according to which consumers always try to maximize their benefit [30]. In DCEs, consumers must choose from a set of different products offered at determined prices. The products differ regarding the tested product attributes (e.g., share of local feed, price, etc.). According to micro-economic theory, participants will choose the product with the highest benefit. By means of DCEs, consumers' benefit for each tested product attribute can thus be revealed, as well as the influence of each product attribute on the probability of purchasing/choosing the product. In the DCE of this study, the products varied by six attributes: plant-based protein share, EU organic label, origin label for the protein source, environmental claim, nutritional label, and price (see Table 2). The EU organic label was included since previous studies had shown the importance of this aspect to consumers. The five price levels used in the choice experiment were within the price range that encompassed observed market prices at food retailers in Germany during the winter of 2018/1019. The reported attributes and attribute levels were used for generating the experimental design of the choice experiment. The DCE was carried out for four product categories (meatballs, mortadella, salami, chicken nuggets) on the basis of the same underling experimental design structure.

Table 2. Attributes and attribute levels used in the Discrete Choice Experiment (DCE).

Attributes	Levels
plant-based protein share	100% (vegetarian), 50%, 35%, 20%, 20, 0% (meat)
EU organic label	yes, no
origin label prot.source	locally produced produced in Ger/Belg no indicated origin
environmental claim	20% reduced carbon foot print no indicated claim
nutritional label	high content of non-saturated fatty acids high in fibre no indicated label
price	high, middle, low

In each choice set, consumers had a choice between four product alternatives and a no-choice option. The no-choice option was included to get a more realistic purchase situation and thus raise the validity of the data [31]. Furthermore, there was always one

100% meat option and one vegetarian option in the sets, whereas for two options, the plant-based protein share varied between 50% and 20%.

A D-efficient unlabelled design (0.949) was generated using the software Ngene [32], and for each product category, eight choice sets were generated. Thus, in total, there were 32 choice sets. The priors used were based on expert judgement and the literature.

Each participant received two choice sets from each product category and thus had to answer in total eight choice sets. The survey order of the choice sets of the alternatives was randomised to prevent ordering effects [33]. The products, respectively the characteristics, are depicted in photographs (see Table 3).

Table 3. Choice set example—meat balls.

	Meat Ball 1	**Meat Ball 2**	**Meat Ball 3**	**Meat Ball 4**
	300 g	300 g	300 g	300 g
Ingredients (plant-based protein share)	100% pork	50% pork + 50% plant-based protein seed	65% pork + 35% plant-based protein seed	100% vegetarian
Organic label				
Price	3.29€	2.29€	2.29€	2.29€
Origin of meat, resp. plant-based protein source	Locally produced			Produced in Germany
Environmental claim		20% reduced carbon foot print		

2.2. Multinomial Logistic Regression

Multinomial logistic regression is the regression analysis to conduct when the dependent variable is nominal with more than two levels. It is used to model nominal outcome variables, in which the log odds of the outcomes are modelled as a linear combination of the predictor variables. The multinomial logistic model belongs to the family of generalized linear models and as mentioned is used when the response variable is a categorical variable. Suppose that variable Y_i represents the offered alternatives in a choice experiment (e.g., the choice between meat and meathybrid), with $i = 1, \ldots, n$, and n is the number of possible product alternatives. In the case that n equals 2, Y has outcomes Y_1 and Y_2. Both the counts of Y_1 and Y_2 follow a binomial distribution. The probability of occurrence of Y_1 is π_1 and that of Y_2 is π_2. Logistic regression relates probability π_1 to a set of predictors using the logit link function:

$$logit(\pi_1) = ln(\frac{\pi_1}{\pi_2}) = ln(\frac{\pi_1}{1-\pi_1}) = \mathbf{x}'\beta \quad (1)$$

where $\mathbf{x}$ is a vector of predictors (e.g., FNS, MEAS or buying frequency of organic meat) and β is a vector of model coefficients that are typically estimated by maximum likelihood. Equation (1) can be rewritten as:

$$(\frac{\pi_1}{1-\pi_1}) = exp(\mathbf{x}'\beta) = exp(\eta) \quad (2)$$

The quotient in Equation (2) is referred to as the odds. From Equation (2), it follows that:

$$\pi_1 = \frac{exp(\eta)}{1 + exp(\eta)} \tag{3}$$

The binomial logistic regression model is easily generalized to the multinomial case. If there are n product alternatives, there are also n variables $Y_1, \ldots, Y_n$ with corresponding probabilities of occurrence $\pi_1, \ldots, \pi_n$. Analogous to binomial logistic regression, the odds $\pi_1/\pi_n, \ldots, \pi_n - 1/\pi_n$ are modelled by means of $exp(\eta_1), \ldots, exp(\eta_{n-1})$. From $\sum_{i=1}^{n} \pi_i = 1$, it follows that:

$$\pi_1 = \frac{exp(\eta_i)}{exp(\eta_1) + exp(\eta_2) + \ldots + exp(\eta_n)} \tag{4}$$

where $exp(\eta_n) = 0$. This model ensures that all probabilities are in the interval [0, 1] and that the probabilities sum to 1.

In this paper, the dependent variable is taken from the DCE where respondents had to indicate if they would buy/choose one out of the four offered options or none of these options. The FNS, HS and MEAS, as well as other parameters (e.g., FAMILIARITY= buying frequency of meat substitutes) entered the regression analysis as independent variables. In addition, all three scales were interacted with the different levels of the attribute "plant-based protein share" for analysing their effect on meat, hybrids and the vegetarian alternative.

Given the theoretical background, an model was built according to the following expression:

$$\begin{aligned}
\mathbf{x}'\beta = {} & \text{meat} * \beta_1 + (\text{meat + 50plant}) * \beta_2 + (\text{meat + 35plant}) * \beta_3 + (\text{meat + 20plant}) * \beta_4 \\
& + \text{reduced CO}_2 * \beta_5 + \text{organic} * \beta_6 + \text{Ger/Bel origin} * \beta_7 + \text{local origin} * \beta_8 \\
& + \text{high in fibre} * \beta_9 + \text{high of nsf.acids} * \beta_{10} + \text{price} * \beta_{11} \\
& + \text{HS} * \text{meat} * \beta_{12} + \text{HS} * (\text{meat + 50plant}) * \beta_{13} + \text{HS} * (\text{meat + 35plant}) * \beta_{14} \\
& + \text{HS} * (\text{meat + 20plant}) * \beta_{15} + \text{FNS} * \text{meat} * \beta_{16} + \text{FNS} * (\text{meat + 50plant}) * \beta_{17} \\
& + \text{FNS} * (\text{meat + 35plant}) * \beta_{18} + \text{FNS} * (\text{meat + 20plant}) * \beta_{19} + \text{MEAS} * \text{meat} * \beta_{20} \\
& + \text{MEAS} * (\text{meat + 50plant}) * \beta_{21} + \text{MEAS} * (\text{meat + 35plant}) * \beta_{22} + \text{MEAS} * (\text{meat + 20plant}) * \beta_{23} \\
& + \text{FAMILIARITY} * \text{meat} * \beta_{24} + \text{FAMILIARITY} * (\text{meat + 50plant}) * \beta_{25} \\
& + \text{FAMILIARITY} * (\text{meat + 35plant}) * \beta_{26} + \text{FAMILIARITY} * (\text{meat + 20plant}) * \beta_{27} + \text{no-option}
\end{aligned} \tag{5}$$

From the estimation results, odds ratios are calculated. Odds ratios in logistic regression can be interpreted as the effect of one unit of change in X in the predicted odds ratio with the other variables in the model held constant.

In this study, for estimating the specified model, the software R [34] and the package mlogit [35] were used. For the visualisation of the odds ratios from the estimated model, the package sjplot [36] was applied.

3. Results

3.1. General Buying Behaviour

At the beginning of the questionnaire, the participants had to indicate where they buy most of their meat products. The classical retailer took the first position (48.6%) followed by discount shops (38.6%). Butcheries were in third position (10.2%). All other options were only of minor importance (see Figure 1).

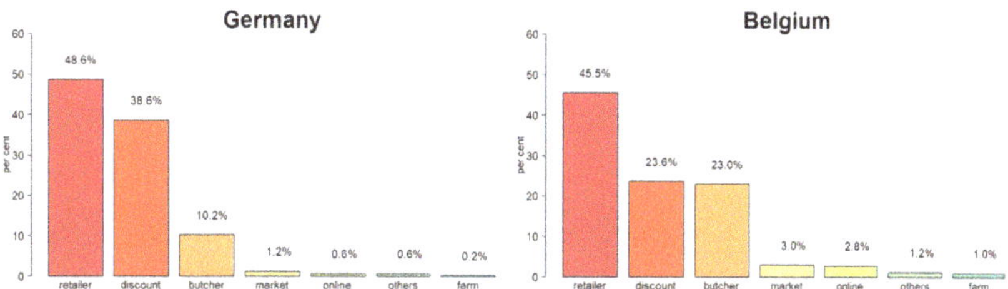

Figure 1. Preferred buying location of meat/meat products.

Furthermore, respondents were asked for their buying frequency of organic, respectively free-range, meat. About 22% of the participants indicated buying such products often (18.2%) or always (4.2%) (see Table 4). In 2019, a survey was conducted by Kitchen Stories investigating the purchasing behaviour towards organic food in Germany. In the mentioned study, somewhat higher values were found with 13.2% buying mostly organic products, while for 18.6% of the respondents, organic food made up more than half of the shopping cart.

Table 4. Buying frequency of organic/free-range meat.

	Germany	Belgium
never	24.4%	15.4%
sometimes	57.7%	62.2%
often	15.2%	18.2%
always	3.0%	4.2%

3.2. *Scales: Meat Attachment Scale, Neophobia Food Scale and Health Scale*

3.2.1. Meat Attachment Scale

In Germany, due to the confirmatory factor analysis, the item "I would feel fine with a meatless diet" was deleted from the scale because in the four-factor solution, this item had a similar loading on different factors and its deletion increased the calculated indices. The reliability analysis for the global MEAS showed in Germany a high internal consistency with a standardized Cronbach α of 0.86. The Comparative Fit Index (CFI = 0.962), the Tucker–Lewis Index (TLI = 0.952) and the Root Mean Squared Error of Approximation (RMSEA = 0.060) showed acceptable values.

In Belgium, likewise due to the confirmatory factor analysis, the item "I would feel fine with a meatless diet" was deleted from the scale because in the four-factor solution, this item had a similar loading on different factors. The reliability analysis for the global MEAS showed a high internal consistency with a standardized Cronbach α of 0.86. The Comparative Fit Index (CFI = 0.959), the Tucker–Lewis Index (TLI = 0.947) and the Root Mean Squared Error of Approximation (RMSEA = 0.067) showed acceptable values.

In comparison to Graça et al. [15], in both countries, we received higher values for the non-reversed items and lower values for the reversed item, which was due to the fact that vegans and vegetarians were not part of this study (see Table 5). On average, respondents agreed to all of the statements. The highest means were received for the statements "I love meals with meat" (3.94) and the reverse-coded item "Meat reminds me of diseases". The MEAS findings demonstrates that on average, German and Belgium respondents considered meat not as an unhealthy product, but as an essential part of their diet.

Table 5. Meat Attachment Scale (MEAS) questionnaire.

	Germany			Belgium		
Statement	**std. α**	$\bar{x}$	σ	**std. α**	$\bar{x}$	σ
I love meals with meat.	0.84	3.94	1.00	0.84	3.69	1.03
To eat meat is one of the good pleasures in life.	0.85	3.38	1.08	0.84	3.36	1.10
I'm a big fan of meat.	0.84	3.58	1.07	0.84	3.58	1.07
A good steak is without comparison.	0.84	3.76	1.12	0.85	3.43	1.13
By eating meat I'm reminded of the death and suffering of animals. (r)	0.86	3.50	1.19	0.87	3.42	1.25
To eat meat is disrespectful towards life and the environment. (r)	0.86	3.30	1.19	0.87	3.23	1.12
Meat reminds me of diseases. (r)	0.86	3.86	1.18	0.87	3.70	1.15
To eat meat is an unquestionable right of every person.	0.86	3.57	1.12	0.86	3.60	1.05
According to our position in the food chain, we have the right to eat meat.	0.86	3.68	1.13	0.87	3.70	1.15
Eating meat is a natural and indisputable practice.	0.85	3.75	0.98	0.85	3.58	1.00
I don't picture myself without eating meat regularly.	0.85	3.56	1.14	0.85	3.44	1.10
If I couldn't eat meat I would feel weak.	0.85	3.12	1.19	0.85	3.07	1.07
I would feel fine with a meatless diet.		3.32	1.14		3.11	1.11
If I was forced to stop eating meat I would feel sad.	0.85	3.38	1.15	0.85	3.35	1.14
Meat is irreplaceable in my diet.	0.84	3.43	1.11	0.85	3.29	1.02

Note: 5-point Likert scale: 1 = strongly disagree, 2 = disagree, 3 = neither disagree nor agree, 4 = agree, 5 = strongly agree.

3.2.2. Food Neophobia Scale

For Germany, after deleting two items from the original FNS list due to low item correlations in the reliability analysis and one item due to the confirmatory factor analysis, the FNS showed an acceptable internal consistency with a standardized Cronbach α of 0.76 (see Table 6).

Table 6. Food Neophobia Scale (FNS). (r), inversely re-coded.

	Germany			Belgium		
Statement	**std. α**	$\bar{x}$	σ	**std. α**	$\bar{x}$	σ
I am constantly sampling new and different food. (r)	0.74	2.75	1.17	0.74	2.79	1.14
I do not trust new (different or innovative) food.		2.93	1.11	0.73	2.81	1.05
If I don't know what a food is, I won't try it.		3.85	1.00	0.74	3.16	1.08
I prefer food from different cultures. (r)	0.72	2.59	1.07	0.75	2.92	1.03
I am reluctant to eat foreign food that I see for the first time.	0.75	2.96	1.21	0.71	2.86	1.17
If I go to a buffet, meetings or parties, I'll eat new food. (r)	0.73	2.32	1.09	0.73	2.45	0.99
I'm afraid to eat food that I did not eat before.	0.74	2.49	1.23	0.71	2.66	1.18
I am very particular about the food I eat.		2.94	1.13	0.74	3.00	1.26
I will eat almost anything. (r)	0.76	2.32	1.13		2.65	1.20
I like to try new ethnic restaurants. (r)	0.70	2.36	1.10	0.73	2.61	1.07

Note: 5-point Likert scale: 1 = totally disagree, 2 = disagree, 3 = neither disagree nor agree, 4 = agree, 5 = totally agree.

The confirmatory factor analysis (two-factor solution) produced acceptable values for the three considered indices (CFI = 0.961, TLI = 0.937 and RMSEA = 0.074). The deleted items were: "I do not trust new (different or innovative) food", "If I don't know what a food is, I won't try it", and "I am very particular about the food I eat". For use in the regression analysis, the individual scores, that is the z-standardised mean value across the seven items, were calculated. The higher the FNS is, the higher is the individuals' food neophobia.

For Belgium, Item No. 9 had to be deleted due to the findings of the confirmatory factor analysis, and the FNS showed an acceptable internal consistency with a standardized Cronbach α of 0.75. The Comparative Fit Index (CFI = 0.949) and the Tucker–Lewis Index (TLI) (0.929), as well as the Root Mean Squared Error of Approximation (RMSEA) 0.073 showed acceptable values for the two-factor solution.

3.2.3. Health Scale

For the applied Health Scale (HS) in Germany (α = 0.81) and Belgium (α = 0.88), acceptable internal consistencies could be measured (see Table 7). Because the HS consisted only of three items, a CFAwith one factor has zero degrees of freedom. In this case, the model is saturated, and there are no degrees of freedom left over to assess model fit. Nonetheless, due to the high Cronbach α values and high factor loadings (>0.6) in the factor analyses, the developed scale was used for the subsequent analysis.

Table 7. Health Scale (HS).

	Germany			Belgium		
Statement	**std. α**	$\overline{x}$	σ	**std. α**	$\overline{x}$	σ
I think it is important to eat healthily	0.74	5.78	1.30	0.83	5.52	1.43
My health is dependent on how and what I eat	0.67	5.38	1.40	0.80	5.29	1.49
If one eats healthily, one gets ill less frequently	0.82	5.33	1.37	0.88	5.14	1.53

Note: 5-point Likert scale: 1 = totally disagree, 2 = disagree, 3 = neither disagree nor agree, 4 = agree, 5 = totally agree.

3.3. *Consumption and Perception of Substitutes*

The survey questionnaire comprehended several direct questions about the consumption of meat substitutes. In this context, respondents were asked if they deliberately substitute meat on the days they did not eat meat. In this context, a high proportion of 54.2% of the respondents stated consciously choosing meatless alternatives (see Table 8).

Table 8. Deliberate substitution of meat on the days respondents did not eat meat.

	Germany	Belgium
	%	%
yes	54.2	58.7
no	45.8	41.3

Subsequently, this group had to indicate with which products they concretely substitute meat. For this purpose, they received a list of twelve products, and from that, up to three products could be chosen. The option fish was selected by 48.3% of this segment in Germany and 66.7% in Belgium, followed by cheese (G: 47.6%, B: 29.6%), eggs (G: 41.7%, B: 58.8%), pasta (G: 39.5%, B: 36.7%) and salad (G: 35.4%, B: 16.7%) as the most preferred substitutes (see Table 9). We highlight that the top three on the list were non-vegan alternatives, whereas vegan alternatives like protein-rich lentils, tofu, or seitan were only of minor importance. Furthermore, the findings correspond with the results of De Boer et al. [2], who found a similar ranking (fish, eggs, cheese, etc.) with lentils, nuts, seitan, tempeh and tofu as less often mentioned items (<20%). From a sustainability perspective, the first ranked products do not offer much advantage compared with meat [37].

Additionally, all respondents were asked how often they buy plant-based meat substitutes, such as veggie burgers. Interestingly, only 4.0% (Germany), respectively 4.4% (Belgium), indicated consuming such products frequently, whereas 14.4%, respectively 16.6%, stated doing so at least sometimes (see Table 10). The figures are somewhat lower as those found by De Boer et al. [2] for the Netherlands, where 8% of the respondents reported buying such products frequently. In contrast, similar values as in Germany were observed by Siegrist and Hartmann [25] for Switzerland, where about 23.5% stated consuming substitutes.

Table 9. Ranking list of consumed meat alternatives.

	Germany		Belgium	
nr	**Product**	**%**	**Product**	**%**
1	Fish	48.3	Fish	66.7
2	Cheese	47.6	Egg(s)	58.8
3	Egg(s)	41.7	Pasta	36.7
4	Pasta	39.5	Cheese	29.6
5	Salad	35.4	Salad	16.7
6	Other legumes	15.1	Lentils	10.9
7	Lentils	9.6	Nuts	6.5
8	Nuts	8.9	Other legumes	5.4
9	Tofu	6.3	Tofu	5.1
10	Seitan	1.8	Other	2.3
11	Other	1.1	Tempeh	1.0
12	Tempeh	0.4	Seitan	0.7

Table 10. Frequency of consumption of meat alternatives such as veggie burgers.

	Germany	Belgium
	%	%
never	45.6	41.3
tried it once	16.0	14.6
rarely	20.0	23.2
sometimes	14.4	16.6
frequently	4.0	4.4

In the study, respondents had to indicate if they considered either meathybrids or meat as tastier. Furthermore, they had to decide which of the alternatives was better for the environment, better for animal welfare and healthier. Concerning the parameters environment and animal welfare, the meathybrid was evaluated as much better than the meat option (see Table 11). Contrarily, meat was perceived as tastier in comparison to the meathybrid by 62.4% of the respondents in Germany and 62.7% in Belgium. Concerning the perceived healthiness, the findings differed between the countries. Whereas in Germany, the hybrid was perceived as healthier, the opposite held for Belgium. We highlight that contrary to the reported literature for the perception of meat substitutes, at least in Germany, meathybrids were on average seen as healthier than meat. This outcome is quite surprising against the background that only respondents that consume meat were interviewed.

Table 11. Perception meat vs. hybrid.

	Germany			Belgium		
	Meat	**Neither/Nor**	**Hybrid**	**Meat**	**Neither/Nor**	**Hybrid**
tastier	62.4%	20.8%	16.8%	62.7%	14.0%	23.4%
healthier	31.0%	27.2%	41.8%	45.3%	14.4%	40.3%
better for environment	15.8%	31.0%	53.2%	22.6%	24.2%	53.3%
better for animal welfare	15.6%	26.8%	57.6%	20.2%	28.9%	50.9%

3.4. Multinomial Logit Regression Analysis

In the multinomial regression analysis, it was explored whether the MEAS, the FNS, the HS and all other analysed parameters had an impact on the decision in the DCE (see Section 3.3 for the applied model). In the DCE, the respondents were directly asked if they would choose one out of the four offered product alternatives or none of these products. On the basis of the choice experiments carried out, four logistic regression models for the product categories mortadella, chicken nuggets, salami and meat balls were calculated for Germany and four models for Belgium (see Table 12). In the estimation models, the vegetarian option was set as the reference category for the estimation against the alternatives meat, meat + 50%-plant-based protein seed, meat + 35% plant-based protein seed and meat + 20%-plant-based protein seed. As expected, the price parameter was negative and significant in all models and with one exemption, whereas the organic label predominantly exerted a positive effect on product choice.

The coefficients for the meat options were all significant and revealed the highest positive values in comparison to all other parameters. Thus, the fact that the product was a pure meat product had the highest relevance in the analysed sample of meat eaters. Nonetheless, we highlight that all coefficients for the meathybrids were positive as well, and out of the 24 coefficients for hybrids, fifteen were significant. That is, the vegetarian product was the least preferred in the experiment, whereas the pure meat products had the highest consumer preference, followed by the hybrids. For the plant-based shares of 50%, 35% and 20% in the meathybrids, no real preference order can be stated. Dependent on the country and the product category, sometimes, the plant-based share of 50% was the most preferred option (e.g., nuggets in Germany = 0.930 ***) and, sometimes, the lowest share (e.g., meat balls in Belgium = 0.837 *). Nonetheless, it can be generalised that the hybrids performed better in the DCE compared to the vegetarian alternative. Furthermore, the previous use of meat substitutes had a positive impact on the choice of hybrids in particular for a 50% plant-based protein share (seven out of eight cases). Contrarily, this parameter had a negative impact on the choice of the meat alternatives in the product categories mortadella, salami and meat balls.

Concerning the environmental label "reduced CO_2", six out of the eight coefficients were positive and significant. Thus, the use of such a label on the product packaging for hybrids can be recommended. For the applied health labels, this holds only for the product category chicken nuggets and the claim "high of non-saturated acids", whereas for the other products, no such effect could be measured. Across all products and countries, the local origin had a positive effect on product choice (six out of eight parameters were significant), whereas these held for the national labels only for Germany. As expected, the MEAS exerted in both countries a positive impact on the preference of the meat alternatives in all product categories, whereas for the hybrids, there were only a few significant parameters (three out of 24). Therefore, this psychological construct represents a barrier for the consumption of hybrids because it directly increases the preference for the default option "pure" meat. This finding is in line with Graça et al. [38], who found negative significant associations from meat attachment to meat substitution.

Concerning the HS, it can be stated that the lower the health consciousness was, the lower was the preference for meat. Interestingly, for most of the hybrid variants across product categories and countries (19 out of 24 parameters), there were no significant differences between the hybrid variants and the vegetarian alternatives. Thus, it can be concluded that on average, the health conscious segment saw no serious differences in the health characteristics of these options. That is, the vegetarian and hybrids alternatives were seen both as healthier compared to the meat alternative from this segment. For the impact of the FNS on the preference of hybrids, the results were quite mixed. Only nine out of 24 parameters were significant, and no real order or systemic behaviour can be identified. Therefore, the hypothesis, that food neophobia is a barrier for the choice of hybrids, cannot be affirmed. In this context, we point out that in Belgium, the FNS even reduced the choice probability of the pure meat alternatives (three out four cases significant), whereas no such effect can be found in Germany. In Figures 2 and 3, the odds ratios of the estimations are graphically displayed. The figures clearly show that the sample had a far distance from the highest preference for the pure meat alternatives.

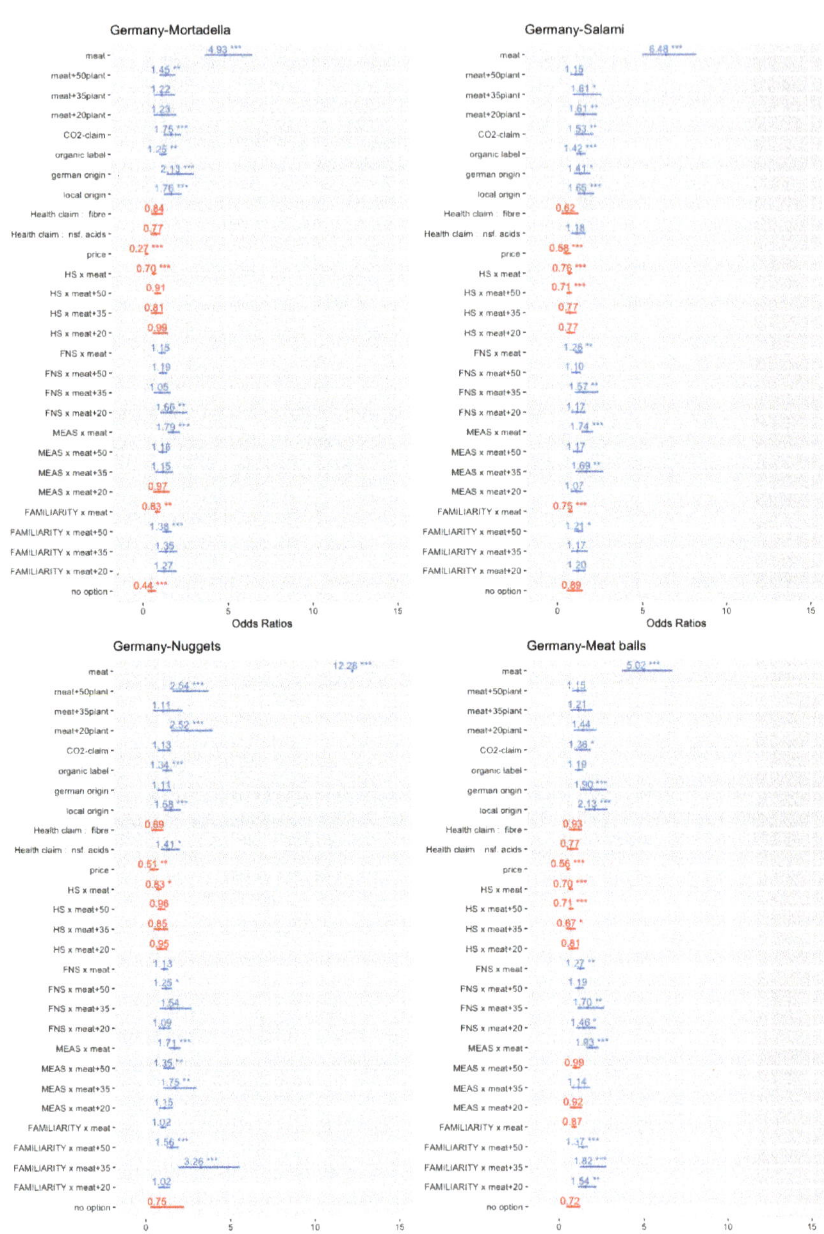

Figure 2. Odds ratios—estimations for Germany. * $p < 0.1$; ** $p < 0.05$; *** $p < 0.01$.

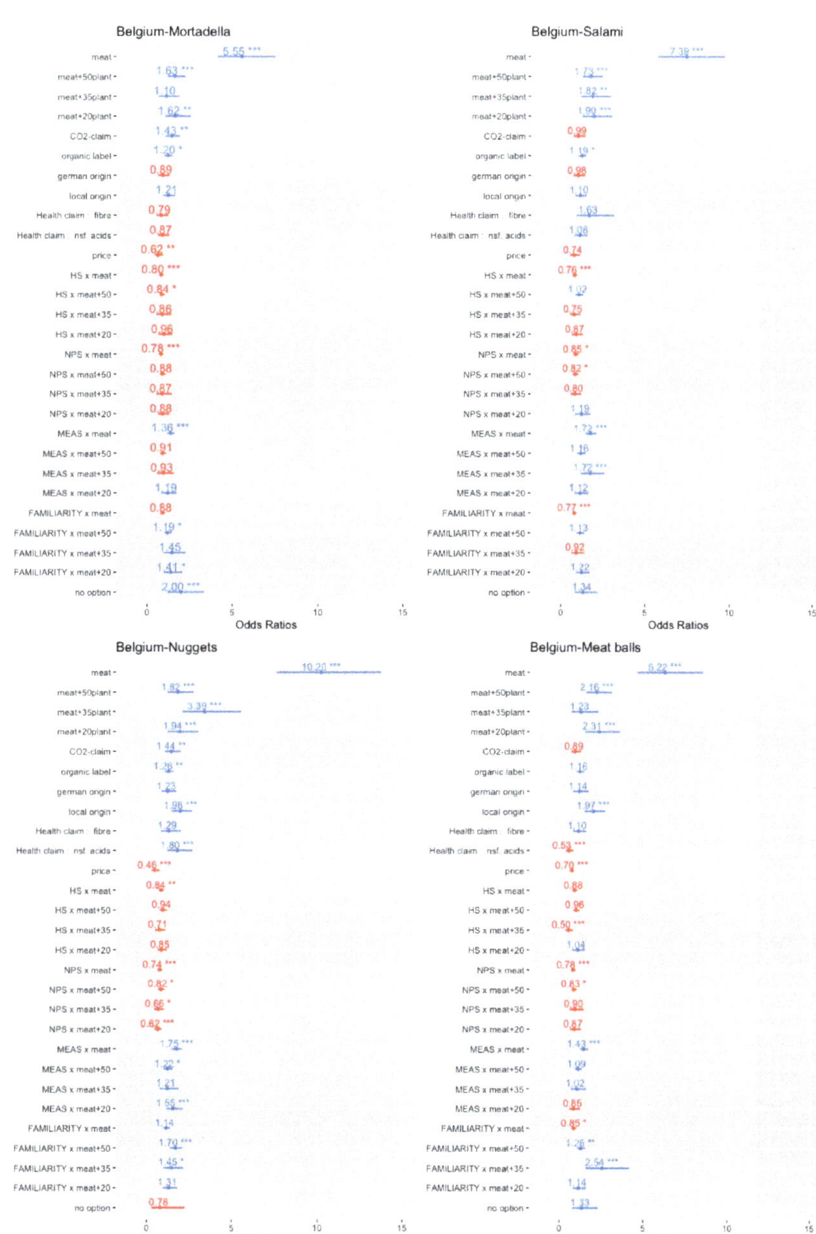

Figure 3. Odds ratios—estimations for Belgium. * $p < 0.1$; ** $p < 0.05$; *** $p < 0.01$.

Table 12. Estimation results.

	Mortadella		Salami		Nuggets		Meat Balls	
	GER (1)	BEL (2)	GER (3)	BEL (4)	GER (5)	BEL (6)	GER (7)	BEL (8)
meat	1.596 *** (0.143)	1.713 *** (0.154)	1.869 *** (0.126)	2.001 *** (0.134)	2.508 *** (0.158)	2.322 *** (0.152)	1.614 *** (0.152)	1.828 *** (0.158)
meat + 50plant	0.372 ** (0.167)	0.486 *** (0.170)	0.137 (0.178)	0.549 *** (0.171)	0.930 *** (0.208)	0.601 *** (0.210)	0.138 (0.187)	0.770 *** (0.178)
meat + 35plant	0.202 (0.244)	0.100 (0.274)	0.473 * (0.260)	0.598 ** (0.241)	0.107 (0.366)	1.222 *** (0.250)	0.187 (0.265)	0.203 (0.314)
meat + 20plant	0.205 (0.269)	0.482 ** (0.236)	0.473 ** (0.208)	0.643 *** (0.227)	0.925 *** (0.241)	0.662 *** (0.233)	0.366 (0.236)	0.837 *** (0.222)
reduced CO_2	0.557 *** (0.150)	0.355 ** (0.159)	0.426 ** (0.181)	−0.014 (0.185)	0.120 (0.175)	0.361 ** (0.171)	0.311 * (0.176)	−0.113 (0.168)
organic	0.221 ** (0.099)	0.186 * (0.102)	0.348 *** (0.104)	0.175 * (0.106)	0.289 *** (0.111)	0.234 ** (0.109)	0.171 (0.107)	0.151 (0.105)
Ger/Bel origin	0.756 *** (0.186)	−0.119 (0.187)	0.344 * (0.178)	−0.019 (0.175)	0.102 (0.183)	0.205 (0.180)	0.644 *** (0.210)	0.132 (0.207)
local origin	0.563 *** (0.150)	0.190 (0.150)	0.502 *** (0.154)	0.094 (0.153)	0.457 *** (0.160)	0.685 *** (0.156)	0.758 *** (0.158)	0.678 *** (0.155)
High in fibre	−0.170 (0.211)	−0.231 (0.220)	−0.483 (0.378)	0.490 (0.326)	−0.371 (0.252)	0.256 (0.222)	−0.069 (0.208)	0.099 (0.182)
High of nsf.acids	−0.265 (0.188)	−0.136 (0.193)	0.167 (0.185)	0.073 (0.186)	0.347 * (0.199)	0.590 *** (0.207)	−0.257 (0.212)	−0.633 *** (0.216)
price	−1.293 *** (0.205)	−0.484 ** (0.190)	−0.544 *** (0.197)	−0.308 (0.194)	−0.672 ** (0.265)	−0.786 *** (0.256)	−0.576 *** (0.100)	−0.358 *** (0.093)
HS * meat	−0.362 *** (0.099)	−0.227 *** (0.082)	−0.281 *** (0.100)	−0.273 *** (0.093)	−0.181 * (0.098)	−0.178 ** (0.089)	−0.351 *** (0.098)	−0.132 (0.089)
HS * meat + 50plant	−0.089 (0.113)	−0.175 * (0.099)	−0.343 *** (0.121)	0.022 (0.117)	−0.045 (0.121)	−0.060 (0.117)	−0.345 *** (0.114)	−0.039 (0.104)
HS * meat + 35plant	−0.211 (0.215)	−0.149 (0.253)	−0.266 (0.215)	−0.287 (0.200)	−0.163 (0.247)	−0.336 (0.209)	−0.404 * (0.207)	−0.701 *** (0.223)
HS * meat + 20plant	−0.011 (0.224)	−0.038 (0.212)	−0.266 (0.173)	−0.144 (0.192)	−0.056 (0.168)	−0.167 (0.161)	−0.214 (0.183)	0.040 (0.187)
FNS * meat	0.137 (0.099)	−0.243 *** (0.080)	0.233 ** (0.101)	−0.164 * (0.093)	0.126 (0.101)	−0.301 *** (0.092)	0.241 ** (0.095)	−0.245 *** (0.089)
FNS * meat + 50plant	0.175 (0.114)	−0.126 (0.099)	0.091 (0.130)	−0.204 * (0.112)	0.221 * (0.128)	−0.201 * (0.118)	0.173 (0.118)	−0.188 * (0.102)
FNS * meat + 35plant	0.051 (0.232)	−0.141 (0.257)	0.450 ** (0.227)	−0.222 (0.197)	0.433 (0.293)	−0.423 * (0.223)	0.531 ** (0.233)	−0.104 (0.232)
FNS * meat + 20plant	0.504 ** (0.236)	−0.128 (0.193)	0.159 (0.182)	0.170 (0.193)	0.085 (0.172)	−0.486 *** (0.163)	0.375 * (0.205)	−0.140 (0.192)
MEAS * meat	0.580 *** (0.103)	0.307 *** (0.081)	0.554 *** (0.102)	0.544 *** (0.094)	0.535 *** (0.103)	0.560 *** (0.091)	0.658 *** (0.101)	0.359 *** (0.089)
MEAS * meat + 50plant	0.151 (0.115)	−0.097 (0.098)	0.156 (0.126)	0.144 (0.113)	0.303 ** (0.126)	0.200 * (0.117)	−0.013 (0.118)	0.083 (0.101)
MEAS * meat + 35plant	0.142 (0.224)	−0.068 (0.259)	0.523 ** (0.245)	0.543 *** (0.203)	0.560 ** (0.274)	0.190 (0.225)	0.132 (0.242)	0.023 (0.226)

Table 12. *Cont.*

	Mortadella		Salami		Nuggets		Meat Balls	
	GER (1)	BEL (2)	GER (3)	BEL (4)	GER (5)	BEL (6)	GER (7)	BEL (8)
MEAS * meat + 20plant	−0.032 (0.251)	0.170 (0.196)	0.067 (0.178)	0.114 (0.192)	0.139 (0.181)	0.439 *** (0.162)	−0.087 (0.195)	−0.163 (0.192)
FAMILIARITY * meat	−0.188 ** (0.096)	−0.125 (0.082)	−0.290 *** (0.094)	−0.256 *** (0.091)	0.016 (0.100)	0.135 (0.093)	−0.135 (0.092)	−0.157 * (0.092)
FAMILIARITY * meat + 50pl.	0.321 *** (0.106)	0.173 * (0.096)	0.195 * (0.116)	0.124 (0.108)	0.442 *** (0.120)	0.529 *** (0.116)	0.316 *** (0.108)	0.233 ** (0.102)
FAMILIARITY * meat + 35pl.	0.299 (0.204)	0.369 (0.231)	0.160 (0.216)	−0.086 (0.209)	1.181 *** (0.272)	0.371 * (0.205)	0.601 *** (0.215)	0.931 *** (0.254)
FAMILIARITY * meat + 20pl.	0.242 (0.224)	0.343 * (0.199)	0.185 (0.165)	0.201 (0.174)	0.019 (0.173)	0.267 (0.167)	0.429 ** (0.180)	0.128 (0.175)
no option	−0.826 *** (0.281)	0.692 *** (0.265)	−0.113 (0.257)	0.290 (0.248)	−0.292 (0.563)	−0.243 (0.535)	−0.330 (0.277)	0.285 (0.277)
Observations	1000	1002	1000	1002	1000	1002	1000	1002
Log Likelihood	−1220.917	−1366.974	−1201.299	−1272.771	−1177.140	−1261.003	−1227.343	−1310.656

* $p < 0.1$; ** $p < 0.05$; *** $p < 0.01$.

4. Discussion and Conclusions

The results show that more than fifty percent of consumers substitute meat at least occasionally. Thus, about half of the respondents revealed an eligible consumption behaviour with respect to sustainability and healthiness, at least sometimes. Furthermore, about a fifth indicated sometimes consuming, respectively frequently, meat alternatives such as veggie burgers. However, most of the consumed meat alternatives had an animal origin (dairy products, fish, eggs), which, like meat production, come along with an environmental burden. In this context, we highlight, that the findings of this study demonstrate that at least a substantial amount of consumers are open-minded to the "meathybrid" concept. Even a higher share believes that this new alternative is better for the environment and the animals in comparison to meat. In the DCE, the tested meathybrid variants with different shares of plant-based proteins took the second position, followed by the vegetarian-based alternative. Therefore, meathybrids could facilitate the diet transition of meat-eaters in the direction toward a more sustainable consumption. The analysed consumer segment was more open-minded to the meathybrid concept in comparison to the vegetarian substitutes. Thus, there is chance that hybrids could serve as a low-threshold option for a transition in the direction toward a more sustainable diet.

Nonetheless, on the road-map of this transition, some major problems and issues have to be tackled. Other research [39,40] suggested that the current meathybrids and meat replacers are still relatively unfamiliar and that their image in relation to the expected taste as shown in this study is quite mixed. While meathybrids have a plant-based protein share, they are not necessarily optimal from an environmental perspective, because their processing stage can require a considerable input of energy, and they often contain a high share of egg protein [41,42]. Although technological development has led to improvements in the quality of meathybrids in recent years, it is still important to develop improved meathybrids and meat substitutes, which are significantly better and superior compared to meat in several ways, such as taste, texture and environmental performance with a lower input of energy and egg whites [43,44].

Despite the technological challenges, there are cultural, respectively socio-cultural challenges as well. As shown, meat attachment as a psychological construct represents a barrier for diet change and transition. Future research should address this topic in more detail and analyse how to overcome this attitude.

Concerning the impact factors on choosing either meathybrids or meat, it becomes obvious that familiarity with meat substitutes, respectively their former use, plays a great role in preference formation. Therefore, it can be recommended to increase the share of meat substitutes/meathybrids in school/public canteens and to financially support other canteens that replace meat with plant substitutes or hybrids. Herewith, consumers are confronted more often with meat alternatives, and familiarity with such products can be built up on a mid- to long-term time horizon.

Author Contributions: Conceptualization, N.T., J.W., G.V.R. and V.H.; methodology, A.P.; formal analysis, A.P.; writing, original draft preparation, A.P., M.-C.B., S.S. and K.B.; writing, review and editing, A.P., M.-C.B. and S.S.; visualization, A.P.; project administration, M.-C.B.; funding acquisition, N.T., V.H. and G.V.R. All authors read and agreed to the published version of the manuscript.

Funding: This IGF Project of the FEI is/was supported via AiF within the programme for promoting the Industrial Collective Research (IGF) of the German Ministry of Economic Affairs and Energy (BMWi), based on a resolution of the German Parliament. Project AiF 196 EN https://www.fei-bonn.de/gefoerderte-projekte/projektdatenbank/cornet-aif-196-en.projekt.

Institutional Review Board Statement: Not applicable.

Informed Consent Statement: Informed consent was obtained from all subjects involved in the study.

Data Availability Statement: The data presented in this study are available on request from the corresponding author. The data will be made publicly to a later stage.

Conflicts of Interest: The authors declare no conflict of interest.

References

1. Max Roser, H.R.; Ortiz-Ospina, E. *World Population Growth*. 2013. Available online: https://ourworldindata.org/world-population-growth (accessed on 10 November 2020)
2. De Boer, J.; Schösler, H.; Aiking, H. "Meatless days" or "less but better"? Exploring strategies to adapt Western meat consumption to health and sustainability challenges. *Appetite* **2014**, *76*, 120–128. [CrossRef] [PubMed]
3. Hallström, E.; Röös, E.; Börjesson, P. Sustainable meat consumption: A quantitative analysis of nutritional intake, greenhouse gas emissions and land use from a Swedish perspective. *Food Policy* **2014**, *47*, 81–90. [CrossRef]
4. Richi, E.B.; Baumer, B.; Conrad, B.; Darioli, R.; Schmid, A.; Keller, U. Health risks associated with meat consumption: A review of epidemiological studies. *Int. J. Vitam. Nutr. Res.* **2015**, *85*, 70–78. [CrossRef] [PubMed]
5. Kroeze, C.; Bouwman, L. The Role of Nitrogen in Climate Change. *Curr. Opin. Environ. Sustain.* **2011**, *3*, 279–280.
6. De Vries, W.; Kros, J.; Reinds, G.J.; Butterbach-Bahl, K. Quantifying impacts of nitrogen use in European agriculture on global warming potential. *Curr. Opin. Environ. Sustain.* **2011**, *3*, 291–302. [CrossRef]
7. Erisman, J.W.; Galloway, J.; Seitzinger, S.; Bleeker, A.; Butterbach-Bahl, K. Reactive nitrogen in the environment and its effect on climate change. *Curr. Opin. Environ. Sustain.* **2011**, *3*, 281–290. [CrossRef]
8. Profeta, A.; Hamm, U. Do consumers care about local feedstuffs in local food? Results from a German consumer study. *NJAS Wagening. J. Life Sci.* **2019**, *88*, 21–30. [CrossRef]
9. Godfray, H.C.J.; Aveyard, P.; Garnett, T.; Hall, J.W.; Key, T.J.; Lorimer, J.; Pierrehumbert, R.T.; Scarborough, P.; Springmann, M.; Jebb, S.A. Meat consumption, health, and the environment. *Science* **2018**, *361*. [CrossRef] [PubMed]
10. McAuliffe, G. A thematic review of life cycle assessment (LCA) applied to pig production. *Environ. Impact Assess. Rev.* **2015**, *56*, 12–22. [CrossRef]
11. Zhu, X.; van Ierland, E.C. Protein Chains and Environmental Pressures: A Comparison of Pork and Novel Protein Foods. *Environ. Sci.* **2004**, *1*, 254–276. [CrossRef]
12. Grandin, T. Welfare Problems in Cattle, Pigs, and Sheep that Persist Even Though Scientific Research Clearly Shows How to Prevent Them. *Animals* **2018**, *8*, 124. [CrossRef] [PubMed]
13. Dawkins, M. *Animal Suffering*, 1st ed.; Springer: Dordrecht, The Netherlands, 1980; p. 150. [CrossRef]
14. Stoll-Kleemann, S.; Schmidt, U.J. Reducing meat consumption in developed and transition countries to counter climate change and biodiversity loss: a review of influence factors. *Reg. Environ. Chang.* **2017**, *17*, 1261–1277. [CrossRef]
15. Graça, J.; Calheiros, M.M.; Oliveira, A. Attached to meat? (Un)Willingness and intentions to adopt a more plant-based diet. *Appetite* **2015**, *95*, 113–125. [CrossRef] [PubMed]
16. Pliner, P.; Hobden, K. Development of a scale to measure the trait of food neophobia in humans. *Appetite* **1992**, *19*, 105–120. [CrossRef]
17. Apostolidis, C.; McLeay, F. Should we stop meating like this? Reducing meat consumption through substitution. *Food Policy* **2016**, *65*, 74–89. [CrossRef]
18. Neville, M.; Tarrega, A.; Hewson, L.; Foster, T. Consumer-orientated development of hybrid beef burger and sausage analogues. *Food Sci. Nutr.* **2017**, *5*, 852–864. [CrossRef]

19. Hartmann, C.; Siegrist, M. Consumer perception and behaviour regarding sustainable protein consumption: A systematic review. *Trends Food Sci. Technol.* **2017**, *61*, 11–25. [CrossRef]
20. Tucker, C.A. The significance of sensory appeal for reduced meat consumption. *Appetite* **2014**, *81*, 168–179. [CrossRef]
21. Kreuter, F.; Presser, S.; Tourangeau, R. Social Desirability Bias in CATI, IVR, and Web Surveys: The Effects of Mode and Question Sensitivity. *Public Opin. Q.* **2008**, *72*, 847–865. [CrossRef]
22. Phillips, D.L.; Clancy, K.J. Some Effects of "Social Desirability" in Survey Studies. *Am. J. Sociol.* **1972**, *77*, 921–940. [CrossRef]
23. Hair, J.; Black, W.; Babin, B.; Anderson, R. *Multivariate Data Analysis: A Global Perspective*; Pearson Education: Upper Saddle River, NJ, USA, 2010; Volume 7.
24. Revelle, W. *Procedures for Psychological, Psychometric, and Personality Research*; Technical Report; Northwestern University: Evanston, IL, USA, 2019.
25. Siegrist, M.; Hartmann, C. Impact of sustainability perception on consumption of organic meat and meat substitutes. *Appetite* **2019**, *132*, 196–202. [CrossRef] [PubMed]
26. Hartmann, C.; Shi, J.; Giusto, A.; Siegrist, M. The psychology of eating insects: A cross-cultural comparison between Germany and China. *Food Qual. Prefer.* **2015**, *44*, 148–156. [CrossRef]
27. Jackson, D.L.; Gillaspy, J.A.; Purc-Stephenson, R. Reporting Practices in Confirmatory Factor Analysis: An Overview and Some Recommendations. *Psychol. Methods* **2009**, *14*, 6–23. [CrossRef] [PubMed]
28. Graça, J.; Oliveira, A.; Calheiros, M.M. Meat, beyond the plate. Data-driven hypotheses for understanding consumer willingness to adopt a more plant-based diet. *Appetite* **2015**, *90*, 80–90. [CrossRef]
29. Visschers, V.H.; Hartmann, C.; Leins-Hess, R.; Dohle, S.; Siegrist, M. A consumer segmentation of nutrition information use and its relation to food consumption behaviour. *Food Policy* **2013**, *42*, 71–80. [CrossRef]
30. McFadden, D. The measurement of urban travel demand. *J. Public Econ.* **1974**, *3*, 303–328. [CrossRef]
31. Hensher, D.A. Hypothetical bias, choice experiments and willingness to pay. *Transp. Res. Part B Methodol.* **2010**, *44*, 735–752. [CrossRef]
32. Choicemetrics. *Ngene 1.2 User Manual & Reference Guide*; Technical Report; Choicemetrics: Sydney, Australia, 2018.
33. Loureiro, M.L.; Umberger, W.J. A choice experiment model for beef: What US consumer responses tell us about relative preferences for food safety, country-of-origin labeling and traceability. *Food Policy* **2007**, *32*, 496–514. [CrossRef]
34. R Core Team. *R: A Language and Environment for Statistical Computing*; R Core Team: Vienna, Austria, 2020.
35. Croissant, Y. mlogit: Multinomial Logit Models. 2020. Available online: https://cran.r-project.org/web/packages/mlogit/index.html (accessed on 10 November 2020)
36. Lüdecke, D. sjPlot: Data Visualization for Statistics in Social Science. 2020. Available online: https://cran.r-project.org/package=sjPlot (accessed on 10 November 2020)
37. Schösler, H.; de Boer, J.; Boersema, J.J. Can we cut out the meat of the dish? Constructing consumer-oriented pathways towards meat substitution. *Appetite* **2012**, *58*, 39–47. [CrossRef]
38. Graça, J.; Calheiros, M.M.; Oliveira, A. Situating moral disengagement: Motivated reasoning in meat consumption and substitution. *Personal. Individ. Differ.* **2016**, *90*, 353–364. [CrossRef]
39. Elzerman, J.E.; van Boekel, M.A.; Luning, P.A. Exploring meat substitutes: Consumer experiences and contextual factors. *Br. Food J.* **2013**, *115*, 700–710. [CrossRef]
40. Hoek, A.C.; Luning, P.A.; Weijzen, P.; Engels, W.; Kok, F.J.; de Graaf, C. Replacement of meat by meat substitutes. A survey on person- and product-related factors in consumer acceptance. *Appetite* **2011**, *56*, 662–673. [CrossRef] [PubMed]
41. Aiking, H. Future protein supply. *Trends Food Sci. Technol.* **2011**, *22*, 112–120. [CrossRef]
42. Davis, J.; Sonesson, U.; Baumgartner, D.U.; Nemecek, T. Environmental impact of four meals with different protein sources: Case studies in Spain and Sweden. *Food Res. Int.* **2010**, *43*, 1874–1884. [CrossRef]
43. Aiking, H.; de Boer, J. Background, Aims and Scope. In *Sustainable Protein Production and Consumption: Pigs or Peas?* Aiking, H., de Boer, J., Vereijken, J., Eds.; Springer: Dordrecht, The Netherlands, 2006; pp. 1–21. [CrossRef]
44. Boland, M.J.; Rae, A.N.; Vereijken, J.M.; Meuwissen, M.P.; Fischer, A.R.; van Boekel, M.A.; Rutherfurd, S.M.; Gruppen, H.; Moughan, P.J.; Hendriks, W.H. The future supply of animal-derived protein for human consumption. *Trends Food Sci. Technol.* **2013**, *29*, 62–73. [CrossRef]

Article

A Cross-Cultural Evaluation of Liking and Perception of Salted Butter Produced from Different Feed Systems

Emer C. Garvey [1,2], Thorsten Sander [3,4], Tom F. O'Callaghan [5], MaryAnne Drake [6], Shelley Fox [7], Maurice G. O'Sullivan [2], Joseph P. Kerry [8] and Kieran N. Kilcawley [1,*]

1 Food Quality & Sensory Department, Teagasc Food Research Centre, Moorepark, Fermoy, P61 C996 Co. Cork, Ireland; emer.garvey@teagasc.ie
2 Sensory Group, School of Food and Nutritional Science, University College Cork, T12 R220 Cork, Ireland; maurice.osullivan@ucc.ie
3 Department of Food, Nutrition, Facilities, FH Münster, Corrensstraße 25, D-48149 Münster, Germany; tsander@fh-muenster.de
4 Innovationsmanagement, Sensorische Produktevaluation und Consumer Trends, Marie-Jahn-Str.20, 30177 Hannover, Germany
5 School of Food and Nutritional Sciences, University College Cork, T12 Y337 Cork, Ireland; tom_ocallaghan@ucc.ie
6 Department of Food, Bioprocessing and Nutrition Sciences, Southeast Dairy Foods Research Center, North Carolina State University, Raleigh, NC 27695, USA; maryanne_drake@ncsu.edu
7 St. Angela's Food Technology Centre, Lough Gill, 999928 Sligo, Ireland; sfox@stangelas.nuigalway.ie
8 Food Packaging Group, School of Food and Nutritional Science, University College Cork, T12 R220 Cork, Ireland; Joe.Kerry@ucc.ie
* Correspondence: kieran.kilcawley@teagasc.ie

Received: 3 November 2020; Accepted: 25 November 2020; Published: 28 November 2020

Abstract: Perception and liking among Irish, German and USA consumers of salted butter produced from different feed systems—outdoor grass (FS-GRSS), grass/clover (FS-CLVR), and indoor concentrate (FS-TMR)—was investigated. A consumer study was conducted in all three countries. Irish and German assessors participated in ranking descriptive analysis (RDA), whereas descriptive analysis (DA) was carried out by a trained panel in the USA. Volatile analysis was conducted to identify differences in aroma compounds related to cow diet. Overall, there was no significant difference in overall liking of the butters, among USA, German and Irish consumers, although cross-cultural preferences were evident. Sensory attribute differences based on cow diet were evident across the three countries, as identified by German and Irish assessors and trained USA panelists, which are likely influenced by familiarity. The abundance of specific volatile aromatic compounds, especially some aldehydes and ketones, were significantly impacted by the feed system and may also contribute to some of the perceived sensory attribute differences in these butters.

Keywords: dairy; diet; butter preference; sensory; volatiles

1. Introduction

Globally, consumers are increasingly aware of their food choices, with respect to country of origin, production practices, sustainability, and potential health-promoting properties, prior to purchase [1]. Satisfaction of these extrinsic aspects can influence overall liking, and thus purchase intent and even willingness to pay a premium, particularly for meat and dairy products [2–6]. There has been substantial interest in exploring consumer's perception towards meat and dairy products produced from a pasture/grazing diet. A recent review by Stampa, Schipmann-Schwarze et al. (2020) [7], mainly focusing on studies undertaken in Europe and the United States of America (USA), outlined that attitudes towards environmental practices and health benefits of consuming pasture-raised livestock products were the main drivers for consumption and willingness to pay premium prices for these types of products.

Farming practices in Ireland consist of fresh pasture for the majority of lactation, allowing for utilization of a low-cost, readily available feed source to produce high-quality milk products [8,9]. Since the abolishment of milk quotas in Europe in 2015, milk production in Ireland has increased substantially, with a 13% increase in milk intake from 2015 to 2017 [10–12]. In 2019, Ireland's dairy sector grew in value by 11%, with butter being the largest export category [13] and further opportunities exist to expand current markets and develop new markets.

The most apparent differences in dairy products produced from cows on a pasture-rich diet, versus concentrates, are changes to the fatty acid (FA) profile and color. Inclusion of fresh pasture significantly increases levels of unsaturated FAs in milk [14–16], and β-carotene levels, enhancing a yellow color, particularly obvious in butter, as β-carotene is fat soluble [17]. Although milk and dairy products are not regarded as a dietary source of omega-3 FAs, higher amounts of α-linoleic acid and conjugated linoleic acid (CLA) are present in bovine milk from pasture based diets [14,15,18–20]. β-carotene derived from fresh pasture gives a yellow hue to butter, with the intensity of yellow positively correlating to the amount of fresh pasture grazed [21,22]. β-carotene is also a precursor for fat-soluble vitamin A, and a powerful antioxidant, and therefore also beneficial in the diet [23]. Butter hardness and spreadability are also dictated by the FA profile and thus impacted by the cow's diet, with higher numbers of unsaturated FAs lowering the melting point [24].

Butter is coveted for its rich sensory attributes, with butter flavor being a significant driver for liking [25]. Although flavor is mainly dictated by the milk fat itself and added salt, volatile aroma compounds also play an important role in the sensory perception of butter. Previous studies have identified a range of potentially important volatiles including lactones, ketones, acids, esters, aldehydes, pyrroles and sulfur compounds, thought to influence the sensory perception of butter or sweet cream butter [26–28] and further work is required to determine factors that may influence the generation of these volatiles, such as cow diet. Although the influence of diet on the sensory properties of butter has been previously investigated [22,29], studies are limited and no studies have been published exploring the cross-cultural liking of salted butter. Therefore, an objective of this study was to investigate the liking and perception of salted butters, produced from cows outdoors on two pasture-based diets—perennial ryegrass, or perennial ryegrass/white clover—and cows indoors on a concentrate diet (total mixed rations) by consumers (Ireland, Germany and the USA), untrained assessors (Ireland and Germany) and trained panelists (USA). In addition, volatile analysis was performed to elucidate potential differences in sensory perception. The information generated should result in an improved understanding of the cross-cultural perception of Irish dairy products beneficial for the export markets.

2. Materials and Methods

2.1. Experimental Diets and Milk Production

Fifty-four spring calving Friesian cows were selected from the general herd at the Teagasc Animal and Grassland Research and Innovation Centre, Moorepark, Fermoy, Co. Cork. Teagasc has both an animal welfare body and animal ethics committee. The animal welfare body is a legal requirement of Article 26 of Directive 2010/63/EU and Regulation 50 of S.I. No. 543 of 2012. The Health Products Regulatory Authority provided project authorization, and the Health Products Regulatory Authority License number for this project is AE19132/P019. The cows were randomized based on milk yield, milk solids yield, calving date and lactation number, and allocated to one of three experimental feed systems (FS) (n = 18) (18 cows in each FS)—outdoor pasture grazing on perennial ryegrass (*Lolium perenne* L.) (FS-GRSS), outdoor pasture grazing on perennial ryegrass supplemented with white clover (*Trifolium repens* L.) (FS-CLVR) or housed indoor provided with a diet of TMR (FS-TMR). In-depth details of the three diets were outlined by O'Callaghan et al. (2016) [22]. Morning and evening milks from each of the experimental herds (FS-GRSS, FS-CLVR and FS-TMR) were collected and assigned as per feed system to 5000 L refrigerated tanks. Combined milk was kept at 4 °C prior to sample collection, which took place within 24 h after milking.

2.2. Butter Manufacture

Butter production took place on 3 separate occasions over a 3 week period. Butter produced from each experimental feed system was produced in triplicate (producing 3 batches per feed system). The procedure was identical to that as outlined by O'Callaghan, et al. (2016) [22].

The butter was packed into 200 ± 20 g sticks using an extruder and wrapped in grease-proof paper followed by an outer wrapping of aluminum foil. Butter was vacuum packed and stored at −20 °C until subsequent sensory and volatile analysis. Butter was defrosted in a refrigerator 24 h before the relevant analysis. Prior to each sensory study, the butter was tempered at room temperature for 1 h.

2.3. Consumer Study

2.3.1. Consumer Selection

Three consumer sensory panels from three countries—Ireland, Germany and USA—were created for the purpose of this study. Panels were comprised of 108 German (70% female, 30 male, age = 18–68), 103 USA (79% female, 21% male, age = 21–64) and 96 Irish consumers (68% female, 32% male = 18–60), who regularly consumed butter. The German consumers consisted of a mixture of faculty and students, recruited in the University of Applied Sciences, Muenster, Germany. The USA consumers were recruited by the Southeast Dairy Foods Research Center, North Carolina State University, Raleigh, North Carolina, USA. The Irish consumers consisted of students and staff from Teagasc Food Research Centre, Moorepark, Fermoy, Co. Cork, Ireland and St. Angela's Food Technology Centre, Sligo, Ireland.

2.3.2. Product Evaluation by Hedonic, Intensity and Just-About-Right Scales

Butter evaluation took place in accordance with international standards [30], in each country. The ballots were identical in all three countries except that salt liking was not included in the USA ballot. The consumer study evaluation was designed to collect information on attribute liking, intensity perception and optimum levels of each attribute by applying a just-about-right scale (JAR). Liking was assessed for overall appearance, color, flavor, and texture attributes, intensity for color, flavor, saltiness, freshness, and firmness, and JAR scale evaluation was used to assess color, flavor, saltiness and texture. Each attribute was evaluated for liking and intensity on a 9-point hedonic scale; 9 = "extremely like" and 1 = "extremely dislike" and 9 = "high intensity" and 1 = "low intensity", respectively (Table S1). Five-point JAR scales were employed with extremes 1 = "much too little" and 5 = "much too much". Since 9 butters were produced in total from the 3 experimental diets (3 butters produced in triplicate), the samples were presented to panelists in an incomplete block design, randomly allocated and single blinded using three-digit codes, allowing for even distribution of each treatment and butter trial. Therefore, the order of presentation was balanced, not randomized, and the ballot order for each product was identical as per standard consumer tests. Panelists were presented with three individual 5 ± 0.02 g servings of butter alongside water crackers and water. Crackers were provided for palatability purposes, the butter was presented on the crackers, but consumers were clearly instructed that the sensory evaluation was to be carried out solely on the sensory attributes of the butter. Panelists were encouraged to rinse their palate thoroughly after each sample and a 1 min rest between each sample was enforced. Consumers from the University of Applied Sciences, Muenster and Teagasc Food Research Centre completed the ballot questionnaire on paper. Consumers at North Carolina State University and St. Angela's College undertook the questionaire via Compusense Cloud (Guelph, Canada) and FIZZ (Biosystems, Couternon, France), respectively.

2.4. Ranking Descriptive Analysis

Ranking descriptive analysis (RDA) [31], a modification of flash profiling, was performed by untrained assessors at the University of Applied Sciences, Muenster, Germany, and at the Teagasc Food Research Centre, Ireland. However, as stated, assessors were recruited on their frequency to consume butter, and had previous sensory analysis experience. Each panel was comprised of 20 German and 20 Irish assessors, respectively. Attributes were generated by a focus group consisting of 7 people, comprised of members from the Food Quality and Sensory Science Department at the Teagasc Food Research Centre, Ireland. The established list of attributes was chosen on their ability to best describe the different butter samples produced by each treatment. Irish and German assessors were asked to assess the attributes relative to color (yellow color), aroma (buttery, milky, grassy and rancid), flavor (salty, sweet, creamy, sour, stale) and texture (melt in the mouth) (Table S2). Similar to the consumer study, each butter trial sample was presented to assessors on a water cracker for evaluation. The presentation for the RDA was a complete block design. Assessors were briefly coached on the explanation of each attribute in relevance to butter and asked to evaluate the intensity of each on a 9 cm continuous scale. Sensory analysis was conducted in duplicate over two separate occasions.

2.5. Descriptive Analysis Evaluation

Butter flavor was evaluated by a trained descriptive sensory panel using an established flavor language for butter [25] (Cooked/Nutty, Milkfat, Grassy, Mothball, Stale, Salty Taste and Color Intensity). All sensory testing was conducted in accordance with the North Carolina State University Institutional Review Board for Human Subjects guidelines. Panelists (n = 6) had more than 100 h of previous experience with the sensory analysis of dairy products. Prior to this study, panelists participated in 20 h of additional training on the three butters, FS-GRSS, FS-CLVR and FS-TMR to calibrate and confirm sensory attributes. Samples were prepared 24 h in advance and refrigerated at 4 °C. Prior to evaluation, butters were tempered to 15 °C. A cube of butter (~20 g) was placed in 3-digit-coded, 60 mL lidded cups (Sweetheart Cup Company, Owings Mills, MD, USA). Samples were evaluated on a 15-point intensity scale, in duplicate, on paper ballots by each panelist in a randomized balanced block design.

2.6. Volatile Analysis by HS-SPME-GC-MS

Volatile analysis was carried out by headspace solid-phase microextraction gas chromatography mass spectrometry (HS-SPME-GC-MS) utilizing a Gerstel MultiPurpose Sampler (GMPS) rail system (Anatune, Cambridge CB3 0NA, UK) connected to a Shimadzu GP2010 plus GC (Mason Technology Ltd., Dublin, Ireland). A 50/30 µm divinylbenzene/carboxen/polydimethylsiloxane (DVB/CAR/PDMS) SPME fiber was employed for analysis (Agilent Technologies Ireland Ltd., Cork, Ireland). The chosen HS-SPME parameters were as described by O'Callaghan [22], with modifications. More sample was used and a longer extraction time was applied in an attempt to recover a higher number compounds. Butter was thawed overnight at room temperature and 3 g was added to an La-Pha-Pack amber 20 mL screw-capped SPME vial with magnetic caps and silicone/polytetrafluoroethylene 1.3 mm 45° shore A septa (Apex Scientific Ltd., Co. Kildare, Ireland) and equilibrated for 10 min while exposed to a temperature of 40 °C, with pulsed agitation for 5 s at 350 rpm using the GMPS agitator/heater. The SPME fiber was exposed to the headspace above the samples, at a depth of 21 mm, for an extraction time of 60 min at 40 °C. The fiber was retracted, injected into the GC inlet with a merlin microseal (Merck, Arklow, Ireland) and desorbed for 3 min at 250 °C, followed by 3 min at 270 °C in the GMPS fiber bakeout station, to minimize carryover of compounds. Each butter trial was analyzed in triplicate. An external standard solution (1-butanol, dimethyl disulfide, butyl acetate, cyclohexanone) (Merck, Arklow, Ireland) at 1000 ppm in methanol (Merck, Arklow, Ireland) was also analyzed at the start and end of each batch, and levels of each external standard were quantified and compared to reference values to ensure that both the SPME extraction and MS detection were performing within specification.

The GC analysis was performed on a Shimadzu 2010 Plus GC (Mason Technology Ltd., Dublin, Ireland), equipped with a split/splitless injector, operating in the splitless mode. The carrier gas was helium held at a pressure of 43.8 psi and a flow rate of 1.2 mL/min. The volatile compounds were separated on a DB-624 UI (60 m × 0.32 mm × 1.80 μm) column (Agilent Technologies Ireland Ltd., Cork, Ireland). The temperature of the column oven was set at 40 °C, held for 5 min, increased at 5 °C/min to 230 °C then increased at 15 °C/min to 260 °C. The total GC run time was 65 min. Compound identification was carried out by a Shimadzu TQ8030 mass spectrometry detector (Mason Technologies Ltd., Dublin, Ireland) ran in the single-quad mode. The ion source temperature was 220 °C and the interface temperature was set at 260 °C. The MS mode was electronic ionization (70 eV) with the mass range scanned between m/z 35 and 250 amu. Compounds were identified using mass spectra comparisons to the NIST 2014 mass spectral library, the Shimadzu commercial library FFNSC (Flavor and Fragrance Natural and Synthetic Compounds library) version 2 and an in-house library created in GCMS Solutions software (Shimadzu, Japan) created with standards (where possible), and with target and qualifier ions and linear retention indices for each compound [32]. Spectral deconvolution was also performed to confirm identification of compounds using AMDIS (Automated Mass Spectral Deconvolution and Identification System, www.amdis.net).

2.7. Statistical Analysis

Data analysis was handled accordingly based on the normality of the data. Hedonic data from the sensory evaluation was analyzed using a non-parametric Kruskal–Wallis test ($\alpha = 0.05$), with post hoc Mann-Whitney to identify the significant differences between samples. Bonferroni adjustment was applied to account for type 1 error, therefore working at an alpha level of 0.017. Analysis of variance (ANOVA) with post hoc Tukey significant test was applied to RDA and descriptive analysis data, working at an alpha level of 0.05. Just-about-right (JAR) data was assessed using chi-square statistic. A combination of parametric and non-parametric tests was used to evaluate the volatile data. All parametric and non-parametric tests were performed using SPSS IBM SPSS Statistics 24 for windows (SPSS Inc., IBM Corporation, NY, USA).

3. Results and Discussion

3.1. Consumer Evaluation

The average results of the sensory evaluation of consumers liking towards FS-GRSS, FS-CLVR and FS-TMR butters are presented in Table 1. Overall, there were no significant ($p < 0.05$) differences in overall liking of all three butters within each consumer cultural group; however, cross-cultural differences were evident.

3.1.1. Irish Consumers

There was no significant ($p > 0.05$) difference among Irish consumers liking of the sensory attributes (overall appearance, color, flavor, saltiness, texture) or overall liking of the three butters (Hedonics Table 1). These results contradict a previous study by O'Callaghan et al. 2016 [22], who found that Irish consumers preferred the appearance and flavor of butters produced from grass and clover diets, compared to butter produced from TMR. However, in this study, Irish consumers did perceive the color intensity of FS-GRSS and FS-CLVR butters significantly ($p < 0.017$) higher than FS-TMR butter (Intensity Scale Evaluation Table 1), presumably due to higher β-carotene levels, with 41.6% of panelists grouping the FS-TMR butter as 'not yellow enough' (JAR Evaluation Table 1). However, this did not negatively influence their liking of the FS-TMR butter.

Table 1. Cross-cultural comparison of liking, intensity rating and just-about-right scale evaluation by Irish, German and United States of America consumers, of butters produced by different diets—FS-GRSS, FS-CLVR and FS-TMR.

		Irish Consumers			German Consumers			USA Consumers		
		FS-CLVR	FS-GRSS	FS-TMR	FS-CLVR	FS-GRSS	FS-TMR	FS-CLVR	FS-GRSS	FS-TMR
Hedonics										
	Overall appearance	6.51 ± 1.56^{x}	6.46 ± 1.65^{x}	5.99 ± 1.76^{y}	5.62 ± 1.59^{y}	5.69 ± 1.65^{y}	5.44 ± 1.69^{y}	6.46 ± 1.63^{abx}	6.33 ± 1.57^{bx}	6.98 ± 1.42^{ax}
	Color	6.44 ± 1.61^{x}	6.19 ± 1.80	5.80 ± 1.80^{y}	5.66 ± 1.50^{y}	5.88 ± 1.56	5.56 ± 1.38^{y}	6.43 ± 1.64^{abx}	6.18 ± 1.72^{b}	6.91 ± 1.57^{ax}
	Flavor	6.46 ± 1.74	6.44 ± 1.67^{y}	6.28 ± 1.68^{x}	6.19 ± 1.65	6.38 ± 1.66^{y}	5.89 ± 1.69^{y}	6.52 ± 1.75^{b}	7.10 ± 1.55^{ax}	6.76 ± 1.79^{abx}
	Salt	6.00 ± 1.63	5.73 ± 1.79	5.78 ± 1.68	5.83 ± 0.75	5.87 ± 0.73	5.52 ± 0.72			
	Texture (firmness for USA)	6.54 ± 1.58	6.35 ± 1.72^{x}	6.05 ± 1.75^{x}	6.40 ± 1.84	6.52 ± 1.51^{x}	6.17 ± 1.84^{x}	6.26 ± 1.67^{a}	5.95 ± 1.69^{aby}	5.50 ± 1.81^{by}
	Overall liking	6.46 ± 1.67^{x}	6.36 ± 1.76^{y}	6.27 ± 1.66^{y}	6.04 ± 1.74^{y}	6.23 ± 1.61^{y}	5.89 ± 1.90^{z}	6.59 ± 1.65^{x}	7.13 ± 1.47^{xy}	6.85 ± 1.67^{x}
Intensity Scale Evaluation										
	Color	6.14 ± 1.89^{a}	5.68 ± 2.41^{ab}	5.22 ± 2.40^{bx}	6.31 ± 0.80^{a}	6.02 ± 0.76^{a}	3.27 ± 0.81^{ay}			
	Flavor	6.34 ± 1.78	5.73 ± 2.01	6.21 ± 1.60^{x}	6.38 ± 1.79^{a}	6.14 ± 1.87^{a}	5.29 ± 2.23^{by}			
	Salt	5.31 ± 2.00^{y}	5.15 ± 2.15^{y}	5.06 ± 1.96	6.17 ± 1.88^{ax}	5.95 ± 1.81^{ax}	5.28 ± 1.89^{b}			
	Freshness	6.34 ± 1.73^{x}	6.10 ± 1.93	5.95 ± 1.82	5.56 ± 0.89^{y}	5.78 ± 0.82	5.80 ± 0.97			
	Firmness	5.46 ± 2.05^{x}	5.34 ± 2.15^{x}	5.61 ± 1.94^{x}	3.62 ± 2.08^{by}	3.42 ± 2.06^{by}	4.53 ± 1.98^{ay}			
JAR Evaluation										
Color	Not Yellow Enough	23.96%	31.25%	41.67%	7.41% ^{b}	9.26% ^{b}	51.85% ^{a}	2.9% ^{b}	4.9% ^{b}	26.2% ^{a}
	Just about Right	62.50%	53.125%	42.71%	37.96%	38.89%	42.59%	57.3% ^{ab}	49.5% ^{b}	72.8% ^{a}
	Too Yellow	13.54%	15.625%	15.63%	54.63%	51.85%	5.56%	39.8%	45.6%	1.0%
Flavor	Not Enough Flavor	18.75%	26.04%	17.71%	18.52%	24.07%	42.59%	26.2%	21.4%	33.0%
	Just about Right	62.50%	60.42%	56.25%	51.85%	53.70%	41.67%	58.3%	70.9%	63.1%
	Too Much Flavor	18.75%	13.54%	26.04%	29.63%	22.22%	15.74%	15.5% ^{a}	7.8% ^{ab}	3.9% ^{b}
Salt	Not Enough Salt	18.75%	27.08%	16.67%	17.59%	21.30%	30.56%			
	Just about Right	62.50%	51.04%	59.38%	38.89%	41.67%	37.04%			
	Too Much Salt	18.75%	21.88%	23.96%	43.52%	34.26%	32.41%			
Texture	Not Firm Enough	2.08%	10.42%	10.42%	42.59%	38.89%	19.44%	0.0%	0.0%	1.0%
	Just about Right	76.04	63.54	52.08	55.56%	57.41%	64.81%	58.3% ^{a}	48.5% ^{ab}	35.9% ^{b}
	Much Too Firm	21.88	26.04	37.50	1.85% ^{b}	3.70% ^{b}	15.74% ^{a}	41.7% ^{b}	51.5% ^{ab}	63.1% ^{a}

Within each consumer cultural group: values in the same row not sharing the same superscript (a, b) indicate significant difference (identified using Kruskal–Wallis and Mann–Whitney test for multiple comparison, $\alpha = 0.017$). Cross-cultural comparison: values in the same row not sharing the same superscript (x, y, z) indicate significant difference (identified using Kruskal–Wallis and Mann–Whitney test for multiple comparison, $\alpha = 0.017$). Values provided after ± are standard deviations.

3.1.2. German Consumers

Similar to Irish consumers, there was no significant ($p > 0.05$) difference in liking by German consumers for all three butters for sensory attributes, or for overall liking (Hedonics Table 1). In agreement with Irish consumers, German consumers rated the color of both FS-GRSS and FS-CLVR butter significantly ($p < 0.017$) more intense than FS-TMR butter (Intensity Scale Evaluation Table 1) and showed a significant ($p < 0.05$) higher score for 'not yellow enough' for FS-TMR butter (JAR Evaluation Table 1). German consumers also found the salt intensity of FS-GRSS and FS-CLVR butters, significantly ($p < 0.017$) higher than FS-TMR butter, even though identical levels of salt were added to each batch. Additionally, there was a significant ($p < 0.017$) difference in the perception of firmness intensity, with FS-TMR butter perceived as firmer than FS-GRSS and FS-CLVR butters. Although FA analysis was not undertaken in this study, butters from an identical experimental trial [22], and other studies [33,34], have characterized butter produced from pasture diets higher in unsaturated FAs, therefore the pasture-derived butter is likely to be softer, due to a lower melting point. The higher salty intensity perception of FS-GRSS and FS-CLVR butter as perceived by the German consumers may also relate to the softer texture of these butters and their behavior in the mouth—more rapid melting compared to FS-TMR butter (Figure 1). This would also appear to be confirmed by the JAR Evaluation of texture for FS TMR butter, which was deemed significantly ($p < 0.05$) higher for 'much too firm'. Dadalı and Elmacı (2019) [35] found that margarine with the lowest score for hardness was perceived as the saltiest, despite having identical salt contents.

Figure 1. Average results ($n = 20$) from ranking descriptive analysis evaluation of FS-GRSS, FS-CLVR and FS-TMR butters by Irish (**a**) and German (**b**) assessors. * $p < 0.05$.

3.1.3. USA Consumers

For overall appearance and color liking, USA consumers scored FS-CLVR and FS-TMR butters significantly ($p < 0.017$) higher than FS-GRSS butter (Hedonics Table 1). When attempting to elucidate the drivers for liking of butter, Krause et al. 2007 [25] identified from a focus group that USA consumers found a light yellow color desirable in butter, which likely explains why USA consumers in this study rated their liking of appearance and color of FS-TMR butter the highest (Hedonics Table 1), promoting the theory of familiarity dictating preference [36,37]. However, it is difficult to interpret why consumers liked the FS-CLVR butter similarly to FS-TMR butter for appearance and color, yet rated the FS-GRSS butter significantly lower for these same attributes. In the JAR Evaluation, USA consumers also scored FS-CLVR and FS-TMR butters significantly ($p < 0.05$) higher for 'just about right' for yellow color. However, these same consumers also scored FS-TMR butter significantly ($p < 0.05$) higher for 'not enough yellow', compared to FS-GRSS and FS-CLVR butter. Interestingly, FS-GRSS butter flavor was perceived as the most favorable by USA consumers. However, it was not significantly ($p > 0.017$) different from FS-TMR butter. This may reflect the trend in the USA for value added grass fed milk and other dairy products [38]. In terms of texture liking, USA consumers liked the FS-CLVR butter significantly ($p < 0.017$) more than FS-TMR butter. Krause et al. 2007 [25] also identified a cluster of USA consumers, referred to as 'margarine lovers', who were also butter users, but preferred the sensory attributes of margarines, i.e., the soft texture. This result corresponds to that of the JAR Evaluation, where consumers ranked FS-CLVR and FS-GRSS butters higher for 'just-about-right' texture, with FS-CLVR butter significantly ($p < 0.05$) higher compared to FS-TMR butter. In addition, USA consumers ranked FS-TMR butter significantly ($p < 0.05$) higher for 'much too firm'.

3.2. Cross-Cultural Perceptions of Butters

Irish and USA consumers scored the overall appearance liking of FS-GRSS and FS-CLVR butters significantly ($p < 0.017$) higher compared to German consumers (Hedonics Table 1). However, Irish and German consumers both scored FS-TMR butter statistically ($p < 0.017$) lower for liking of appearance. Cross-cultural liking of color corresponds to liking of appearance, with Irish and USA consumers liking FS-CLVR butter significantly ($p < 0.017$) more than German consumers. Irish consumers are accustomed to yellow butter and previously showed highest liking for butter produced from grass and clover, compared to TMR [22]. Referring to the study by Krause et al. 2007 [25], the same cluster of USA consumers who liked softer 'margarine like' butters also preferred those butters and spreads which were darker in color, in agreement with this study. Similarly, USA consumers scored the FS-GRSS butter significantly ($p < 0.017$) higher for liking of flavor than Irish and German consumers. Both Irish and German consumers had similar liking for the texture of FS-GRSS butter, which was significantly ($p < 0.017$) higher compared to USA consumers. Overall acceptability of FS-GRSS butter was rated significantly ($p < 0.017$) higher by USA consumers compared to Irish and German consumers, which may be driven by their high liking for flavor. Both USA and Irish consumers considered FS-CLVR butter to be significantly ($p < 0.017$) higher for overall acceptability compared to German consumers. Salt intensity was perceived significantly ($p < 0.017$) higher by German consumers compared to Irish consumers for both FS-GRSS and FS-CLVR butter (Intensity Scale Evaluation Table 1). Irish consumers are familiar with soft butter, and did not perceive a significant difference in salt taste in agreement with a previous study [22]. Butter sold in Germany, however, is typically unsalted, which may explain the higher perceived salt intensity in the pasture butters, although texture, as discussed earlier, is also likely a contributory factor. Perception of freshness intensity was significantly lower ($p < 0.017$) by German consumers compared to Irish consumers for FS-CLVR butters. Irish consumers found FS-GRSS butter to be significantly ($p < 0.017$) firmer compared to German consumers, which again, is likely to be related to familiarity [22].

Significant ($p < 0.017$) differences were identified for liking of FS-TMR butter, with USA consumers rating the overall appearance, color and overall liking significantly ($p < 0.017$) higher, compared to Irish and German consumers, with a similar trend identified by Krause et al. 2007 [25]. There was no significant ($p > 0.017$) difference in liking of flavor of FS-TMR butter between Irish and USA consumers, which differed to O' Callaghan et al. 2016 [22], where Irish consumers rated liking of flavor of TMR butter significantly lower than butter produced from milk from grass and clover diets. There was a significant ($p < 0.017$) difference in the liking of texture of FS-TMR butter by Irish and German consumers compared to USA consumers. For intensity ratings, Irish consumers ranked color, flavor and firmness of FS-TMR butter significantly ($p < 0.05$) higher than German consumers. No significant differences were evident in relation to JAR Evaluation for color, flavor, salt (USA consumers did not assess salt) or texture between the cultural groups. Overall, the cross-cultural perception of butter attributes by the three consumer groups was within a similar range on the hedonic scale, signifying a liking of butter among all three consumer groups.

3.3. Ranking Descriptive Analysis

The average results of Irish and German assessors perception of butters are portrayed in Figure 1a,b and in Table 2. Corresponding to the results for yellow intensity (Table 1), both Irish and German assessors rated FS-TMR butter significantly ($p < 0.05$) lower for yellow color. Both German and Irish assessors rated FS-CLVR butter as having a more intense yellow color (Figure 1a,b), compared to FS-GRSS butter, although not significant, this may suggest that FS-CLVR butter contained higher amounts of β-carotene (β-carotene levels vary with intake of grass and clover outdoors). O'Callaghan et al. 2016 [22] identified grass butters as having the highest levels of β-carotene. However, Panthi, Sundekilde et al. (2019) [39] identified Massdam cheeses produced from cows supplemented with white clover as having higher amounts of β-carotene compared to cheese produced from only grass. As mentioned, β-carotene content in milk will vary due to the levels of grass and clover ingested by cows due to differences in availability within the pasture. German assessors perceived FS-TMR butter as significantly ($p < 0.05$) darker than Irish assessors; however, this does not match results from the intensity scale portion of the consumer study (Table 1). This result may be due to the fact that in Ireland only butter derived from pasture is commercially available, while in Germany butter's from both pasture or concentrate are widely available, and therefore Irish consumers may have scored TMR butter lower for color intensity due to a lack of familiarity.

Table 2. Cross-cultural comparison of RDA evaluation by Irish and German assessors of butters produced by different diets—FS-GRSS, FS-CLVR and FS-TMR.

	Irish Assessors			German Assessors		
	FS-GRSS	**FS-CLVR**	**FS-TMR**	**FS-GRSS**	**FS-CLVR**	**FS-TMR**
Color						
Yellow color	6.81 ± 0.9 ^{a}	6.33 ± 1.37 ^{a}	2.41 ± 0.85 ^{by}	6.63 ± 1.12 ^{a}	6.41 ± 1.46 ^{a}	3.72 ± 1.37 ^{bx}
Aroma						
Buttery	4.66 ± 1.53 ^{a}	5.13 ± 1.51 ^{a}	3.48 ± 1.22 ^{b}	5.48 ± 1.86	5.93 ± 1.99	5.20 ± 1.93
Milky	3.89 ± 1.45 ^{y}	4.05 ± 1.29	3.40 ± 1.24	5.03 ± 2.56 ^{x}	4.66 ± 2.25	3.93 ± 2.08
Grassy	2.41 ± 1.17	2.54 ± 1.34	1.71 ± 0.72	2.50 ± 1.92	2.42 ± 1.97	1.69 ± 1.11
Rancid	1.39 ± 0.69	1.76 ± 1.40	1.31 ± 0.73	2.11 ± 1.76	2.05 ± 1.64	1.49 ± 1.08
Flavor						
Salty	5.71 ± 1.13 ^{a}	5.33 ± 1.66 ^{a}	4.16 ± 1.77 ^{b}	6.24 ± 2.00 ^{a}	6.23 ± 1.47 ^{a}	4.81 ± 1.77 ^{b}
Sweet	3.64 ± 1.50	3.99 ± 1.45	3.34 ± 1.41	3.92 ± 1.95	3.50 ± 1.76	3.61 ± 1.84
Creamy	6.29 ± 1.25 ^{a}	6.24 ± 1.37 ^{a}	4.91 ± 1.46 ^{b}	5.27 ± 1.90	5.10 ± 1.90	4.60 ± 2.03
Sour	1.64 ± 0.57 ^{y}	1.76 ± 1.15	1.83 ± 0.79	2.85 ± 1.67 ^{x}	2.97 ± 2.20	2.79 ± 1.82
Stale	1.51 ± 0.81	1.76 ± 1.07	1.61 ± 0.74	1.59 ± 0.78	1.80 ± 1.49	1.83 ± 1.44
Off flavor	1.44 ± 0.63	2.16 ± 1.84	1.56 ± 1.01	1.90 ± 1.33	1.81 ± 1.25	1.66 ± 1.33
Texture						
Melt in the mouth	7.19 ± 0.85 ^{a}	6.79 ± 1.26 ^{a}	4.19 ± 1.73 ^{by}	7.36 ± 1.20	6.86 ± 1.46	5.96 ± 2.21 ^{x}

Values are the average results ±standard deviations. Within consumer group: values in the same row not sharing the same superscript (a, b) indicate significant difference (Confidence level 5%, identified using ANOVA and Tukey post). Cross-cultural comparison: values in the same row not sharing the same superscript (x, y) indicate significant difference (confidence level 5%, identified using student *t*-test).

Irish assessors perceived the aroma of FS-GRSS and FS-CLVR butters as significantly ($p < 0.05$) more buttery compared to FS-TMR butter, similar to O'Callaghan et al. 2016 [22], where consumers rated butter produced from grass significantly higher for diacetyl aroma compared to TMR butter. There was no significant difference ($p < 0.05$) detected among the German assessors for buttery aroma. German assessors perceived FS-GRSS butters to be significantly more milky compared to Irish assessors. Grassy and rancid aroma were not identified as significantly ($p > 0.05$) different by either German or Irish assessors. Both Irish and German assessors perceived significant ($p < 0.05$) differences in saltines perception, with values for FS-GRSS and FS-CLVR butters significantly higher, compared to FS-TMR butter. This is likely linked to the texture—melt in the mouth attribute—where Irish assessors perceived FS-GRSS and FS-CLVR butters to melt much more rapidly ($p < 0.05$) in the mouth. As previously mentioned, this is likely due to changes in fatty acid profile, which influence spreadability and melting properties of butter [40]. Similarly, Irish assessors perceived the flavor attribute creamy as significantly ($p < 0.05$) higher in both FS-GRSS and FS-CLVR butters, than in FS-TMR butter, which is also likely directly related to the texture, and in turn to the unsaturated fatty acid profile of these butters. Irish assessors have previously perceived pasture butters as significantly more creamy [22]. Irish assessors in this study also found that FS-GRSS and FS-CLVR butters were significantly higher ($p < 0.05$) for melt in the mouth than FS-TMR butter. German assessors scored melt in the mouth statistically higher ($p < 0.05$) for TS-TMR butter than Irish assessors. German assessors also found FS-CLVR butter significantly ($p < 0.05$) more sour compared to Irish assessors, and this may relate to the lower rating of freshness as perceived in the consumer study (Intensity Scale Evaluation Table 1).

3.4. Descriptive Analysis Evaluation of FS-GRSS, FS-CLVR and FS-TMR Butters by Trained USA Panelists

The results of the descriptive analysis (DA) undertaken by trained USA panelists are listed in Table 3. There was no significant ($p > 0.05$) difference perceived for cooked/nutty, milkfat, and salty taste, between all three butters. Grassy was rated significantly ($p < 0.05$) higher in FS-GRSS butter, followed by FS-CLVR butter. However, it was not detected in FS-TMR butter, corresponding to results from Cheng et al. (2020) [41], who reported significant difference in grassy/hay perception of bovine skim milk powder, produced from pasture versus indoor TMR diets, as assessed by a descriptive panel. Villeneuve et al. 2013 [42] identified that grassy intensity was higher from milk produced by pasture and correlated to the higher levels of the aldehyde pentanal. The attribute stale was not perceived in any of the butters, and the levels of saltiness perception were similar (Table 3). Panelists found a significant ($p < 0.05$) difference in the color intensity of the three butters, in the following order: FS-CLVR > FS-GRSS > FS-TMR. Mothball flavor was noted in FS-CLVR butter but not in FS-GRSS or FS-TMR butters. This is a feed specific flavor that has been documented in previous studies (butter, cheese, dried ingredients) manufactured from pasture feeding, and appears to be specific to certain types of pastures [25,43].

Table 3. Sensory attribute means from trained USA panel evaluation of FS-GRSS, FS-CLVR and FS-TMR butters. Table presents DA averages and standard deviations.

	Feed System		
Sensory Attribute	**FS-CLVR**	**FS-GRSS**	**FS-TMR**
Cooked/Nutty	3.1 ± 0.1	3.06 ± 0.2	3.3 ± 0.1
Milkfat	3.1 ± 0.1	3.1 ± 0.2	3.2 ± 0.1
Grassy	1.2 ± 0.1 ^{b}	1.4 ± 0.1 ^{a}	ND ^{c}
Mothball	1.3 ± 0.1 ^{a}	ND ^{b}	ND ^{b}
Stale	ND	ND	ND
Salty Taste	11.1 ± 0.1	10.9 ± 0.7	11.2 ± 0.1
Color Intensity	4.2 ± 0.1 ^{a}	3.4 ± 0.3 ^{b}	1.8 ± 0.1 ^{c}

Means represent duplicate evaluations from three experimental replications by six highly trained panelists. Attributes were scored using a 0 to 15 point universal intensity scale consistent with the Spectrum descriptive analysis method. ND—not detected. Means in a row followed by different superscript letters are different ($p < 0.05$). Values provided after ± are standard deviations.

3.5. Volatile Compounds

HS-SPME-GC-MS analysis of the butters identified a total of 30 volatile compounds across the three feeding systems (Table 4). Aldehydes, ketones, acids, terpenes and lactones were the main chemical classes contributing to the volatile profile of all three butters. We have only discussed compounds where the abundances are significantly different with respect to the feed systems.

The aldehyde compounds—pentanal, hexanal, heptanal and decanal—were most influenced by the different feed systems. All of these compounds are associated with lipid oxidation [44]. Levels of pentanal were significantly ($p < 0.05$) more abundant in FS-CLVR butter compared to FS-GRSS and FS-TMR butters in agreement with previous studies on butter [22], and pasteurized bovine milk [45].

Pentanal is derived from the fatty acids arachidonic and linoleic acid has the potential to adversely impact sensory perception by yielding a paint-like, cardboard aroma [45]. Hexanal, derived from linoleic acid [46], can confer a grassy off flavor in butter [47], and was significantly more abundant in FS-CLVR and FS-TMR butter, compared to FS-GRSS butter (Table 4). Heptanal was significantly ($p < 0.017$) more abundant in FS-CLVR and FS-GRSS butters in comparison to FS-TMR butter, and is also a product of linoleic acid, which has a green sweet aroma in dairy products [48]. Decanal, a compound of oleic acid degradation, was significantly ($p < 0.05$) more abundant in FS-CLVR butter compared to FS-GRSS butter and has been identified as having a green, fatty aroma in sweet cream butter [27]. Faulkner et al. 2018 [49] found a similar trend in decanal levels from raw milk produced from grass, grass/clover and TMR diets. Although the relative abundance of precursor unsaturated fatty acids is important, other factors such as the presence of natural pro- and anti-oxidants is also important.

Six ketones were identified in the three butter samples; 2,3-butanedione (diacetyl), a very odor-active compound with a characteristic buttery aroma [50,51], was significantly ($p < 0.05$) more abundant in FS-CLVR and FS-GRSS butters compared to FS-TMR butter (Table 4). Diacetyl was not detected in any of the butters by O'Callaghan et al. (2016) [22], but grass derived butter was rated higher for diacetyl aroma. It is difficult to discern why more diacetyl would be present in the butters produced from milk derived from pasture, but it could be that precursors of diacetyl are higher in those milks, due to different microbial activities in the rumen. Diacetyl is derived from pyruvate where α-acetolactate synthase converts it to α-acetolactate, which is subsequently converted to diacetyl by non-enzymatic oxidative decarboxylation [52]. FS-CLVR butter also had significantly ($p < 0.05$) more 2-butanone compared to FS-GRSS butter, with similar results seen in previous studies on butter [22] and milk [49]. This methyl ketone also derives from pyruvate metabolism [53], similarly to diacetyl. Acetone was significantly ($p < 0.05$) higher in FS-GRSS and FS-CLVR butters than in FS-TMR butter, and has also been identified as a product from concentrate feed [54]. Previous studies on milk and skim milk powder produced from pasture and TMR diets did not find differences in the abundance of acetone based on diet [40,49].

Both butanoic and nonanoic acid were significantly ($p < 0.05$) more abundant in FS-CLVR butter in comparison to FS-TMR and FS-GRSS butter. The presence and concentration of these acids are potentially important in flavor perception, with butanoic acid described as dirty sock in butter, and quite odor active [27]. Bovolenta et al. (2014) [55] found that when cows were exposed to high levels of pasture, there was a significant influence on nonanoic acid in Montasio cheese, with this compound not detected in cheese produced from milk derived from a low pasture diet.

Although not recognized to play such a significant role in odor and sensory quality of butter, toluene was significantly more abundant ($p < 0.05$) in both FS-CLVR and FS-GRSS butter than in FS-TMR butter. Toluene is a by-product of β-carotene metabolism [56], demonstrating its potential as a biomarker for pasture-derived dairy products [44]. As FS-CLVR butter was rated most yellow in this study, this may also suggest higher levels of β-carotene and hence the abundance of toluene.

The greater abundance of ethyl acetate in FS-CLVR butter is difficult to discern except that it is a product of ethanol and acetic acid. The sources of both precursors may be influenced by cow diet [45].

Table 4. Average (n = 9) peak area values (×10^6) of volatile compounds identified in FS-GRSS, FS-CLVR and FS-TMR butters.

Volatile Compound	CAS NUMBER #	Odor Descriptors	RI	REF RI	Feeding System		
					FS-GRSS	FS-CLVR	FS-TMR
Aldehyde							
Pentanal [2]	110-62-3	Pungent, almond like, chemical, malty, apple [22]	731	733	0.054 ± 0.027 [b]	0.489 ± 0.394 [a]	0.049 ± 0.018 [b]
Hexanal [3]	66-25-1	Green, slightly fruity, lemon, herbal, grassy [22]	836	839	0.032 ± 0.013 [b]	0.060 ± 0.035 [a]	0.049 ± 0.018 [a]
Heptanal [3]	111-71-7	Slightly fruity (balsam), fatty, oily [22]	937	943	0.020 ± 0.008 [a]	0.035± 0.027 [a]	0.010 ± 0.004 [b]
Benzaldehyde	100-52-7	Bitter, almond, sweet cherry [22]	1026	1028.9	0.020 ± 0.012	0.019 ± 0.011	0.028 ± 0.011
Nonanal	124-19-6	Green, citrus, fatty, floral [22]	1143	1150	0.039 ± 0.040	0.035 ± 0.028	0.024 ± 0.017
Decanal [3]	112-31-2	Green [28]	1246	-	0.001 ± 0.001 [b]	0.003± 0.001 [a]	0.002 ± 0.002 [ab]
Ketone							
Acetone[2]	67-64-1	Earthy, strong fruity, wood pulp, hay [22]	529	533	0.224 ± 0.125 [ab]	0.216 ± 0.052 [a]	0.124 ± 0.027 [b]
Diacetyl (2,3-Butanedione) [2]	431-03-8	Buttery [50]	628	-	0.044 ± 0.015 [a]	0.079 ± 0.057 [ab]	0.022 ± 0.009 [b]
2-Butanone [2]	78-93-3	Buttery, sour milk, etheric [22]	635	639	0.106 ± 0.019 [b]	0.286 ± 0.156 [a]	0.170 ± 0.082 [ab]
2-Heptanone	113-43-0	Blue cheese, spicy, Roquefort [22]	929	936	0.062 ± 0.011	0.077 ± 0.034	0.060 ± 0.007
2-Nonanone	821-55-6	Malty, fruity, hot milk, smoked cheese [22]	1133	1140	0.017 ± 0.003	0.037 ± 0.033	0.013 ± 0.005
2-Pentanone	107-87-9	Orange peel, sweet, Fruity [22]	725	-	0.029 ± 0.012	0.032 ± 0.010	0.031 ± 0.014
Acid							
Butanoic Acid [1]	107-92-6	Sweaty, butter, cheese, Strong, acid, fecal, rancid [22]	860	864	0.015 ± 0.007 [b]	0.024 ± 0.004 [a]	0.017 ± 0.008 [ab]
Hexanoic Acid [1]	142-62-1	Acidic, sweaty, cheesy, sharp, goaty, bad breath [22]	1045	1052	0.023 ± 0.009	0.041 ± 0.02	0.026 ± 0.010
Nonanoic Acid [2]	112-05-0	Waxy, dirty and cheesy with a cultured dairy nuance **	22.7	-	0.014 ± 0.007 [b]	0.026 ± 0.005 [a]	0.014 ± 0.008 [b]

Table 4. *Cont.*

Volatile Compound	CAS NUMBER #	Odor Descriptors	RI	REF RI	Feeding System		
					FS-GRSS	FS-CLVR	FS-TMR
Hydrocarbons							
Toluene[2]	108-88-3	Nutty, bitter, almond, Plastic [22]	789	794	0.794 ± 0.26^{b}	1.793 ± 0.708^{a}	0.139 ± 0.031^{c}
* o-Xylene	108-38-3	Geranium **	895	-	0.371 ± 0.304	0.303 ± 0.386	0.572 ± 0.449
* *p*-Xylene	106-42-3	Not listed **	923	-	0.177 ± 0.187	0.087 ± 0.143	0.222 ± 0.136
Lactone							
δ-Hexalactone	823-22-3	Creamy, chocolate, sweet aromatic [50]	1215	-	0.129 ± 0.033	0.114 ± 0.026	0.117± 0.022
δ-Octalactone	698-76-0	Coconut like, peach [50]	1413	-	0.026 ± 0.006	0.024 ± 0.007	0.022 ± 0.005
δ-Decalactone	705-86-2	Coconut like, peach [50]	1691	1620.9	0.020 ± 0.007	0.021 ± 0.010	0.019 ± 0.008
Sulfide							
Dimethyl Sulfide	75-18-3	Corn like, fresh pumpkin [50]	534	538	0.011 ± 0.010	0.018 ± 0.016	0.009 ± 0.002
Carbon Disulfide	75-15-0	Sulfury, onion, sweet corn, vegetable, cabbage, tomato, green, radish **	542	548	0.057 ± 0.018	0.132 ± 0.115	0.088 ± 0.088
Ester							
Ethyl Acetate[2]	141-78-6	Solvent, pineapple, Fruity, fruit gum [22]	639	-	0.013 ± 0.010^{ab}	0.033 ± 0.019^{a}	0.010 ± 0.009^{b}
Ethyl Benzene	100-41-4	Not listed **	887	-	0.072 ± 0.046	0.103 ± 0.160	0.206 ± 0.213
Diethyl Ether	60-29-7	Ethereal **	512	-	0.020 ± 0.022	0.024 ± 0.025	0.024 ± 0.026
Other							
Ethanol	64-17-5	Alcoholic, ethereal, medicinal **	503	506	0.038 ± 0.020	0.032 ± 0.018	0.028 ± 0.021
1-Pentene	109-67-1	Not listed **	565	-	0.061 ± 0.024	0.047 ± 0.024	0.076 ± 0.077
α-Pinene	80-56-8	Mint, pine oil [27]	950	951	0.038 ± 0.032	0.014 ± 0.010	0.036 ± 0.026
Dodecane	112-40-3	Alkane **	1193	-	0.006 ± 0.004	0.007 ± 0.002	0.005 ± 0.002

RI: Retention index. REF RI: Reference retention index. # CAS: Chemical Abstracts Service Number. Values in the same row not sharing the same superscript (a, b) specify significant difference in peak area value average. Volatile compound annotated with a [1] indicate statistical analysis carried out by ANOVA and Tukey post hoc test, $\alpha = 0.05$. Volatile compounds annotated with a [2] indicate statistical analysis carried out by Welch test and Games-Howell post hoc test, $\alpha = 0.05$. Volatile compounds annotated with a [3] indicate statistical analysis carried out by Kruskal–Wallis and Mann–Whitney, $\alpha = 0.017$. * Tentative identification due to isomers. Values provided after ± are standard deviations. ** Odor descriptors sourced from http://www.thegoodscentscompany.com/.- No REF RI available.

4. Conclusions

This research assessed the cross-cultural perception and liking of butters produced by three different feed systems. Overall, Irish, German, and USA consumers did not discriminate their overall liking of butters based on feed system, although some clear cultural differences were evident. Familiarity was postulated to contribute to differences in appearance liking and color liking by USA consumers, where the indoor TMR feed system scored highest. However, this did not impact overall liking of butter produced from pasture (FS-GRSS and FS-CLVR) by USA consumers. Both Irish and German consumers rated the color intensity of the pasture butters (FS-GRSS and FS-CLVR) higher than FS-TMR butter. German consumers also found that pasture (FS-GRSS and FS-CLVR) butter was more salty and that FS-TMR butter was firmer. It seems plausible that the texture and salty differences noted by German consumers are likely related to differences in their FA profile, which directly impacts on texture, but also indirectly on salty perception as the greater unsaturated FAs lower the melting point in the pasture butters. RDA by Irish and German assessors also identified important cross-cultural differences—German assessors perceived FS-GRSS butter as significantly more sour and milky, and scored FS-TMR butter higher for melt in the mouth, in comparison to Irish assessors. Both Irish and German assessors found that FS-TMR butter had less yellow color in agreement with the consumers, and German assessors found it darker than Irish assessors. Both Irish and German assessors found that FS-CLVR butter was the most intense yellow. USA panelists also found that color intensity ranged from FS-CLVR > FS-GRSS > FS-TMR. These results correlate with previous studies highlighting the impact of carotenoids, specifically β-carotene on the yellow color of pasture-derived dairy products. The trained USA panelists also found that a grassy flavor was highest in FS-GRSS butter but absent in FS-TMR butter, and that FS-CLV butter had a mothball attribute absent in FS-GRSS and FS-TMR butter. Volatile analysis identified a number of compounds that were statistically different based on diet—aldehydes, ketones, acids, toluene and ethyl acetate. The aldehydes, ketones and acids in the butter from the pasture diets may be influencing the grassy, milky and sour flavors, as perceived by USA and German consumers. Diacetyl may also be influencing enhanced buttery aroma, as perceived by Irish assessors in pasture (FS-GRSS and FS-CLVR) butter.

This study demonstrates that different feed systems affect cross-cultural perception of butter and that familiarity of products from specific feeds systems is a factor, but that it does not adversely impact butter acceptance in terms of overall liking within these cultural groups.

Supplementary Materials: The following are available online at http://www.mdpi.com/2304-8158/9/12/1767/s1, Table S1: Sensory terms for consumer study; in relation to Hedonics, Intensity Scales and for Just About Right assessment; Table S2: Attribute list presented to panelists for Ranked Descriptive Analysis by assessors.

Author Contributions: Funding acquisition, M.G.O. and K.N.K.; conceptualization, K.N.K. and E.C.G.; methodology, T.S., M.D., M.G.O. and K.N.K.; resources, T.S., M.D., S.F., M.G.O., J.P.K. and K.N.K.; project administration, M.G.O., J.P.K. and K.N.K.; formal analysis, E.C.G., T.S., M.D., M.G.O., J.P.K. and K.N.K.; investigation, E.C.G. and T.F.O.; data curation, E.C.G., T.S., T.F.O., M.D. and S.F.; writing—original draft preparation, E.C.G.; writing—review and editing, M.G.O., J.P.K. and K.N.K.; supervision, M.G.O., J.P.K. and K.N.K. All authors have read and agreed to the published version of the manuscript.

Funding: Department of Agriculture, Food and the Marine, Ireland: 14F 812.

Conflicts of Interest: The authors declare no conflict of interest.

References

1. Aschemann-Witzel, J.; Varela, P.; Peschel, A.O. Consumers' categorization of food ingredients: Do consumers perceive them as 'clean label' producers expect? An exploration with projective mapping. *Food Qual. Prefer.* **2019**, *71*, 117–128. [CrossRef]
2. Bir, C.L.; Widmar, N.J.O.; Thompson, N.M.; Townsend, J.; Wolf, C. US respondents' willingness to pay for Cheddar cheese from dairy cattle with different pasture access, antibiotic use, and dehorning practices. *J. Dairy Sci.* **2020**, *103*, 3234–3249. [CrossRef]

3. García-Torres, S.; López-Gajardo, A.; Mesias, F. Intensive vs. free-range organic beef. A preference study through consumer liking and conjoint analysis. *Meat Sci.* **2016**, *114*, 114–120. [CrossRef] [PubMed]
4. Font i Furnols, M.; Realini, C.; Montossi, F.; Sañudo, C.; Campo, M.M.; Oliver, M.A.; Nute, G.R.; Guerrero, L. Consumer's purchasing intention for lamb meat affected by country of origin, feeding system and meat price: A conjoint study in Spain, France and United Kingdom. *Food Qual. Prefer.* **2011**, *22*, 443–451. [CrossRef]
5. Napolitano, F.; Braghieri, A.; Piasentier, E.; Favotto, S.; Naspetti, S.; Zanoli, R. Effect of information about organic production on beef liking and consumer willingness to pay. *Food Qual. Prefer.* **2010**, *21*, 207–212. [CrossRef]
6. Scozzafava, G.; Gerini, F.; Boncinelli, F.; Contini, C.; Marone, E.; Casini, L. Organic milk preference: Is it a matter of information? *Appetite* **2020**, *144*, 104477. [CrossRef] [PubMed]
7. Stampa, E.; Schipmann-Schwarze, C.; Hamm, U. Consumer perceptions, preferences, and behavior regarding pasture-raised livestock products: A review. *Food Qual. Prefer.* **2020**, *82*, 103872. [CrossRef]
8. O'Brien, B.; Dillon, P.; Murphy, J.; Mehra, R.K.; Guinee, T.P.; Connolly, J.F.; Kelly, A.L.; Joyce, P. Effects of stocking density and concentrate supplementation of grazing dairy cows on milk production, composition and processing characteristics. *J. Dairy Res.* **1999**, *66*, 165–176. [CrossRef] [PubMed]
9. Whelan, S.; Carey, W.; Boland, T.; Lynch, M.; Kelly, A.; Rajauria, G.; Pierce, K. The effect of by-product inclusion level on milk production, nutrient digestibility and excretion, and rumen fermentation parameters in lactating dairy cows offered a pasture-based diet. *J. Dairy Sci.* **2017**, *100*, 1055–1062. [CrossRef]
10. CSO. Central Statistics Office, Milk Statistics December 2015. 2015. Available online: https://www.cso.ie/en/releasesandpublications/er/ms/milkstatisticsdecember2015/ (accessed on 9 June 2020).
11. CSO. Central Statistics Office, Milk Statistics December 2015. 2017. Available online: https://www.cso.ie/en/releasesandpublications/er/ms/milkstatisticsdecember2017/ (accessed on 9 June 2020).
12. McKay, Z.; Lynch, M.; Mulligan, F.; Rajauria, G.; Miller, C.; Pierce, K. The effect of concentrate supplementation type on milk production, dry matter intake, rumen fermentation, and nitrogen excretion in late-lactation, spring-calving grazing dairy cows. *J. Dairy Sci.* **2019**, *102*, 5042–5053. [CrossRef]
13. Bord Bia. Export Performance and Prospects. Irish Food, Drink & Horticulture 2019–2020. 2019, p. 10. Available online: https://www.bordbia.ie/globalassets/bordbia.ie/industry/performance-and-prospects/2019-pdf/performance-and-prospects-2019-2020.pdf (accessed on 9 June 2020).
14. Croissant, A.; Washburn, S.; Dean, L.; Drake, M. Chemical Properties and Consumer Perception of Fluid Milk from Conventional and Pasture-Based Production Systems. *J. Dairy Sci.* **2007**, *90*, 4942–4953. [CrossRef] [PubMed]
15. O'Callaghan, T.F.; Hennessy, D.; McAuliffe, S.; Kilcawley, K.N.; O'Donovan, M.; Dillon, P.; Ross, R.; Stanton, C. Effect of pasture versus indoor feeding systems on raw milk composition and quality over an entire lactation. *J. Dairy Sci.* **2016**, *99*, 9424–9440. [CrossRef] [PubMed]
16. White, S.L.; Bertrand, J.A.; Wade, M.R.; Washburn, S.P.; Green, J.T., Jr.; Jenkins, T.C. Comparison of fatty acid content of milk from Jersey and Holstein cows consuming pasture or a total mixed ration. *J. Dairy Sci.* **2001**, *84*, 2295–2301. [CrossRef]
17. Martin, B.; Verdier-Metz, I.; Buchin, S.; Hurtaud, C.; Coulon, J.-B. How do the nature of forages and pasture diversity influence the sensory quality of dairy livestock products? *Anim. Sci.* **2005**, *81*, 205–212. [CrossRef]
18. Khanal, R.; Dhiman, T.R.; Ure, A.; Brennand, C.; Boman, R.; McMahon, D. Consumer Acceptability of Conjugated Linoleic Acid-Enriched Milk and Cheddar Cheese from Cows Grazing on Pasture. *J. Dairy Sci.* **2005**, *88*, 1837–1847. [CrossRef]
19. Liu, N.; Pustjens, A.M.; Erasmus, S.W.; Yang, Y.; Hettinga, K.A.; Van Ruth, S.M. Dairy farming system markers: The correlation of forage and milk fatty acid profiles from organic, pasture and conventional systems in the Netherlands. *Food Chem.* **2020**, *314*, 126153. [CrossRef] [PubMed]
20. Van Valenberg, H.J.F.; Hettinga, K.; Dijkstra, J.; Bovenhuis, H.; Feskens, E.F. Concentrations of n-3 and n-6 fatty acids in Dutch bovine milk fat and their contribution to human dietary intake. *J. Dairy Sci.* **2013**, *96*, 4173–4181. [CrossRef] [PubMed]
21. Agabriel, C.; Cornu, A.; Sibra, C.; Grolier, P.; Martin, B. Tanker Milk Variability According to Farm Feeding Practices: Vitamins A and E, Carotenoids, Color, and Terpenoids. *J. Dairy Sci.* **2007**, *90*, 4884–4896. [CrossRef] [PubMed]

22. O'Callaghan, T.F.; Faulkner, H.; McAuliffe, S.; Sullivan, M.G.O.; Hennessy, D.; Dillon, P.; Kilcawley, K.N.; Stanton, C.; Ross, R.P. Quality characteristics, chemical composition, and sensory properties of butter from cows on pasture versus indoor feeding systems. *J. Dairy Sci.* **2016**, *99*, 9441–9460. [CrossRef]
23. Jayedi, A.; Rashidy-Pour, A.; Parohan, M.; Zargar, M.S.; Shab-Bidar, S. Dietary and circulating vitamin C, vitamin E, β-carotene and risk of total cardiovascular mortality: A systematic review and dose–response meta-analysis of prospective observational studies. *Public Health Nutr.* **2019**, *22*, 1872–1887. [CrossRef]
24. Walstra, P. Physical chemistry of milk fat globules. *Adv. Dairy Chem.* **1995**, *2*, 131–178.
25. Krause, A.; Lopetcharat, K.; Drake, M. Identification of the Characteristics That Drive Consumer Liking of Butter. *J. Dairy Sci.* **2007**, *90*, 2091–2102. [CrossRef] [PubMed]
26. Li, Y.; Wang, Y.; Yuan, D.; Zhang, L. Comparison of SDE and SPME for the analysis of volatile compounds in butters. *Food Sci. Biotechnol.* **2020**, *29*, 55–62. [CrossRef] [PubMed]
27. Lozano, P.R.; Miracle, E.R.; Krause, A.J.; Drake, M.; Cadwallader, K.R. Effect of Cold Storage and Packaging Material on the Major Aroma Components of Sweet Cream Butter. *J. Agric. Food Chem.* **2007**, *55*, 7840–7846. [CrossRef] [PubMed]
28. Mallia, S.; Escher, F.; Dubois, S.; Schieberle, P.; Schlichtherle-Cerny, H. Characterization and Quantification of Odor-Active Compounds in Unsaturated Fatty Acid/Conjugated Linoleic Acid (UFA/CLA)-Enriched Butter and in Conventional Butter during Storage and Induced Oxidation. *J. Agric. Food Chem.* **2009**, *57*, 7464–7472. [CrossRef] [PubMed]
29. Hurtaud, C.; Faucon, F.; Couvreur, S.; Peyraud, J.-L. Linear relationship between increasing amounts of extruded linseed in dairy cow diet and milk fatty acid composition and butter properties. *J. Dairy Sci.* **2010**, *93*, 1429–1443. [CrossRef]
30. ISO. 2014 ISO 11136. 2014. Available online: https://www.iso.org/ics/67.240/x/ (accessed on 16 October 2020).
31. Richter, V.B.; De Almeida, T.C.A.; Prudencio, S.H.; Benassi, M.D.T. Proposing a ranking descriptive sensory method. *Food Qual. Prefer.* **2010**, *21*, 611–620. [CrossRef]
32. Van Den Dool, H.; Kratz, P.D. A generalization of the retention index system including linear temperature programmed gas—liquid partition chromatography. *J. Chromatogr. A* **1963**, *11*, 463–471. [CrossRef]
33. Couvreur, S.; Hurtaud, C.; Lopez, C.; Delaby, L.; Peyraud, J. The Linear Relationship between the Proportion of Fresh Grass in the Cow Diet, Milk Fatty Acid Composition, and Butter Properties. *J. Dairy Sci.* **2006**, *89*, 1956–1969. [CrossRef]
34. Silva, C.; Silva, S.P.; Prates, J.A.; Bessa, R.J.B.; Rosa, H.J.D.; Rego, O.A. Physicochemical traits and sensory quality of commercial butter produced in the Azores. *Int. Dairy J.* **2019**, *88*, 10–17. [CrossRef]
35. Dadalı, C.; Elmacı, Y. Characterization of volatile release and sensory properties of model margarines by changing fat and emulsifier content. *Eur. J. Lipid Sci. Technol.* **2019**, *121*, 1900003. [CrossRef]
36. Kim, Y.-K.; Jombart, L.; Valentin, D.; Kim, K.-O. Familiarity and liking playing a role on the perception of trained panelists: A cross-cultural study on teas. *Food Res. Int.* **2015**, *71*, 155–164. [CrossRef]
37. Pagès, J.; Bertrand, C.; Ali, R.; Husson, F.; Lê, S. SENSORY ANALYSIS COMPARISON OF EIGHT BISCUITS BY FRENCH AND PAKISTANI PANELS. *J. Sens. Stud.* **2007**, *22*, 665–686. [CrossRef]
38. Harwood, W.; Drake, M. Identification and characterization of fluid milk consumer groups. *J. Dairy Sci.* **2018**, *101*, 8860–8874. [CrossRef] [PubMed]
39. Panthi, R.R.; Sundekilde, U.K.; Kelly, A.L.; Hennessy, D.; Kilcawley, K.N.; Mannion, D.; Fenelon, M.A.; Sheehan, J.J. Influence of herd diet on the metabolome of Maasdam cheeses. *Food Res. Int.* **2019**, *123*, 722–731. [CrossRef] [PubMed]
40. Hurtaud, C.; Peyraud, J.L. Effects of Feeding Camelina (Seeds or Meal) on Milk Fatty Acid Composition and Butter Spreadability. *J. Dairy Sci.* **2007**, *90*, 5134–5145. [CrossRef]
41. Cheng, Z.; Sullivan, M.G.O.; Kerry, J.; Drake, M.; Miao, S.; Kaibo, D.; Kilcawley, K.N. A cross-cultural sensory analysis of skim powdered milk produced from pasture and non-pasture diets. *Food Res. Int.* **2020**, *109749*, 109749. [CrossRef]
42. Villeneuve, M.-P.; Lebeuf, Y.; Gervais, R.; Tremblay, G.; Vuillemard, J.; Fortin, J.; Chouinard, P. Milk volatile organic compounds and fatty acid profile in cows fed timothy as hay, pasture, or silage. *J. Dairy Sci.* **2013**, *96*, 7181–7194. [CrossRef]
43. Drake, M.; Yates, M.; Gerard, P.; Delahunty, C.; Sheehan, E.; Turnbull, R.; Dodds, T. Comparison of differences between lexicons for descriptive analysis of Cheddar cheese flavour in Ireland, New Zealand, and the United States of America. *Int. Dairy J.* **2005**, *15*, 473–483. [CrossRef]

44. Clarke, H.J.; O'Sullivan, M.G.; Kerry, J.P.; Kilcawley, K.N. Correlating Volatile Lipid Oxidation Compounds with Consumer Sensory Data in Dairy Based Powders during Storage. *Antioxidants* **2020**, *9*, 338. [CrossRef]
45. Kilcawley, K.N.; Faulkner, H.; Clarke, H.J.; Sullivan, M.G.O.; Kerry, J.P. Factors Influencing the Flavour of Bovine Milk and Cheese from Grass Based versus Non-Grass Based Milk Production Systems. *Foods* **2018**, *7*, 37. [CrossRef] [PubMed]
46. Fujisaki, M.; Endo, Y.; Fujimoto, K. Retardation of volatile aldehyde formation in the exhaust of frying oil by heating under low oxygen atmospheres. *J. Am. Oil Chem. Soc.* **2002**, *79*, 909–914. [CrossRef]
47. Panseri, S.; Soncin, S.; Chiesa, L.; Biondi, P.A. A headspace solid-phase microextraction gas-chromatographic mass-spectrometric method (HS-SPME-GC/MS) to quantify hexanal in butter during storage as marker of lipid oxidation. *Food Chem.* **2011**, *127*, 886–889. [CrossRef] [PubMed]
48. Friedrich, J.E.; Acree, T.E. Gas Chromatography Olfactometry (GC/O) of Dairy Products. *Int. Dairy J.* **1998**, *8*, 235–241. [CrossRef]
49. Faulkner, H.; O'Callaghan, T.F.; McAuliffe, S.; Hennessy, D.; Stanton, C.; Sullivan, M.G.O.; Kerry, J.P.; Kilcawley, K.N. Effect of different forage types on the volatile and sensory properties of bovine milk. *J. Dairy Sci.* **2018**, *101*, 1034–1047. [CrossRef] [PubMed]
50. Mallia, S.; Escher, F.; Schlichtherle-Cerny, H. Aroma-active compounds of butter: A review. *Eur. Food Res. Technol.* **2007**, *226*, 315–325. [CrossRef]
51. Schieberle, P.; Gassenmeier, K.; Guth, H.; Sen, A.; Grosch, W. Character Impact Odour Compounds of Different Kinds of Butter. *LWT* **1993**, *26*, 347–356. [CrossRef]
52. Liu, J.; Chen, L.; Dorau, R.; Lillevang, S.K.; Jensen, P.R.; Solem, C. From Waste to Taste—Efficient Production of the Butter Aroma Compound Acetoin from Low-Value Dairy Side Streams Using a Natural (Nonengineered) Lactococcus lactis Dairy Isolate. *J. Agric. Food Chem.* **2020**, *68*, 5891–5899. [CrossRef] [PubMed]
53. Valero, E.; Villamiel, M.; Miralles, B.; Sanz, J.; Martínez-Castro, I. Changes in flavour and volatile components during storage of whole and skimmed UHT milk. *Food Chem.* **2001**, *72*, 51–58. [CrossRef]
54. Contarini, G.; Povolo, M.; Leardi, R.; Toppino, P.M. Influence of Heat Treatment on the Volatile Compounds of Milk. *J. Agric. Food Chem.* **1997**, *45*, 3171–3177. [CrossRef]
55. Bovolenta, S.; Romanzin, A.; Corazzin, M.; Spanghero, M.; Aprea, E.; Gasperi, F.; Piasentier, E. Volatile compounds and sensory properties of Montasio cheese made from the milk of Simmental cows grazing on alpine pastures. *J. Dairy Sci.* **2014**, *97*, 7373–7385. [CrossRef] [PubMed]
56. Clarke, H.J.; Griffin, C.; Rai, D.K.; O'Callaghan, T.F.; O'Sullivan, M.G.; Kerry, J.P.; Kilcawley, K.N. Dietary Compounds Influencing the Sensorial, Volatile and Phytochemical Properties of Bovine Milk. *Molecules* **2019**, *25*, 26. [CrossRef] [PubMed]

Publisher's Note: MDPI stays neutral with regard to jurisdictional claims in published maps and institutional affiliations.

Article

Using Cross-Cultural Consumer Liking Data to Explore Acceptability of PGI Bread—Waterford Blaa

Rachel Kelly [1,2], Tracey Hollowood [3], Anne Hasted [4], Nikos Pagidas [5], Anne Markey [1] and Amalia G. M. Scannell [1,2,6,*]

1 UCD School of Agriculture and Food Science, University College Dublin, Belfield, Dublin D04 V1W8, Ireland; rachel.kelly.3@ucdconnect.ie (R.K.); anne.markey@ucd.ie (A.M.)
2 Sensory Food Network, Dublin D15 DY05, Ireland
3 Sensory Dimensions, Unit F1-F2, Cowlairs, Nottingham, Nottinghamshire NG5 9RA, UK; tracey@hollowood.myzen.co.uk
4 QI Statistics, Penhales, Ruscombe Lane, Ruscombe, Reading RG10 9JN, UK; anne@qistatistics.co.uk
5 Kerry Europe & Russia, Millennium Park, Naas, Co., Kildare W91 W923, Ireland; nikos.pagidas@kerry.com
6 UCD Institute of Food and Health, University College Dublin, Belfield, Dublin D04 V1W8, Ireland
* Correspondence: amalia.scannell@ucd.ie

Received: 30 July 2020; Accepted: 31 August 2020; Published: 1 September 2020

Abstract: Waterford Blaa is one of only four Irish food products granted protected geographical (PGI) status by the European Commission. This study aimed to determine whether cultural background/product familiarity, gender, and/or age impacted consumer liking of three Waterford Blaa products and explored product acceptability between product-familiar and product-unfamiliar consumer cohorts in Ireland and the UK, respectively. Familiarity with Blaa impacted consumer liking, particularly with respect to characteristic flour dusting, which is a unique property of Waterford Blaa. UK consumers felt that all Blaas had too much flour. Blaa A had the heaviest amount of flouring and was the least preferred for UK consumers, who liked it significantly less than Irish consumers ($p < 0.05$). Flavour was also important for UK consumers. Blaa C delivered a stronger oven baked odour/flavour compared to Blaa A and was the most preferred by UK consumers. Irish consumer liking was more influenced by the harder texture of Blaa B, which was their least preferred product. Age and gender did not impact liking for Blaas within Irish consumers, but gender differences were observed among UK consumers, males liking the appearance significantly more than females. This is the first paper comparing Waterford Blaa liking of naïve UK consumers with Irish consumers familiar with the product.

Keywords: Waterford Blaa; cross-cultural consumer differences; consumer liking; sensory attributes; gender differences; age differences; PGI status

---

1. Introduction

In Ireland, bread and bread products markets were estimated to be worth €605 million in 2018, with 29% of Republic of Ireland consumers buying bread rolls/baps from traditional artisan bakeries. This was the second most popular type of bread purchased from these bakeries after baguettes/bagels [1]. Waterford Blaa is a white bread product produced specifically in traditional bakeries in county Waterford and a region of south Kilkenny in Ireland. Waterford Blaa obtained its Protected Geographical Indication (PGI) status in 2013 and is one of only four food products in Ireland with this status [2].

Age, gender, and cultural background have been studied in relation to consumer liking of numerous food and beverage products. Older consumers rated wines higher than middle aged and young consumers for evoking positive emotions, e.g., active, enthusiastic, and good natured [3]. Older consumers also liked jam-filled cakes more [4] and rated berries higher for usage, familiarity, and liking than younger participants [5]. The impact of gender on sensory perception has also been studied by a number of authors [6–8], and product dependent gender differences [3,5] have been associated with several food products, including spicy food [9] and wine [3,10].

Cross-cultural differences have been seen when comparing the perceptions of populations from different ethnic backgrounds. Japanese consumers were found to be more tolerant of changes in taste intensity of orange juice and grapefruit juice than Australian consumers [11], Norwegian consumers liked whole grain breads more than consumers from Estonia, Scotland, and Czech Republic [12], and more recently, the reported sensory perception of Balsamic vinegar among product–familiar consumers in Italy was different to those reported by Korean consumers unfamiliar with the product [13]. Geographically close, the populations of the UK and Ireland have been shown to be quite similar in terms of consumer behaviour. In a cross-cultural study [14], results from the Food Related Lifestyle (FRL) survey found that UK and Irish consumers had identical response styles. That said, there is anecdotal evidence of differences in food liking between the UK and Ireland, e.g., the Irish palate tends to prefer saltier potato chips (crisps) and sweeter breakfast cereal, however, published research comparing the food preferences of Irish and UK consumers is limited.

Cross-cultural differences in opinions of pre-sliced white and brown bread have previously been studied comparing Chinese-Malaysian and Australian consumers [15]. Authors found that, whilst both cultures were affected by product information, Australian consumers were more receptive to information on fibre content than the Chinese-Malaysian consumers. The impact of age on consumer perception of rye bread [16] determined that younger and older Swedish consumers differed in their liking of rye bread but held similar perceptions of its healthiness. Age (18–44 vs. 45–80 years) and gender were also found to impact the selection of descriptive attributes and perceived health effects of bread, and consumers with a lower education level struggled to identify healthy bread compared to more educated Swedish consumers [17].

Liking of food as informed by consumer perception of its sensory attributes has a considerable impact on food choice and self-reported purchase intent [18]. There are, of course, many other "non-sensory" factors that drive consumers' liking in bread, for example, information about the product, its provenance/origin [18,19], salt content [20], perceived healthfulness, e.g., gluten-free status [21], as well as the type of flour [21–23]. Bread is a staple food across the world and is generally well liked with reported average hedonic scores of between 6.4 and 8.0 (using a 9-point hedonic scale). Freshness, flavour, and texture have been shown to impact liking among bread consumers [24,25].

European food exports and imports have grown by 20% and 18%, respectively, in the five years 2012–2017 [26]. Whilst some geographically protected products have become household names on a global scale, e.g., champagne and Parma ham, most products awarded Protected Geographical Indication(PGI), and Protected Designation of Origin(PDO) status are artisanal with small-scale production, and familiarity with the product tends to have a local focus. The key challenge when exporting such food products to markets with different cultural product expectations is to understand how the sensory perception of product-unfamiliar consumers maps to that of the product-familiar local population. This information enables the producer to develop sensory based marketing strategies using appropriate sensory language. This not only manages consumer expectation but improves the likelihood of commercial success [13,27,28].

Previous research using a trained sensory panel established a Quantitative Descriptive Analysis (QDA) profile for Waterford Blaa. Results have shown that Waterford Blaa has a unique sensory profile that distinguishes it from other white bread products, particularly by its characteristic thick flour dusting and strong toasted odour [29]. Furthermore, the sensory profile of Blaa differs between the three artisan bakeries producing it; Blaa A was deemed to have a sour odour, high amount of flour dusting appearance, and an intense, floury mouthcoating texture. Blaa B's had a dense crumb structure appearance, strong oven baked flavour, and springy and firm texture, whereas Blaa C was characterised as having a strong oven baked odour and a thick exposed crumb appearance [29]. This study aims to examine whether cultural background/product familiarity, gender, and/or age impact consumers' perceived liking of Waterford Blaa products and to determine whether it is possible to identify sensory characteristics of Waterford Blaa that influence liking, comparing product-familiar and product-unfamiliar consumer cohorts in Ireland and the UK, respectively.

2. Materials and Methods

2.1. Consumer Test

Two consumer tests were run with support from Sensory Dimensions Nottingham, one consumer study in Nottingham, United Kingdom (product-unfamiliar cohort) and the other consumer study in Waterford, Ireland (product-familiar cohort). All participants were recruited and screened using the following criteria: (a) over 18 years of age, (b) in good health, (c) no food related allergies, (d) UK only: consumption of white bread rolls at least once a week, (e) Ireland only: consumption of Blaa at least once a week), (f) signed written consent to take part in this study.

In the UK trial, n = 115 UK bread consumers (57 males; 68 females) were recruited from Sensory Dimensions volunteer database, while in the Irish study, n = 122 participants (54 males; 68 females) were recruited at The Waterford Food Festival in County Waterford, Ireland. Both studies were completed within one day, however, the Irish and the UK trials were undertaken on separate occasions 12 months apart. In each study, consumers participated in a 30 min tasting session in which they were asked to rate the overall liking of the products as well as their liking for each of the key sensory modalities (appearance, odour, flavour, texture). These questions were asked using a 9-point hedonic category scale that ranged from 1 (Dislike Extremely) to 9 (Like Extremely). In addition, consumers were asked to assess nine sensory attributes characteristic of Waterford Blaa: darkness of the crust; amount of flour dusting; sour and oven baked odour; salty, sour, and oven-baked flavours; firmness; and degree of mouthcoating. These were selected from descriptive sensory profiles of Blaa generated by a trained sensory panel during a previous research project [29]. Consumers used a 5-point Just About Right (JAR) category scale in which 1 represented "too little", 3 represented "just about right", and 5 represented "too much" for each of the selected characteristics. All subjects gave their informed consent for inclusion before they participated in the study. The study was conducted in accordance with the Declaration of Helsinki and GDPR regulations.

Because no novel ingredients were tested, specific ethical approval was not required. However, this study followed marketing and consumer research protocols, ISO 11136: 2017 Sensory analysis—Methodology—General guidance for conducting hedonic tests with consumers in a controlled area and IFST Professional Sensory Science Group guidelines.

2.2. Sample Preparation and Experimental Design

Waterford Blaa was produced by each of the three authorised artisanal producers using the ten stages outlined in Table 1. Blaa samples were produced and shipped by courier to arrive at the testing venue by the following morning for the consumer test. Each participant assessed Waterford Blaa products producers according to a balanced randomised experimental design. Each Waterford Blaa was presented sequentially to the consumers as a whole bread roll on a white paper plate with a plastic knife. For each question, participants were provided with instructions for assessment including the need to look at, smell, and bite into the Waterford Blaa for assessment of appearance, odour, flavour, and texture. Still water was used as a palate cleanser between samples, and participants were given a break of at least 5 min between samples to avoid sensory fatigue.

Table 1. Stages of Waterford Blaa production.

Stage	Production Stage	Comment
1	Mixing	All ingredients mixed and divided into large pieces of dough
2	Resting	Dough is rested for 10–20 min
3	Pinning	Gives the dough characteristic round shape before baking
4	Sub-Dividing and resting (flour addition)	Dough subdivided into smaller pieces and rested and flour added
5	Flattening and rolling (Flour addition)	Dough flattened and placed on trays, rolled into prover for 1 h, more flour added and visual assessment
6	Proving (flour addition)	More flour added in one or two hand movements and proved
7	Baking	Dough baked in oven at 220–230 °C for 25–30 min
8	Assessment	Baker assesses dough before and after baking
9	Cooling	Blaa cooled for 1 h in ambient temperature
10	Packing	Blaa packaged into boxes or plastic wrapping

2.3. Data Analysis

Hedonic and JAR data from UK and Irish cohorts were analysed separately using XLSTAT Sensory (version 2018.5).

Hedonic data for overall liking and liking for each modality were analysed using a two factor ANOVA model with Tukey's multiple comparison test comparing the three Blaa products to determine if consumers differed significantly in their liking. Combined hedonic data from UK and Irish consumers were further analysed for each Blaa separately using a three factor ANOVA model to determine if age, gender, or culture impacted liking. For this analysis, data were split into three age categories: 18–34, 35–54, and >55 years old.

JAR data were collapsed into three categories (too much, too little, and JAR), and the frequency of response (%) was calculated in each category. Penalty analysis was applied to UK and Irish data to determine the relative impact of specific product characteristics on overall liking.

3. Results and Discussion

Bread in its many forms is a world-wide staple. Typical overall liking scores for bread (generated using a 9-point hedonic scale) have been reported as 6.02 and 6.28 [15,30]. This study measured overall consumer liking for each modality as well as judgements about key sensory attributes to determine whether diversity exists between the perceptions of Irish consumers who are familiar with Waterford Blaa and UK consumer cohorts who are not.

3.1. The Impact of Cultural Differences on Liking for Waterford Blaa

UK and Irish consumers differed in how much they liked the three Blaa products (Table 2). UK consumers liked Blaa C and B the most and Blaa A the least, whereas Irish consumers liked Blaa C and A the most and Blaa B the least. Liking for appearance, odour, texture, and flavour of the three Blaas followed the same trend as their results for overall liking.

Table 2. Mean consumer liking of three Waterford Blaas for UK and Irish consumers. Data analysed using ANOVA with Tukey's post hoc test.

	Overall Liking	Overall Liking		Appearance Liking		Odour Liking		Flavour Liking		Texture Liking	
BLAA	UK+IRISH	UK	IRISH	UK	IRISH	UK	IRISH	UK	IRISH	UK	IRISH
A	5.8 ^{b}	5.2 $^{b;B}$	6.4 $^{ab:A}$	4.6 $^{b;B}$	6.5 $^{ab;A}$	6.0 $^{b;B}$	6.8 $^{a;A}$	5.2 $^{b;B}$	6.2 $^{ab;A}$	5.1 $^{b;B}$	6.1 $^{a;A}$
B	6.1 ^{b}	6.3 ^{a}	5.8 ^{b}	6.3 ^{a}	6.0 ^{b}	6.3 ^{ab}	6.3 ^{b}	6.2 ^{a}	5.7 ^{b}	6.4 $^{a;A}$	5.2 $^{b;B}$
C	6.6 ^{a}	6.6 ^{a}	6.6 ^{a}	6.7 ^{a}	6.7 ^{a}	6.7 ^{a}	6.9 ^{a}	6.7 ^{a}	6.7 ^{a}	6.7 ^{a}	6.4 ^{a}

Lowercase letters indicate significant differences from the Tukey's post hoc between samples for each modality ($p < 0.05$). Uppercase letters indicate the significant differences between the cohorts for each Blaa within the one modality.

UK consumers liked the appearance, the flavour, and the texture of Blaa C and B significantly more than Blaa A ($p < 0.05$) and the odour of Blaa C significantly more than Blaa A ($p < 0.05$); there were no significant differences in liking between Blaa C and Blaa B for any modality ($p > 0.05$). The mean scores for overall liking and each modality were very consistent for Blaa C (range: 6.6–6.7) and Blaa B (range: 6.2–6.4) and, therefore, it was not possible to determine the relative impact of each modality on overall liking for these two products. By contrast, UK consumers found appearance, flavour, and texture of Blaa A much less acceptable, particularly the appearance (mean score = 4.6), although they liked the odour (mean score = 6.0). It is probable that the sensory qualities affecting appearance and, to a lesser extent, flavour and texture, negatively impacted the result for overall liking.

Appearance attributes have been shown to be key in influencing liking and the decision to eat, as they shape a consumer's first impression of a food product. For example, UK consumers liked the appearance of white bread significantly more ($p < 0.0001$) than red beetroot bread [19]. Norwegian consumers considered darker bread rolls to have more rye and therefore to be healthier than white bread, whilst Scottish, Estonian, and Czech consumers preferred the whitest bread [12]. In Ireland, a recent study showed 57% of the participants preferred white bread with males consuming more white bread than females [31]. In this study, Irish consumers liked the odour and the texture of Blaa C and A significantly more than Blaa B ($p < 0.05$) and the appearance and the flavour of Blaa C significantly more than Blaa B ($p < 0.05$); there was no significant difference in liking between Blaa C and Blaa A for any modality ($p > 0.05$). The mean scores for overall liking and each modality were >6 for Blaa C and Blaa A and, as with the UK study, it was not possible to determine the relative impact of each modality on overall liking. Irish consumers found appearance, flavour, and particularly texture (mean score = 5.2) of Blaa B less acceptable, and it is likely that these modalities negatively impacted overall liking.

Whilst mean scores for the most liked Blaa C were similar for UK and Irish consumers, mean scores for the least liked Blaa A were markedly lower in the UK and below the 6.02 liking score previously reported [30]. Further analysis comparing data from both cohorts confirmed that Irish consumers liked Blaa A significantly more than UK consumers ($p < 0.05$). In this study, samples were presented blind with no additional information and, for UK consumers naïve to the product, a standard white floury bap would have been their most similar product experience. Irish consumers in Waterford were selected on the basis of having previously eaten Blaa. This meant that the Irish cohort was more familiar with the product and thus their expectations could be expected to follow the unique qualities of the Blaa product traditionally purchased within their family. It is likely, therefore, based on their prior knowledge and expectation, that UK and Irish consumers differed in their tolerance of and response to the unique qualities of Blaa A. This observation is consistent with other cross-cultural studies that found a positive correlation between product familiarity and product liking in Asian and Western panels—the rationale being that consumers who already knew a product reported feelings of safeness while, conversely, consumers not familiar with a product did not have this security, allowing the potential to give enhanced negative judgements when experiencing it for the first time [32].

Products displaying EU recognised PGI/PDO status enable the consumer to feel the product is safe and of superior quality. Awareness of PGI or PDO status among consumers has been linked to increased positive consumer association with the product and a willingness to pay more for products they perceive to be better and of higher quality [27,33,34]. Interestingly, results of the combined data follow the same trend as the UK results and do not mirror the preferences from the Irish cohort (Table 2). Whilst this is not surprising, as acceptability varied more among UK consumers, it highlights the importance of understanding consumer behaviour in each location rather than assuming that one location, or average results, can be used to predict liking in other populations. This is particularly important when marketing unique products from other countries/regions. For example, the unusual sensory qualities of Blaa A might have been more acceptable to UK consumers had they expected a difference and understood the cultural significance of those differences as opposed to treating the Waterford Blaa as just another bread roll.

Comparison of UK and Irish results for each Blaa provided further evidence of cultural differences. Irish consumers liked Blaa A ($p < 0.0001$) significantly more than UK consumers, whereas UK consumers liked Blaa B ($p < 0.05$) significantly more than Irish consumers. Not only was Blaa C the most preferred by both consumer groups, there were no significant differences in liking between UK and Irish consumers ($p > 0.05$).

Whilst Blaa has PGI status and is manufactured to traditional recipes, the three Blaas used in this study had differing sensory profiles, in particular with regard to their appearance (Figure 1). Each Blaa's unique sensory qualities were responsible for the significant differences in liking within each cohort and between UK and Irish consumers. Blaa A was the most polarising; whilst familiar to Irish consumers, its sensory characteristics were furthest away from those of a standard "floury bap" expected by UK consumers.

Figure 1. Appearance of the three Waterford Blaa products used in this study.

It is well established that perceived quality is not the same as expected quality [35]. Branding has a significant impact on consumer hedonic scores [36], and it has been shown that altering packaging information and other external cues can affect sensory perception [37] and influence consumer decisions [38]. Therefore, by positively marketing these differences to the UK consumer as the products' unique selling point prior to launching such a product into the UK market, one could moderate the consumer expectation and consequently improve the sentiment towards the appearance of Blaas A and B.

3.2. The Impact of Age on Liking for Waterford Blaa

The population of developed countries is aging, and research into the differences in sensory perception between young and older consumers have found there to be considerable differences in rating of textural properties of food between the elderly and the young in a number of food products [39,40]. Factors impacting consumer liking have been shown to differ depending on age, younger consumers tending to value hedonics and price point when buying, whereas older consumers prioritise health benefits of a product [16,17,22].

In the context of Waterford Blaa, age had no impact on consumer liking within either Irish or UK consumers; there was no significant difference in overall liking nor liking for each modality between the three age categories ($p > 0.05$). There was also no significant interaction between age and Blaa for either the UK or Irish consumers.

3.3. The Impact of Gender on Liking for Waterford Blaa

Gender-stereotyped food associations exist in every culture [41], for example, in modern western culture, meat is often associated with men while salad is considered a more female associated dish [42]. A gender effect has been previously reported in consumer studies using orange juices and jam filled cakes [7], where males rated these products higher in likability than females.

In this study, the gender of Irish consumers did not impact liking for Waterford Blaa; there was no significant difference between males and females for overall liking nor liking for any modality ($p > 0.05$). For UK consumers, females liked the appearance of Blaa significantly less than males ($p < 0.05$). It is possible that UK females paid more attention to the appearance of the Blaas, or were more negatively impacted by its qualities, than UK males. This could be associated with females being more health conscious than men, as gluten products such as bread have been considered "not as healthy" as other carbohydrate sources [43].

3.4. Impact of Cultural Differences on Perception of Flour Associated Attributes

Despite all three Blaas originating from Waterford and having PGI status, each Blaa had distinct qualities relating back to long-standing family recipes, particularly with respect to levels of flour coating. In this study, for example, anecdotal evidence suggested that Irish consumers easily identified the distinctive flour dusting on the top of Blaa A, the darker in colour of Blaa B, and the uniform shape of Blaa C (Figure 2).

Figure 2. Deviation from ideal/just about right (JAR) for Blaa A from Irish and UK consumers. Data expressed as percent of responses from the collapsed JAR scale in which: Too little (blue) = rating of 1+2, Too much (red) = rating of 4+5, and Just right (green) = rating of 3.

Diagnostic assessment of the products using JAR scales highlighted differences in consumer judgements relating to the sensory attributes that characterise Blaa. Differences were not only evident when comparing between the products, they were also dependent on the geographic location of the consumer groups. Blaa A was considered to be the furthest from "just about right" for both UK and Irish consumers (Figure 2). Both cohorts judged the product to have too much flour dusting and therefore to be too pale; to have too little oven baked odour, salt, flavour, and oven baked flavour; to be too soft; and to have too much floury mouth-coating. Excessive flour dusting and mouth-coating as well as being too pale were the attributes that had the most impact with >50% of Irish and >70% of UK consumers expressing this view. It is possible that the excessive flour coating also reduced the intensity of salt and baked flavours. The judgements were more pronounced in UK consumers for whom a smaller proportion felt it was "just about right" for each sensory characteristic compared to Irish consumers.

Blaa B yielded different opinions from UK and Irish consumers (Figure 3). Judgements regarding flour dusting and mouthcoating were polarising; 47% of UK consumers felt there was too much, whereas 38% of Irish consumers felt there was too little. Irish consumers (42%) considered Blaa B to be too firm, whilst 37% of UK consumers felt that it was too dark. Results were similar for odour and flavour attributes.

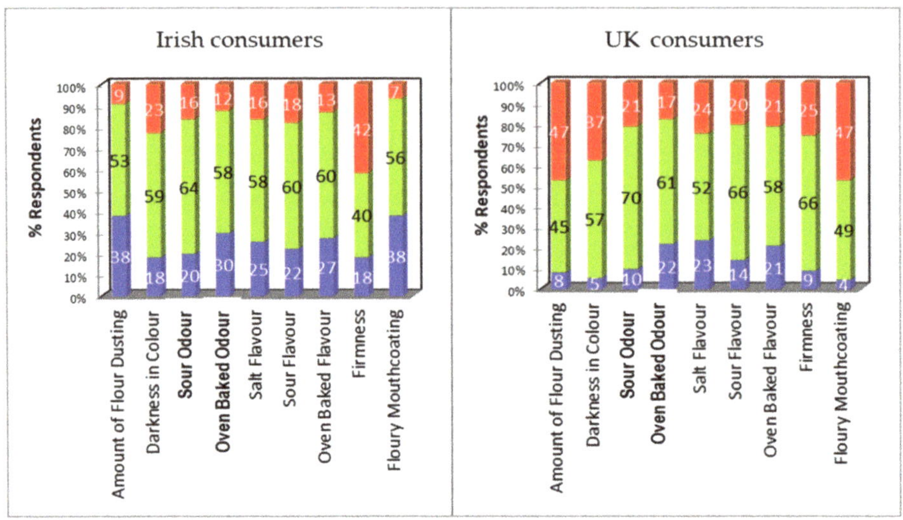

Figure 3. Deviation from ideal/just about right (JAR) for Blaa B from Irish and UK consumers. Data expressed as percent of responses from the collapsed JAR scale in which: Too little (blue) = rating of 1+2, Too much (red) = rating of 4+5, and Just right (green) = rating of 3.

Comparing the three products, Blaa C had the highest proportion of UK and Irish consumers who felt that it was "just about right" with regard to all odour and flavour attributes, darkness of colour, and firmness (Figure 4). Once again, consumers were less satisfied with the amount of flour dusting and mouthcoating and, consistent with Blaa B, their opinion was polarising; ~62% of UK consumers felt there was too much, whereas ~34% of Irish consumers felt there was too little.

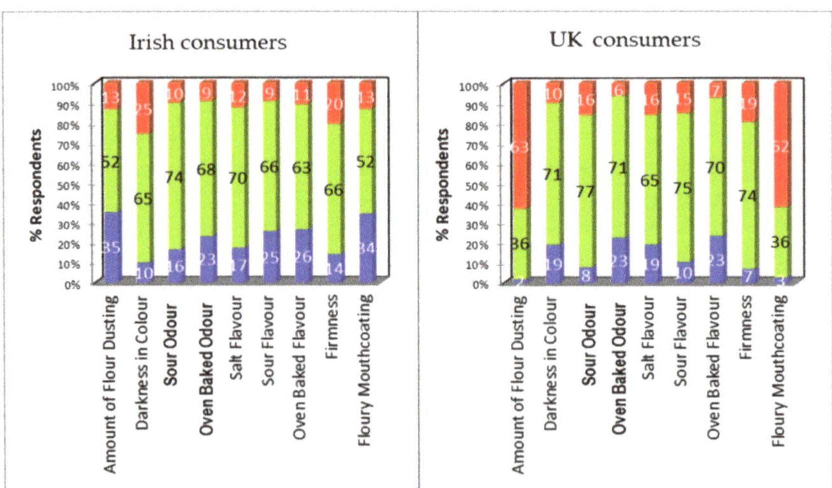

Figure 4. Deviation from ideal/just about right (JAR) for Blaa C from Irish (**left**) and UK (**right**) consumers. Data expressed as percent of responses from the collapsed JAR scale in which: Too little (blue) = rating of 1+2, Too much (red) = rating of 4+5, and Just right (green) = rating of 3.

Penalty analysis applied to JAR data identifies the relative impact of each attribute judgement on the degree of overall liking (Table 3). Weighted penalties >0.6 are indicative of attributes that drive disliking; they result from a high proportion of consumers sharing a judgement about a product, a large drop in overall liking because of that judgement (compared to those who felt the product was "just about right"), or a combination of both.

Table 3. Penalty analysis results for all attributes with a weighted penalty > 0.6 *.

Product	Location	Variable	JAR Group Mean Hedonic Score	Judgement	%	Mean Drop in Overall Liking	Weighted Penalty
Blaa A	Irish	Flour Dusting	7.3	Too much	58.2	1.7	1.0
		Darkness in Colour	7.7	Too little	52.5	2.5	1.3
		Oven Baked Odour	7.2	Too little	41.0	2.1	0.9
		Oven Baked Flavour	7.1	Too little	40.2	1.8	0.7
		Floury Mouthcoating	7.4	Too much	54.1	1.8	1.0
	UK	Flour Dusting	5.9	Too much	70.4	1.1	0.8
		Darkness in Colour	6.7	Too little	75.7	1.9	1.5
		Oven Baked Odour	5.8	Too little	49.6	1.3	0.6
		Oven Baked Flavour	6.2	Too little	63.5	1.5	1.0
		Firmness	6.1	Too little	40.0	2.0	0.8
		Floury Mouthcoating	6.4	Too much	69.6	1.7	1.1
Blaa B	Irish	Firmness	6.6	Too much	41.8	1.5	0.6
	UK	Darkness in Colour	7.1	Too much	37.4	1.9	0.7
Blaa C	UK	Flour Dusting	7.4	Too much	62.6	1.2	0.8
		Floury Mouthcoating	7.4	Too much	61.7	1.3	0.8

* JAR mean score = mean overall liking for those who rated "just about right". Judgement = Too much or Too little from the collapsed scale. % = proportion of consumers expressing that judgement. Mean drop in overall liking = difference in overall liking between consumers with that judgement and those who rated "just about right". Weighted penalty = % x Mean drop.

Based on weighted penalties, judgements related to flour dusting and mouthcoating, pale colour, and oven baked odour and flavour impacted the overall liking score for Blaa A in both UK and Irish consumers. Despite this, UK consumers liked Blaa A significantly ($p < 0.0001$) less than Irish consumers; this was due to two factors a) the mean liking score from Irish consumers who rated Blaa A "just about right" was significantly higher than the equivalent score from UK consumers and b) more Irish consumers judged Blaa A to be "just about right" than UK consumers.

For UK consumers, it is probable that the significantly lower scores for appearance liking (excessive flour dusting and pale colour), flavour liking (weak baked flour flavour), and texture liking (soft texture and floury mouthcoating) were impacted by these attribute judgements.

Whilst judgements regarding flour dusting and mouthcoating in Blaa B had been contrary between Irish and UK consumers, neither view translated into a notable penalty on overall liking. Irish consumers (>40%) considered Blaa B to be too hard, which resulted in its texture being liked significantly ($p < 0.05$) less than Blaa A and Blaa C and being their least liked product overall. UK consumers judged Blaa B to be too dark; however, despite a weighted penalty of 0.7, there was no obvious impact on the mean score for overall liking or appearance, mainly due to the high mean score from the 57% of consumers who judged it to be "just about right".

Blaa C was the most liked product for both Irish and UK consumers and had the highest proportion of consumers who felt that it was "just about right" with regard to all attributes except flour dusting and mouthcoating. There were no notable penalties from Irish consumers. Whilst 60% of UK consumers felt there was too much flour dusting and mouthcoating and did show a reasonable drop in overall liking, this was compensated by a particularly high overall liking mean score from those who judged it to be "just about right".

A high proportion of UK consumers commented that all Blaas had too much flour dusting and mouth-coating, and it seemed the greater the flour dusting was, the more UK consumers disliked the product. The flour dusting and the characteristic floury mouthcoating are characteristic features of the product familiar to Irish Blaa consumers but a new experience for the UK consumers. The Irish consumers were all regular Waterford Blaa consumers who ate Blaa at least once a week, thus when asked at the end of the session whether Blaa A had too much flour -frequently occurring responses included "*Blaa is meant to be floury*" and "*Sure, you can knock off the excess flour on the plate if you don't like it*".

When considering export of local products into international markets heretofore unfamiliar with it, it is critical to understand consumer preferences. This can be used to obtain a direction for improving product performance [44] or to determine an appropriate marketing campaign to manage the market's product expectation. This study achieved this by gathering information about consumers perception of the sensory characteristics of Waterford Blaa concurrently with overall liking scores [45] followed by analysing JAR and hedonic response data using penalty analysis [46]. JAR scales involve a direct comparison of the products with their conceptual "ideal"; by rating the perceived deviation from their ideal, they can deliver useful insights into product characterisation and market appeal [47]. However, this does presuppose that an ideal product exists and that the consumer knows what it is and can express what about the product is ideal/just about right. PGI/PDO producers entering a new market with unfamiliar consumers will gain useful insights into how their product compares with or differs from perceived similar products in that market, as the consumer ideal will perforce come from their eating experience.

4. Conclusions

This study aimed to investigate consumer acceptability for Waterford Blaa and determine if culture, age, and gender impacted liking among UK and Irish consumers. Familiarity with Blaa did impact consumer liking; this was particularly evident from the consumers' responses to the heavy flour dusting, which is a unique property of Waterford Blaa. UK consumers felt that all Blaas had too much flour on top and too much of a floury mouthcoating. Consequently, Blaa A with the heaviest amount of flouring was the least preferred for UK consumers who liked it significantly less than Irish consumers ($p < 0.05$). Results for Blaa A and C suggest that flavour was also a key driver for UK consumers. Whilst Blaa C also had a heavily floured top, it delivered a stronger oven baked odour/flavour compared to Blaa A, which was enough to compensate for the flour coating and resulted in Blaa C being the most preferred by UK consumers. By contrast, whilst Irish consumers were aware of the same qualities in Blaa A and C, it did not negatively impact their results for overall liking or liking for any individual modality. Irish consumers were more impacted by the harder texture of Blaa B, which was their least preferred product. Age and gender did not impact liking for Blaas within Irish consumers, whereas for UK consumers, males liked the appearance significantly more than females.

Understanding how new markets might react to unfamiliar products and identifying the barriers to acceptability is important for the artisan bakers who may wish to expand into different markets. To this end, consumer education in a naïve market to raise consumer awareness and manage expectations is important, particularly for the more unusual product qualities, e.g., in the case of Waterford Blaa, the flour related attributes such as amount of flour dusting (appearance) and floury mouthcoating (texture). Results from this study offer useful feedback to the artisan bakers who could also take a second approach and slightly modify the production to match the consumer expectation. In the case of Blaa A, for example, a little less flour dusting and a slightly longer bake time could provide quick fixes to the pale "colour" and the too little "baking odour" and "baking flavour" attributes. Using sensory profiling, this consumer study underscores the importance of the traditional Research and Development guide to success, i.e., know your product; know your customer.

Author Contributions: Conceptualization, A.G.M.S., T.H., N.P. and R.K.; Methodology, A.G.M.S., T.H., N.P. and R.K.; software, A.H. and N.P.; Consumer testing: R.K., T.H. and A.G.M.S.; Data curation: R.K. and A.G.M.S.; Data analysis, R.K., T.H., A.M., A.G.M.S., and A.H.; Writing—original draft preparation, R.K.; Writing & Visualization —review and editing, R.K., A.M., T.H., N.P. and A.G.M.S. Supervision, Project administration, Funding acquisition: A.G.M.S. All authors have read and agreed to the published version of the manuscript.

Funding: This research was funded by grant aid funding under the Food Institutional Research Measure (FIRM), which is administered by the Irish Department of Agriculture, Food and the Marine, (DAFM), grant number 13 SN 401.

Acknowledgments: The authors would like to thank the Waterford Blaa Bakers Association for their support and for providing samples for the consumer testing. We would also like to sincerely thank the management and staff at Sensory Dimension, Nottingham, United Kingdom for their collaboration on this project, in particular with the additional manpower to run the consumer tests.

Conflicts of Interest: The authors declare no conflict of interest.

References

1. Mintel Bread—Ireland—April 2018. Executive Summary. Available online: http://academic.mintel.com/display/891259/?highlight#hit1 (accessed on 24 April 2018).
2. Department of Agriculture, Food and the Marine. Protected Geographical Food Names. Available online: https://www.agriculture.gov.ie/gi/pdopgitsg-protectedfoodnames/products/ (accessed on 26 April 2018).
3. Mora, M.; Urdaneta, E.; Chaya, C. Emotional response to wine: Sensory properties, age and gender as drivers of consumers' preferences. *Food Qual. Prefer.* **2018**, *66*, 19–28. [CrossRef]
4. Michon, C.; O'Sullivan, M.G.; Sheehan, E.; Delahunty, C.; Kerry, J. Study on the influence of age, gender and familiarity with the product on the acceptance of vegetable soups. *Food Qual. Prefer.* **2010**, *21*, 478–488. [CrossRef]
5. Laaksonen, O.A.; Knaapila, A.; Niva, T.; Deegan, K.C.; Sandell, M.A. Sensory properties and consumer characteristics contributing to liking of berries. *Food Qual. Prefer.* **2016**, *53*, 117–126. [CrossRef]
6. Doty, R.L.; Cameron, E.L. Sex differences and reproductive hormone influences on human odor perception. *Physiol. Behav.* **2009**, *97*, 213–228. [CrossRef] [PubMed]
7. Michon, C.; O'Sullivan, M.G.; Sheehan, E.; Delahunty, C.; Kerry, J. Investigation of the influence of age, gender and consumption habits on the liking of jam-filled cakes. *Food Qual. Prefer.* **2010**, *21*, 553–561. [CrossRef]
8. Katou, Y.; Mori, T.; Ikawa, Y. Effect of age and gender on attitudes towards sweet foods among Japanese. *Food Qual. Prefer.* **2005**, *16*, 171–179. [CrossRef]
9. Byrnes, N.K.; Hayes, J. Gender differences in the influence of personality traits on spicy food liking and intake. *Food Qual. Prefer.* **2015**, *42*, 12–19. [CrossRef]
10. Parra, D.R.; Galmarini, M.; Chirife, J.; Zamora, M.C. Influence of information, gender and emotional status for detecting small differences in the acceptance of a new healthy beverage. *Food Res. Int.* **2015**, *76*, 269–276. [CrossRef]
11. Prescott, J.; Bell, G.; Gillmore, R.; Yoshida, M.; O'Sullivan, M.; Korac, S.; Allen, S.; Yamazaki, K. Cross-cultural comparisons of Japanese and Australian responses to manipulations of sourness, saltiness and bitterness in foods. *Food Qual. Prefer.* **1998**, *9*, 53–66. [CrossRef]
12. Rødbotten, M.; Tomic, O.; Holtekjølen, A.; Grini, I.; Lea, P.; Granli, B.; Grimsby, S.; Sahlstrøm, S. Barley bread with normal and low content of salt; sensory profile and consumer preference in five European countries. *J. Cereal Sci.* **2015**, *64*, 176–182. [CrossRef]
13. Torri, L.; Jeon, S.-Y.; Piochi, M.; Morini, G.; Kim, K.-O. Consumer perception of balsamic vinegar: A cross-cultural study between Korea and Italy. *Food Res. Int.* **2017**, *91*, 148–160. [CrossRef] [PubMed]
14. O'Sullivan, C.A.; Scholderer, J.; Cowan, C. Measurement equivalence of the food related lifestyle instrument (FRL) in Ireland and Great Britain. *Food Qual. Prefer.* **2005**, *16*, 1–12. [CrossRef]

15. Mialon, V.; Clark, M.; Leppard, P.; Cox, D. The effect of dietary fibre information on consumer responses to breads and "English" muffins: A cross-cultural study. *Food Qual. Prefer.* **2002**, *13*, 1–12. [CrossRef]
16. Sandvik, P.; Nydahl, M.; Marklinder, I.; Næs, T.; Kihlberg, I. Different liking but similar healthiness perceptions of rye bread among younger and older consumers in Sweden. *Food Qual. Prefer.* **2017**, *61*, 26–37. [CrossRef]
17. Sandvik, P.; Nydahl, M.; Kihlberg, I.; Marklinder, I. Consumers' health-related perceptions of bread—Implications for labeling and health communication. *Appetite* **2018**, *121*, 285–293. [CrossRef]
18. Kihlberg, I.; Johansson, L.; Langsrud, Ø.; Risvik, E. Effects of information on liking of bread. *Food Qual. Prefer.* **2005**, *16*, 25–35. [CrossRef]
19. Hobbs, D.; Ashouri, A.; George, T.; Lovegrove, J.; Methven, L. The consumer acceptance of novel vegetable-enriched bread products as a potential vehicle to increase vegetable consumption. *Food Res. Int.* **2014**, *58*, 15–22. [CrossRef]
20. Antúnez, L.; Giménez, A.; Ares, G. A consumer-based approach to salt reduction: Case study with bread. *Food Res. Int.* **2016**, *90*, 66–72. [CrossRef]
21. Morais, E.; Cruz, A.G.; Faria, J.; Bolini, H.M.A. Prebiotic gluten-free bread: Sensory profiling and drivers of liking. *LWT* **2014**, *55*, 248–254. [CrossRef]
22. Pohjanheimo, T.; Paasovaara, R.; Luomala, H.; Sandell, M.A. Food choice motives and bread liking of consumers embracing hedonistic and traditional values. *Appetite* **2010**, *54*, 170–180. [CrossRef]
23. Turkut, G.M.; Cakmak, H.; Kumcuoglu, S.; Tavman, S. Effect of quinoa flour on gluten-free bread batter rheology and bread quality. *J. Cereal Sci.* **2016**, *69*, 174–181. [CrossRef]
24. Heenan, S.P.; Dufour, J.P.; Hamid, N.; Harvey, W.; Delahunty, C.M. The sensory quality of fresh bread: Descriptive attributes and consumer perceptions. *Food Res. Int.* **2008**, *41*, 989–997. [CrossRef]
25. Dewettinck, K.; Van Bockstaele, F.; Kuhne, B.; Van De Walle, D.; Courtens, T.; Gellynck, X. Nutritional value of bread: Influence of processing, food interaction and consumer perception. *J. Cereal Sci.* **2008**, *48*, 243–257. [CrossRef]
26. European Commission (Eurostat). EU Trade in Food. 2017. Available online: http://ec.europa.eu/eurostat/web/products-eurostat-news/-/EDN-20171016-1?inheritRedirect=true (accessed on 25 April 2018).
27. Resano, H.; Sanjuán, A.I.; Cilla, I.; Roncalés, P.; Albisu, L. Sensory attributes that drive consumer acceptability of dry-cured ham and convergence with trained sensory data. *Meat Sci.* **2010**, *84*, 344–351. [CrossRef] [PubMed]
28. Vu, T.M.H.; Tu, V.P.; Duerrschmid, K. Gazing behavior reactions of Vietnamese and Austrian consumers to Austrian wafers and their relations to wanting, expected and tasted liking. *Food Res. Int.* **2018**, *107*, 639–648. [CrossRef]
29. Kelly, R.; Hollowood, T.; Scannell, A.G.M. Sensory characterisation of an Irish PGI bread: Waterford Blaa. *Eur. Food Res. Technol.* **2019**, *245*, 1307–1319. [CrossRef]
30. Jimenez-Maroto, L.A.; Sato, T.; Rankin, S.A. Saltiness potentiation in white bread by substituting sodium chloride with a fermented soy ingredient. *J. Cereal Sci.* **2013**, *58*, 313–317. [CrossRef]
31. Irish Universities Nutrition Alliance (IUNA). Report on the Pattern of White and Wholemeal Bread Consumption in Irish Adults and Pre-School Children. 2016. Available online: https://www.fooddrinkireland.ie/Sectors/FDI/FDI.nsf/6cbe469001b758968025778b003b324e/04578a16193e56d9802580f9005d1fe0/$FILE/IUNA%20%20-%20IBBA%20Report%20Oct%202016%20Final.pdf (accessed on 19 January 2020).
32. Torrico, D.D.; Fuentes, S.; Viejo, C.G.; Ashman, H.; Dunshea, F.R. Cross-cultural effects of food product familiarity on sensory acceptability and non-invasive physiological responses of consumers. *Food Res. Int.* **2019**, *115*, 439–450. [CrossRef] [PubMed]
33. Fotopoulos, C.; Krystallis, A.; Anastasios, P. Portrait value questionnaire's (PVQ) usefulness in explaining quality food-related consumer behavior. *Br. Food J.* **2011**, *113*, 248–279. [CrossRef]
34. Grunert, K.G.; Aachmann, K. Consumer reactions to the use of EU quality labels on food products: A review of the literature. *Food Control* **2016**, *59*, 178–187. [CrossRef]
35. Deliza, R.; MacFie, H. The generation of sensory expectation by external cues and its effect on sensory perception and hedonic ratings: A review. *J. Sens. Stud.* **1996**, *11*, 103–128. [CrossRef]
36. Morris, C.; Beresford, P.; Hirst, C. Impact of food retailer branding on expectation generation and liking. *J. Sens. Stud.* **2018**, *33*, e12322. [CrossRef]
37. Shankar, M.U.; Levitan, C.; Prescott, J.; Spence, C. The Influence of Color and Label Information on Flavor Perception. *Chemosens. Percept.* **2009**, *2*, 53–58. [CrossRef]

38. Lopes, M.M.T.; Rodrigues, M.D.C.P.; De Araújo, A.M.S. Influence of Expectation Measure on the Sensory Acceptance of Petit Suisse Product. *J. Food Sci.* **2018**, *83*, 798–803. [CrossRef]
39. Kremer, S.; Mojet, J.; Kroeze, J.H.A. Perception of texture and flavour in soups by elderly and young subjects. *J. Texture Stud.* **2005**, *36*, 255–272. [CrossRef]
40. Kremer, S.; Mojet, J.; Kroeze, J.H. Differences in perception of sweet and savoury waffles between elderly and young subjects. *Food Qual. Prefer.* **2007**, *18*, 106–116. [CrossRef]
41. Counihan, C.M.; Kaplan, S.L. *Food and Gender. Identity and Power*; Routledge: London, UK, 2004; pp. 1–4.
42. Cavazza, N.; Guidetti, M.; Butera, F. Ingredients of gender-based stereotypes about food. Indirect influence of food type, portion size and presentation on gendered intentions to eat. *Appetite* **2015**, *91*, 266–272. [CrossRef]
43. Gellynck, X.; Kühne, B.; Van Bockstaele, F.; Van De Walle, D.; Dewettinck, K. Consumer perception of bread quality. *Appetite* **2009**, *53*, 16–23. [CrossRef]
44. Van Kleef, E.; Van Trijp, H.C.M.; Luning, P. Internal versus external preference analysis: An exploratory study on end-user evaluation. *Food Qual. Prefer.* **2006**, *17*, 387–399. [CrossRef]
45. Lawless, H.T.; Heymann, H. *Sensory Evaluation of Food. Principles and Practices*, 2nd ed.; Springer: New York, NY, USA, 2010.
46. Lesniauskas, R.O.; Carr, B.T. Workshop summary: Data analysis: Getting the most out of just-about-right data. *Food Quality Prefer.* **2004**, *15*, 891–899.
47. Worch, T.; Punter, P. Ideal Profiling. In *Novel Techniques in Sensory Characterization and Consumer Profiling*; Varela, P., Ares, G., Eds.; CRC Press: Boca Raton, FL, USA, 2014; pp. 85–136.

Article

Sensory Characteristics Contributing to Pleasantness of Oat Product Concepts by Finnish and Chinese Consumers

Oskar Laaksonen [1], Xueying Ma [1], Eerika Pasanen [1], Peng Zhou [2], Baoru Yang [1] and Kaisa M. Linderborg [1,*]

[1] Food Chemistry and Food Development, University of Turku, FI-20014 Turku, Finland; oskar.laaksonen@utu.fi (O.L.); xueyma@utu.fi (X.M.); eerika.pasanen@utu.fi (E.P.); baoru.yang@utu.fi (B.Y.)

[2] School of Food Science and Technology, Jiangnan University, Wuxi 214122, China; zhoupeng@jiangnan.edu.cn

* Correspondence: kaisa.linderborg@utu.fi; Tel.: +358-50-439-5535

Received: 14 August 2020; Accepted: 2 September 2020; Published: 4 September 2020

Abstract: Oats are increasingly popular among consumers and the food industry. While data exist on sensory characteristics of oats as such, previous studies focusing on the pleasantness of oats, and especially investigations of a wide range of oat products by European and Asian consumers, are scarce. An online questionnaire was organized in Finland (n = 381; 83.7% Finnish) focusing on the liking and familiarity of oat products, followed by sensory tests in Finland (n = 65 and n = 73) and China (n = 103) using the Check-All-That-Apply method and hedonic ratings. A questionnaire revealed that the Finnish consumers rated the pleasantness and familiarity of several oat product categories, such as breads and porridges, higher compared to participants of other ethnicities. Sensory tests showed both similarities, e.g., porridges were described as "natural", "healthy" and "oat-like", and differences between countries, e.g., sweet biscuits, were described as "crispy" and "hard" by Finnish consumers and "strange" and "musty" by Chinese consumers. Sweet products were unanimously preferred. The ethnicity had an important role affecting the rating of pleasantness and familiarity of oat product categories, whereas food neophobia and health interest status also had an influence. The proved healthiness of oats was a crucial factor affecting the choices of consumers and their acceptance in both countries.

Keywords: oat products; consumers; liking; Check-All-That-Apply; cross-cultural; China; Finland

---

1. Introduction

Foods made of oats are traditionally used in many European countries. The importance of oats is increasing globally due to the global need for a shift to a plant-based diet and public health concerns. Oats provide excellent nutritional value and have substantiated health benefits as oat beta-glucans lower plasma cholesterol levels and attenuate postprandial blood glucose rise, and oat fibers increase fecal bulk [1–9]. One of the most traditional oat dishes is porridge made from rolled oats. However, nowadays, a variety of oat products and fractions are available for both consumers and industry. In Finland, for example, the food use of oats is rapidly increasing [10]. Yet, globally, only about 10% of the oat crop is used for human consumption [11], while the majority of oats are used as fodder. The shift in use from the fodder to human food requires research on the sensory characteristics and usage, whereas the substantiated health effects may also influence the choice of consumers. Oats are typically utilized as whole grain. Whole grain foods are generally recommended by official nutritional recommendations in various countries. However, the use of whole grain cereals may have negative effect on the sensory quality perceived by consumers [12].

The original odor of raw oats is somewhat mild and heat treatment and further processing significantly affects the sensory quality and results in a typical "oat-like" odor [13]. The odor and flavor of heated oats have been described as "nutty", "toasted", "sweet" and "cereal-like" [12,14]. Oats have a high content of lipids which results in oats and oat products being susceptible to oxidation. Rancidity in oats and oat products derives from both endogenous enzyme activity as well as processing and storage [14]. Rancid odors and bitter off-flavors produced during oxidation may result in the limited consumption of oat products [14]. The oxidation of lipids and the use of oats as whole grain may also result in increased bitterness by higher contents of oxidation products or phenolic compounds and their derivatives, respectively, in the oat products [12,15,16]. The origin of the oat may also have a role in the sensory properties of the oat product [17]. In a study by Hu et al. [17] the aroma of the *Avena nuda* L. oat flakes originating from China differed from the flakes of *Avena sativa* L. from USA, Canada or northern European countries Denmark or Sweden. Supplementation of fractions such as those rich in beta-glucan into beverage concepts can increase the oat-like flavor but also result in increased perceptions of rancidity [18]. In addition to odor and flavor characteristics, the appearance and texture have been noted to contribute to the liking and potential usage of oat products. Despite previous investigations and definition of the sensory characteristics of oats as such [10,12], there are only few previous studies focusing on the liking and acceptance of oats and oat products.

The cultural background of the consumers affects how food items are perceived and thus also the liking and usage of the products. A wide range of sensory studies have been conducted comparing different nationalities and countries and, especially, comparing consumers of the Western and Eastern locations of the world. Examples of categories studied include meat products [19], honeys [20] and ciders [21]. The cultural background may also affect the behavior and response style in the consumer tests. Examples of such differences include the use of scales [22,23] and creation of descriptors for foods in rapid sensory characterization tests [21]. However, sensory characteristics and their links to the liking of oats and oat-based products have been previously investigated mainly in areas with a Western diet, and with a narrow selection of products compared to those currently available in Scandinavia [24,25]. Global food brands also have a growing influence on Chinese consumers [26], and the global prevalence of various chronic diseases, such as obesity and cardiovascular diseases, further increases. Oats with several approved health claims in EU and USA, have global potential from a public health point of view. Whole grains in general are recommended in the Chinese Dietary Guidelines from 2016. There are indications that Chinese consumers may not be influenced by the nutritional labels or health claims on the food packages [27], as well as indications that because of the increasing health concerns, Chinese consumers appreciate safe and organic foods and are willing to pay more for them [28,29]. However, it has currently not been investigated how the consumers in Finland and China differ from one another in these perceptions. Additionally, the effect of the fact that oats are gluten free [30] on the consumers is currently unclear.

There is an increasing interest in exporting oat products from producing countries to growing markets in East Asia. Despite these growing interests, there are no previous studies investigating or comparing the acceptance of oat product concepts in a cross-cultural context in European countries, where the oat is traditionally consumed, and Eastern countries, such as China, where cereal grains are commonly used, but oats are not equally popular.

This study aimed to identify the key sensory characteristics contributing to the liking of various oat products by conducting an online questionnaire in Finland and sensory tests in Finland and China. The goal was to study the perceptions and acceptance of oats and oat products by Chinese and Finnish consumers. A special focus was laid on the impact of cultural background of the participants as well as the influence of perceptions to new foods [31] and a general interest in the healthiness of food [32]. The studied oat product concepts in the questionnaire and in the sensory tests were designed to include various traditional (such as breads and porridges), and more novel food items (such as oat-based dairy and meat replacements), aiming to cover a wide range of oat products present in the Finnish market.

The Check-All-That-Apply (CATA) method [33] was chosen for the sensory tests to provide rapid characterization of the product concepts.

2. Materials and Methods

2.1. Participants in the Tests

Four studies were conducted (Table 1). Study 1 was conducted as an online questionnaire in Finland. It was mainly advertised in Turku, Finland, but available for volunteers from other Finnish cities as well. Studies 2 and 3 were sensory trials in Turku, Finland, whereas study 4 was conducted as a sensory trial in Wuxi, China. Turku and Wuxi are both cities with large universities attracting students from throughout the countries and globally and physically locate close to bigger cities (Shanghai is close to Wuxi; Helsinki and Stockholm are close to Turku). Study 1 was available for all volunteers in Finland, and the participation of foreigners, especially the Chinese living in Finland, was encouraged. Participants were mainly students and staff of the universities in Turku. The Chinese participants who took part in Study 1 were encouraged to also take part in studies 2 and 3. All participants in studies 2 and 3 completed the online questionnaire (study 1) prior to taking part in the sensory evaluations if they had not done so already. The participants of studies 2 and 3 were allowed, but not required, to take part in both of these tests in Finland. The participants in study 4 were students and staff members of Jiangnan University located in Wuxi, Jiangsu province, China. Overall, 66.9% of the answers were provided by Finnish participants, 25.6% by Chinese participants and 7.6% by other ethnicities.

Table 1. Characteristics of the four studies conducted.

	Study 1	Study 2	Study 3	Study 4
Location	Finland	Turku, Finland	Turku, Finland	Wuxi, China
Time of data collection	April–June 2017	November 2017, January 2018	November–December 2017, January 2018	June 2018
Test method	Online questionnaire	Sensory evaluation in laboratory	Sensory evaluation in laboratory	Sensory evaluation in laboratory
Number of sample categories or samples	11	9	8	10
Number of participants	381	65	73	103
Females, %	75.9	70.8	60.3	81.6
Age range (median)	18–76 (29)	18–68 (27)	18–68 (27)	18–36 (24)
		Country of origin, %		
Finland	83.7	72.3	68.5	-
China	7.6	20.0	19.2	100
Other	8.7	7.7	12.3	-

2.2. Online Questionnaire

The questionnaire (study 1, Table 1) included demographic questions (such as gender, age, country of origin) and questions about general interest towards healthiness, awareness of the healthiness of food and general usage of products containing oats on the first page and ratings for familiarity and pleasantness of selected oat products on the second page. The interest and awareness were rated on seven-point scales (1 = not interested/aware; 7 = extremely interested/aware), whereas usage was rated on a seven-point scale with labels: 1 = 2–4 times a day; 2 = Once a day; 3 = 2–4 times a week; 4 = Once a week; 5 = 1–3 times a month; 6 = A few times a year; 7 = Never. Familiarity of the oat products was rated on five-point scales with verbal anchors (1 = "not familiar at all" to 5 = "extremely familiar") and pleasantness on nine-point balanced hedonic scales (1 = dislike extremely to 9 = like extremely). On the third page of the questionnaire, the standardized Food Neophobia Scale (FNS) [31] and General Health Interest (GHI) scale [32] were rated on a seven-point scale. For both FNS and GHI, a respondent's task was to indicate their extent of agreement (1, "disagree strongly", to 7, "agree strongly") with 10 or 8 statements, respectively, and the potential ranges of the FNS and GHI scale scores were between

10 and 70 or 8 and 56. At the end of the questionnaire, all participants volunteering to take part in studies 2 and 3 gave their contact details in order to link the questionnaire to the sensory evaluation. The questionnaire was created with the Webropol 2.0 survey software (Webropol Oy, Helsinki, Finland) in both Finnish and English.

2.3. Oat Samples

The samples in studies 2–4 were Finnish oat products from varying product categories (Table 2). The samples were commercially available products with oat as the main ingredient or based on a commercially available product custom made without additional flavoring ingredients. The samples were either obtained from the manufacturers or purchased from a local supermarket. Due to the large variety of oat products and product categories in the market, the sensory evaluations in Finland were divided into two separate parts: dry oat samples and moist oat samples. Bread samples were frozen (−18 °C) immediately after purchase for the test in Finland or immediately after arrival in China to ensure that all subjects received bread from the same batches. The samples tested in China were selected based on the results of the tests in Finland, as well as stability during transportation from Finland to China during the summer period, i.e., samples requiring cooled storage were not selected. Samples were prepared, if needed, based on the instructions on the product packages.

The samples in Turku were presented to the participants in glass cups (dry samples) or in disposable 50 mL transparent plastic cups with watch glasses as lids (moist samples). Only disposable cups were used in study 4 in China. All dry samples were served at ambient room temperature. The moist samples, i.e., the oat meat substitute and the porridges were served freshly prepared and warm (approx. +60 °C), and the drinkable samples (oat drink and oat powder) as well as the yoghurt-alternative products were served refrigerated (approx. +4 °C).

Table 2. Oat samples, their preparations, served amounts and serving conditions in the sensory tests in Finland and China.

Sample	Study	Preparation	Served Amount	Serving Dish	Serving Temperature
Sweet Oat Biscuit	2, 4	Split piece into halves	Two halves	Glass cup	Room temp
Salty Oat Biscuit *	2, 4	as such	Two biscuits	Glass cup	Room temp
Puffed Oat Cereal	2, 4	as such	2 tablespoons	Glass cup	Room temp
Oat Drink	3, 4	Shaken	20 mL	Plastic cup	Refrigerated
Oat Powder	3, 4	1 part oat powder + 4 parts water, mixed until homogenized	20 mL	Plastic cup	Refrigerated
Oat Porridge	3, 4	450 mL water mixed with 200 mL flakes, gently simmered over low heat for 15 min in Finland, 3 min in microwave (full power) in China	1 tablespoon	Glass cup	Warm (60 °C)
Oat Granola **	2, 4	as such	2 tablespoons	Glass cup	Room temp
Oat Bread	2	Defrosted (−18 °C) in microwave oven, edges cut away, cut into 6 pieces	One rectangular piece (approx. 2 × 3 cm)	Glass cup	Room temp
Moist Oat Biscuit	2	as such	One piece	Glass cup	Room temp
Oat Roll	2	Defrosted (−18 °C) in microwave oven, edges cut away, cut into 4 pieces	One square piece (approx. 3 × 3 cm)	Glass cup	Room temp

Table 2. *Cont.*

Sample	Study	Preparation	Served Amount	Serving Dish	Serving Temperature
Oat Granola 2 *	2	as such	2 tablespoons	Glass cup	Room temp
Oat Muesli *	2	as such	2 tablespoons	Glass cup	Room temp
Yoghurt-alternative 1	3	Stirred	2 teaspoons	Plastic cup	Refrigerated
Yoghurt-alternative 2	3	Stirred	2 teaspoons	Plastic cup	Refrigerated
Instant Porridge	3	250 mL hot water from electric kettle mixed with 150 mL flakes, simmered for 5 min	1 tablespoon	Glass cup	Warm (60 °C)
Oat Meat Substitute	3	Heated in microwave oven (full power) for 30 s	1 tablespoon	Glass cup	Warm (60 °C)
Granola in Drink *	3	Oat drink shaken	1 tablespoon granola + 15 mL oat drink	Glass cup	Refrigerated
Oat Chips	4	as such	One piece	Plastic cup	Room temp
Hard Bread	4	Split piece in halves	One half	Plastic cup	Room temp
Oat Bread 2	4	Defrosted (−18 °C) in microwave oven, cut into 6 pieces	One rectangular piece (approx. 2 × 3 cm)	Plastic cup	Room temp

* Modified versions of the commercial products without added flavoring ingredients provided by the manufacturer. Granola in drink was mixed by the participant right before testing the sample. ** Manufactured in Sweden. All other products were manufactured in Finland.

2.4. Sensory Evaluations in Finland and China

Three separate sensory characterization tests (Table 1) were carried out using the Check-All-That-Apply (CATA) method [33] and following nine-point balanced hedonic scales (categories from dislike extremely to like extremely) for appearance, odor, flavor and mouthfeel; seven-point scales for odor and flavor intensities; and a five-point scale for purchase interest of the oat samples shown in Table 2. The CATA attribute list consisted of 34 attributes presented on one page and in fixed order for all samples and panelists including odor, flavor, taste and textural attributes as well as more abstract descriptors (Table S1). Attributes on the CATA list were based on the existing literature related sensory properties of oat and other grains and their products (e.g., reviews by Heiniö et al. [12,14]) as well as the preliminary sensory tests by the research team (data not published).

The questionnaires for the sensory evaluations were created with Compusense five 5.6 software (Compusense Inc., Guelph, ON, Canada), and the data were collected in Finland in Finnish and English using the software, whereas printed paper forms in Chinese were used in China. Samples (8–10/study; shown in Table 2) were presented to the participants all at the same time in randomized order (all possible permutations design) with three-digit random codes on the sample cups. Participants were instructed to examine samples in the given order monadically and select all possible CATA attributes in the sample and finally rate pleasantness, intensities and the purchase interests on the scales before evaluating the next sample. Additionally, participants were instructed to drink water between each sample to rinse their mouths. Sensory tests were organized in controlled laboratory conditions in individual sensory booths at the University of Turku, Finland, or at the Jiangnan University, China.

2.5. Statistical Analysis

Hedonic liking data from the online questionnaire were analyzed using ANOVA, with oat product, country of origin (two classes: from Finland or other country), gender, GHI status and FNS status (Table 1) being the main effects tested along with two-way interactions. Agglomerative hierarchical cluster analysis was carried out on the FNS and GHI scores, and consumer liking data to identify potential clusters and ANOVA together with Tukey's post hoc test were used to identify significant differences in liking between clusters. Principal component analyses (PCA) were used to study the correlations between CATA attributes (frequency by the panels) and samples (separate models created for studies 2–4 in Table 1, n = 8–10 and one model containing all samples from these studies, $n = 27$). Full cross validation was used to the select validated number of components or factors in the models. All ANOVA tests and cluster analyses were performed with IBM SPSS Statistics 25.0 (IBM Corporation, Armonk, NY), and multivariate analyses were carried out with Unscrambler X (version 10.5, Camo Software, Oslo, Norway).

3. Results

3.1. Participant Clusters Based on the Online Questionnaire

The FNS status of the participants in study 1 was used to classify participants as less or more neophobic based on their sum scores using cluster analysis models. Similarly, GHI status was used to classify participants as more or less interested in the healthiness of food. The sum ranges for "less" and "more" neophobic were 10–29 and 30–70, respectively, and for "less" and "more" health interested, 8–35 and 36–56, respectively. Cluster analysis with four class variables (gender, country of origin, FNS status and GHI status) identified five clusters of participants (Table 3). The largest cluster (1) consisted of Finnish female participants with a high interest towards the healthiness of food and low FNS status. Clusters 2 and 3 were less interested in the healthiness of food (i.e., low GHI score). The majority of the Finnish male participants were in cluster 3. Clusters 4 and 5 consisted of more food neophobic, yet health interested participants from Finland (cluster 4) and other countries (cluster 5).

Table 3. Comparison of participant clusters [a] and the rated liking of oat product categories in the online questionnaire.

	Cluster 1	Cluster 2	Cluster 3	Cluster 4	Cluster 5	All
Number of participants	111	77	69	62	62	381
Gender (%) [a]	Female (100)	Female (100)	Male (91.3)	Female (100)	Female (61.3)	
Country of origin (%) [a]	Finland (100)	Finland (100)	Finland (100)	Finland (100)	Other (100)	
Food Neophobia Scale [a]	Less FN	Less FN	Less FN	More FN	More FN	
Less neophobic %	100	57.1	65.2	45.2	0	59.8 (21.2)
More neophobic %	0	42.9	34.8	54.8	100	40.2 (38.0)
General Health Interest [a]	More HI	Less HI	Less HI	More HI	More HI	
Less interested %	0	100	53.6	33.9	0	35.4 (28.4)
More interested %	100	0	46.4	66.1	100	64.6 (43.3)
Interest in healthiness of food [b]	6.23 ± 0.74 a	5.14 ± 0.90 c	5.42 ± 1.10 bc	6.26 ± 0.72 a	5.58 ± 1.35 b	5.76 ± 1.06
Awareness of healthiness of food [b]	5.93 ± 0.93 a	5.55 ± 0.95 ab	5.57 ± 0.93 ab	5.97 ± 0.77 a	5.32 ± 1.44 b	5.69 ± 1.04
Usage of products containing oats [b]	2.27 ± 1.41 b	2.56 ± 1.29 b	2.75 ± 1.46 b	2.63 ± 1.36 b	3.47 ± 1.80 a	2.67 ± 1.50
	Pleasantness rating of oat product categories [b]					
Oat breads	8.32 ± 0.87 a	8.01 ± 1.16 ab	7.54 ± 1.38 b	8.10 ± 1.14 ab	6.34 ± 1.99 c	7.76 ± 1.46 A
Oat porridges	8.05 ± 1.17 a	7.10 ± 1.89 bc	6.83 ± 1.86 cd	7.82 ± 1.68 ab	6.10 ± 2.35 d	7.28 ± 1.89 B
Oat mueslis	6.94 ± 1.60 ab	6.43 ± 1.63 ab	6.38 ± 1.76 ab	6.97 ± 1.76 a	6.11 ± 2.42 b	6.60 ± 1.84 CD
Oat flakes and brans	6.94 ± 1.79 a	6.66 ± 1.66 a	5.83 ± 1.79 b	6.85 ± 1.45 a	5.56 ± 1.78 b	6.44 ± 1.79 DE

Table 3. *Cont.*

	Cluster 1	Cluster 2	Cluster 3	Cluster 4	Cluster 5	All
Oat powders	4.90 ± 1.73	4.81 ± 1.56	4.42 ± 1.58	5.15 ± 1.77	4.71 ± 1.81	4.80 ± 1.70 G
Oat meat substitutes	6.52 ± 1.83 a	6.03 ± 1.75 a	5.71 ± 2.16 a	6.15 ± 2.09 a	4.77 ± 2.16 b	5.93 ± 2.05 F
Snack-type oat biscuits	6.66 ± 1.71	6.95 ± 1.54	6.93 ± 1.52	6.77 ± 1.64	6.79 ± 1.58	6.81 ± 1.61 CD
Coffee-table oat biscuits	6.80 ± 1.87 a	7.26 ± 1.66 a	7.07 ± 1.73 a	6.90 ± 1.46 a	6.37 ± 2.06 b	6.89 ± 1.79 BC
Oat-based yoghurts	6.69 ± 1.91 a	6.08 ± 2.08 ab	5.42 ± 1.97 b	5.71 ± 2.01 ab	6.18 ± 2.28 b	6.09 ± 2.08 EF
Oat drinks	6.57 ± 2.05 a	5.61 ± 2.27 ab	4.96 ± 2.20 b	5.42 ± 2.21 b	5.40 ± 2.28 b	5.71 ± 2.25 F
Oat candies	4.80 ± 1.75	4.66 ± 1.77	4.35 ± 1.96	4.56 ± 1.71	4.97 ± 2.19	4.68 ± 1.86 G
Familiarity rating of oat product categories [b]						
Oat breads	4.41 ± 0.59 a	4.25 ± 0.67 a	4.23 ± 0.77 a	4.26 ± 0.68 a	3.48 ± 1.18 b	4.17 ± 0.83 A
Oat porridges	4.71 ± 0.49 a	4.36 ± 0.72 ab	4.29 ± 0.73 b	4.56 ± 0.64 ab	3.81 ± 1.28 c	4.39 ± 0.83 A
Oat mueslis	3.49 ± 0.97	3.21 ± 1.03	3.42 ± 0.93	3.47 ± 1.02	3.19 ± 1.45	3.37 ± 1.08 CD
Oat flakes and brans	3.87 ± 0.95 a	3.45 ± 1.08 ab	3.38 ± 0.81 b	3.58 ± 0.76 ab	2.73 ± 1.31 c	3.46 ± 1.06 BC
Oat powders	2.07 ± 0.97 a	1.48 ± 0.74 b	1.81 ± 1.03 ab	2.08 ± 1.16 a	2.11 ± 1.01 a	1.91 ± 1.01 F
Oat meat substitutes	3.37 ± 0.96 a	2.78 ± 1.07 b	3.03 ± 1.01 ab	3.03 ± 1.06 ab	2.27 ± 1.28 c	2.96 ± 1.12 E
Snack-type oat biscuits	3.64 ± 0.71	3.58 ± 0.66	3.70 ± 0.71	3.60 ± 0.84	3.56 ± 0.88	3.62 ± 0.75 B
Coffee-table oat biscuits	3.56 ± 0.66 a	3.68 ± 0.52 a	3.77 ± 0.71 a	3.48 ± 0.76 a	3.06 ± 1.21 b	3.53 ± 0.81 BC
Oat-based yoghurts	3.48 ± 1.09 a	3.29 ± 0.99 ab	2.93 ± 0.94 b	2.98 ± 0.88 b	3.03 ± 1.29 ab	3.19 ± 1.07 D
Oat drinks	3.63 ± 1.12 a	3.12 ± 1.21 ab	2.93 ± 1.10 b	3.06 ± 1.08 b	2.77 ± 1.21 b	3.17 ± 1.18 DE
Oat candies	1.49 ± 0.84 ab	1.27 ± 0.70 b	1.49 ± 0.85 ab	1.35 ± 0.63 b	1.77 ± 1.09 a	1.47 ± 0.84 G

[a] Clusters based on four class variables: gender and country of origin (Table 1), Food Neophobia Scale (FNS, [31]) status and General Health Interest (GHI, [32]) status. FNS and GHI scores of all 381 participants were first used to create two subsets (less FN/HI or more FN/HI). Mean FNS or GHI scores for the two subsets among all participants are shown brackets after percentages. [b] Mean ratings (± standard deviations; scale 1–7; 1 = not interested/aware/familiar, 7 = extremely interested/aware/familiar; scale for usage: 1 = 2–4 times a day, 2 = Once a day, 3 = 2–4 times a week, 4 = Once a week, 5 = 1–3 times a month, 6 = A few times a year, 7 = Never; hedonic liking ratings on a balanced scale 1–9) for clusters and all participants. Statistical differences (if found) are based on ANOVA with Tukey's post hoc test ($p < 0.05$) and are shown with lowercase letters a–d between clusters and uppercase letters A–G between samples in general.

Clusters differed in the rated interest in the healthiness of food, awareness of the healthiness of the food they are consuming and the general usage of products containing oats (Table 3). The clusters with a lower GHI score (2–3) were less interested in the healthiness of food, but equally aware of the healthiness of food compared to clusters with a higher GHI score. Cluster 5 with non-Finnish participants was less aware of the healthiness of food than cluster 4 with similar GHI and FNS status. The clusters with Finnish respondents (1–4) used oat products more often than cluster 5 which consisted of non-Finnish participants.

The most common reasons for the Finnish participants in the online questionnaire to select oat products were "flavor", "healthy" and "rich in fiber" (Table S2). Among the non-Finnish participants, the "flavor" was not an equally often selected reason in comparison to the other two. In Study 4 in Wuxi, China, "healthy" and "flavor" were among the most often selected reasons, whereas "easy" was also among the top three reasons.

3.2. Sensory Descriptors of Oat Products

The CATA tests were performed in three separate studies (Table 1). In general, the most often checked attributes were "edible", "light color", "oat-like", "mild", "healthy" and "soft" in the three tests. The three former attributes were the most often selected in both locations and "mild" was among the six most often picked attributes, but there were some differences between panels. In Finland, "healthy" was the fourth most often selected attribute, whereas in Study 4, "healthy" and "soft" were replaced by "chewy" and "roasted". Three separate PCA models were created using the CATA attribute frequencies (Figure 1). The first two Principal components (PCs) illustrated in Figure 1a–c showed 68%, 67% and 64% of the variances among the datasets. In Figure 1a with the dry oat products (Study 2), some of the most often used descriptors were linked to biscuits and cereal on the right side of the plots ("crispy/crunchy") and oat breads on the opposite of the first PC ("soft"). On the second PC, "healthy" and "light color" were linked to oat muesli, whereas the moist oat biscuit was described as "strange" and "sticky" and oat granola as "sweet" on the opposite side to healthy but separated by the first PC.

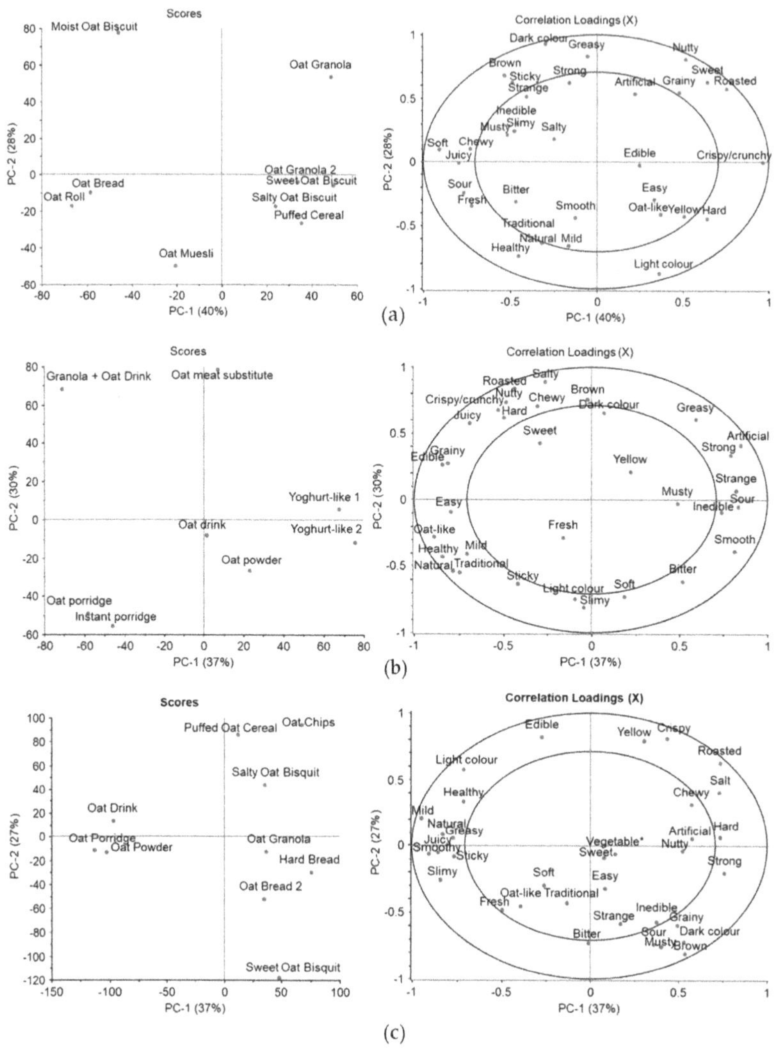

Figure 1. Principal component analysis models based on the Check-All-That-Apply (CATA) attributes used to describe the oat samples in the three sensory tests (Table 1). (**a**) Scores and loadings plots for samples evaluated in study 2; (**b**) samples evaluated in study 3; (**c**) samples evaluated in study 4. Only attributes selected by a minimum of 5% of the panels were used in the models.

Among the "moist" products included in Study 3 (Figure 1b), the unsweetened oat-based yoghurt-like products were described as "strange", "sour", "inedible" and "strong" on the first PC, whereas the oat porridges were more "healthy", "oat-like" and "natural". On the second PC and on the opposite side to the porridges was the meat substitute and oat granola + oat drink (a mixture of the two samples evaluated also separately, Table 2) with "chewy", "salty" and "brown". The latter mixture sample was also described as "crispy/crunchy" and "roasted". In Study 4 conducted in China, the porridge and the drinks (milk and powder drink) were described as "natural" and "healthy", whereas the puffed cereal and oat chips were described as "crispy" and "roasted" on the opposite side of the first PC. The latter two samples were separated from bread samples and sweet biscuit along the second PC with "grainy", "sour" and "inedible". The most frequently used attribute, "edible", located on the opposite side correlated with puffed oat cereal.

In order to compare both test locations, Finland and China, and their descriptors at the same time, additional PCA models were created combining all samples in the tests (n = 27) and their sensory descriptors (Figure 2a) and separate model for the seven samples included in both locations (Figure 2b). Three groups of samples were formed on the first two components with 57% of total variance in the first model. On the right on PC-1 in Figure 2a is the group of samples, such as the granolas and puffed cereals and some of the biscuits, described as "crispy/crunchy", "roasted" and "nutty", but also "hard", "sweet", "grainy" and "edible". On the opposite side of PC-1 are various moist/wet products, such as the porridges, drinks and yoghurts, with "healthy", "natural", "smooth" and "slimy" attributes. Located on the top of PC-2 is the third group of varying samples mostly described as "strange", "inedible" and "musty" and with "dark" and "brown" colors. The most notable difference between the Finnish and Chinese panels was with the sweet oat biscuit. It was associated with the aforementioned third group in the Chinese test but in the first group, in the Finnish test. Similarly, with the second model (Figure 2b), most of the samples were described similarly in both locations except the sweet oat biscuit along the PC-2.

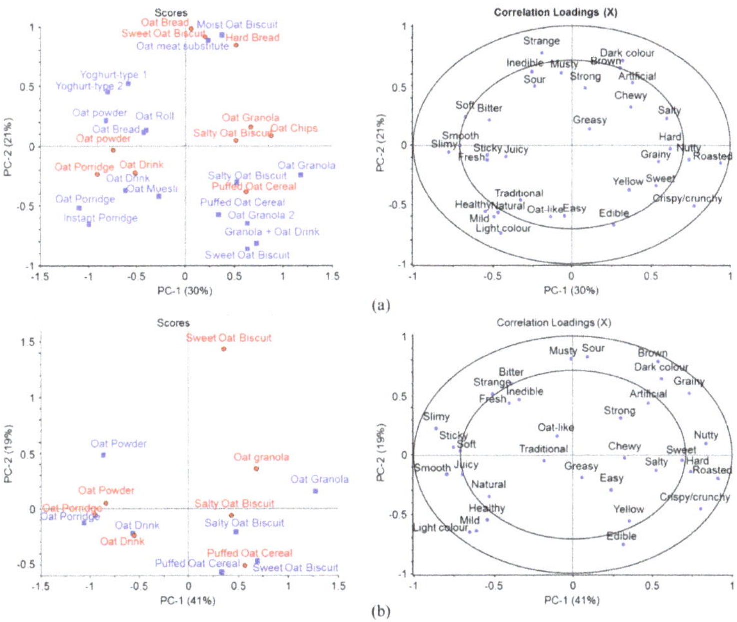

Figure 2. PCA models with (**a**) all samples included in the CATA tests in Finland (blue rectangles) and China (red circles) (n = 27) with their sensory descriptors and (**b**) only seven samples included in both locations. Only attributes selected by a minimum of 5% of the panels were used in the models.

3.3. Comparison of Hedonic Liking Ratings by Finnish and Chinese Participants

Seven of the oat samples shown in Table 2 were included in both test locations. In general, sweeter samples, such as sweet oat biscuit and oat granolas, were the most preferred based on the flavor (Table 4), whereas especially the unsweetened oat powder drink and yoghurt-like products were the most disliked in Finland, and the hard bread, as well as the unsweetened oat powder drink, were the most disliked in China (Table S3). In most cases, the samples were perceived more pleasant by the Finnish in comparison to the Chinese participants. Puffed oat cereal and the oat powder were perceived more pleasant by the Chinese panel. Intensities of oat odor and/or flavor were perceived more intense by the Finnish in sweet biscuit, oat drink and powder drink, and more intense by the Chinese in oat porridge and oat granola.

Table 4. Comparison of the rated pleasantness, intensities oat odor and flavor and interest to purchase for the oat samples.

Samples [a]	Participants	Pleasantness [b]				Intensity [b]		Purchase Interest [b]
		Appearance	Odor	Flavor	MF & T	Oat Odor	Oat Flavor	
Sweet Oat Biscuit	China (n = 103)	6.30 ± 1.2	6.56 ± 1.2	7.17 ± 1.4 b	6.82 ± 1.2 b	3.97 ± 1.4 b	4.24 ± 1.5 b	3.65 ± 1.1 b
	Finland (n = 65)	6.43 ± 1.6	6.38 ± 1.5	7.60 ± 1.6a	7.30 ± 1.3 a	4.46 ± 1.6 a	4.74 ± 1.5 a	4.03 ± 1.0 a
Salty Oat Biscuit	China (n = 103)	5.45 ± 1.3	5.33 ± 1.1	5.64 ± 1.5	5.67 ± 1.5 b	3.52 ± 1.4	3.99 ± 1.5	2.95 ± 1.1
	Finland (n = 65)	5.32 ± 1.5	5.65 ± 1.5	5.91 ± 1.8	6.35 ± 1.6 a	3.62 ± 1.6	4.15 ± 1.4	2.69 ± 1.3
Puffed Oat Cereal	China (n = 103)	6.50 ± 1.4 a	5.84 ± 1.1 a	6.77 ± 1.2 a	6.93 ± 1.3	2.57 ± 1.4	2.94 ± 1.3	3.44 ± 1.1
	Finland (n = 65)	5.79 ± 1.9 b	4.95 ± 1.6 b	6.29 ± 1.7 b	7.03 ± 1.5	2.37 ± 1.4	3.05 ± 1.7	3.06 ± 1.5
Oat Drink	China (n = 103)	5.90 ± 1.2	5.28 ± 0.8 b	6.20 ± 1.4 b	6.33 ± 1.6	1.95 ± 1.1	2.73 ± 1.5 b	3.07 ± 1.2
	Finland (n = 73)	5.97 ± 1.9	5.60 ± 1.0 a	6.60 ± 1.6 a	6.82 ± 1.8	1.79 ± 1.2	3.49 ± 1.7a	3.38 ± 1.4
Oat Powder	China (n = 103)	5.51 ± 1.2	5.00 ± 0.8	4.20 ± 1.7 a	4.52 ± 1.8 a	2.70 ± 1.6 b	3.51 ± 1.8 b	2.08 ± 1.1 a
	Finland (n = 73)	5.15 ± 1.6	5.05 ± 1.6	3.32 ± 1.7 b	3.70 ± 1.9 b	3.63 ± 1.9 a	4.51 ± 1.6 a	1.48 ± 0.8 b
Oat Porridge	China (n = 103)	4.58 ± 1.7 b	5.53 ± 1.6 b	5.19 ± 1.6 b	5.13 ± 1.8 b	4.90 ± 1.6 a	5.13 ± 1.7 a	2.50 ± 1.2 b
	Finland (n = 73)	6.05 ± 1.7 a	6.62 ± 1.3 a	6.33 ± 1.4 a	6.63 ± 1.4 a	4.27 ± 1.5 b	4.36 ± 1.7 b	3.78 ± 1.2 a
Oat Granola	China (n = 103)	6.30 ± 1.4 b	6.25 ± 1.3 a	6.83 ± 1.5 b	6.26 ± 1.4 b	3.77 ± 1.6	4.73 ± 1.5 a	3.56 ± 1.1
	Finland (n = 73)	7.09 ± 1.6 a	5.66 ± 1.8 b	7.77 ± 1.1 a	7.63 ± 1.2 a	3.51 ± 1.8	4.06 ± 1.6 b	3.82 ± 1.3

[a] Samples included both in Finland and China. [b] The evaluation scale was a balanced hedonic 9-point scale for the pleasantness (MF & T mouthfeel and texture), 7-point scale for the intensities and 5-point scale for the purchase interest. Rated attributes presented to panelists as in the table after completion of the Check-All-That-Apply (CATA) test. Sample information presented in Table 2. The letters a–b marked the significant statistical differences ($p < 0.05$) within each comparison.

4. Discussion

Currently, there is little published information concerning the sensory properties of oat products as most of studies have focused mainly on the oat itself or on certain specific food processes. Most of the studies have utilized trained panels and descriptive analysis techniques. The benefits of the rapid methods such as the Check-All-That-Apply (CATA) test used in this study are that they can be constructed to include a larger participant number compared to studies with trained panels. The use of CATA in cross-cultural studies is especially reasonable and feasible, as the influence of a single participant is smaller than in trained panels. While CATA methods may be more difficult to interpret than studies with trained panels [33,34], methods focusing on the similarities and differences among products, such as Projective Mapping used in this study, may overcome these issues [22]. In general, the data collection procedures in cross-cultural sensory studies are recommended to be as similar as possible [22]. However, the use of paper forms in Wuxi, for example, instead of computers as in Turku, or the differences in the sample sets in studies 2–4 (Table 2), may have influenced the outcome.

Here, we provided the CATA list in Finnish or English to participants in Finland and in Chinese in study 4 in China. The translations were conducted by a food scientist, a native Chinese speaker with >5 years of experience in living in Finland, and thus with good understanding of food culture in both countries. However, in any cross-cultural study, the possibility remains for words having a slightly different tone in different languages and cultures. Results of the CATA tests were in accordance with literature [12,14] to some extent as "light color" or "mild" were linked to the less processed oat porridges and plain oat muesli and "strong" and "dark color"' correlated with many of the more processed product samples.

We found ethnicity (country of origin) to have a stronger influence on the clustering of the participants based on the online questionnaire than gender, FNS or GHI status. The participants in cluster 5 consisting largely of non-Finnish responders were less interested or aware of the healthiness of their food compared to the other clusters. The hedonic ratings may also have been influenced by the familiarity of certain foods. Refined and polished grains are traditionally dominant in Chinese food culture, although consumption of whole grain-based products have been an emerging trend with the increasing awareness of the consumers about the health benefits of whole grain products.

This is in contrast to the situation in Finland and other Northern European countries, where whole grain-based cereal products have been part of the traditional food culture, and products such as oat breads and different whole grain porridges are well known. Cluster 5 included participants from a wide range of countries, and with varying time spent in Finland prior to participation to the test (data not shown). Thus, some of them may have become more familiar with the local foods and have started using them regularly, whereas many may still prefer to use foods and ingredients familiar to them. Previous studies have indicated that Chinese people who have moved abroad may be more connected to the original Chinese food in comparison to other ethnicities, and the effect may persist in several generations [26]. The fact that cultural differences may have influenced the answering of the questions (e.g., use of scales) should also be recognized [22].

The acceptance of snack-type oat products did not differ between the clusters indicating that they are universally popular. Moreover, among the most important reasons for selecting oat products (Table S2), the Chinese panel in study 4 chose "Easy". Snack-type foods are also increasing in popularity in Finland, and their consumption may be linked to increasing obesity [35]. Thus, oats have potential as an ingredient in healthier snack-type foods in varying markets.

CATA tests showed similarities and differences between the Finnish and Chinese panelists. As expected, the "sweet" samples, such as granolas and the sweet biscuits, were more preferred in both locations in comparison to the more "sour" or "bitter" unsweetened products (Figures 1 and 2). Interestingly, the sweet biscuit evaluated by the Chinese panel in study 4 was located in the third group of samples (on top of the PC-2 in the scores plot, Figure 2), whereas the corresponding sample evaluated by the Finnish panel was located in the first group on the opposite side of the PC-2. Although being a generally liked sample, it was described differently. Potentially, the descriptor "hard" or characteristics associated with the descriptor "strange" were not familiar to the Chinese panelists. The strangeness can be linked to unfamiliarity and thus decreased liking of the products in the third group. Unfortunately, as a limitation of this study [22], all samples used in studies 2 and 3 in Finland were not studied in China in study 4 due to the required cold storage during transportation. Some of those, e.g., the unsweetened oat yoghurt-like products and the meat substitutes, were among the distinctively described or more disliked samples in study 3. However, the introduction of such products to foreign markets would require costly cooled transportation or local production, and thus is not feasible in comparison to products not requiring cold storage.

Oats are a traditional and important food ingredient in Nordic countries where breads and porridges are part of the traditional daily diet. This is also shown in this study where the breads and porridges were the most familiar product categories in the online questionnaire (Table 3). At the same time, these foods were the most liked. The familiarity of foods plays an important role in liking and in usage of foods [36]. Here, the most novel and thus unfamiliar product categories, such as oat-based meat substitutes, oat powders or oat candies, were more disliked. In China, the dominating ingredients are rice, wheat and soy, and bread is not a traditionally used food item. Thus, oat breads and other oat foods are unfamiliar to Chinese food culture. However, there may be significant regional differences in the preferences. Wheat is grown in the northern parts of China, whereas rice is most popular in the southern regions. Compared to wheat and rice, oat cultivation is limited to few Northern provinces including mainly Shanxi, Shaanxi, Inner Mongolia and Hebei [37]. The oat samples in this study were provided in plastic cups without any additional information about the product (Table 2). For consumers who are familiar to the products, the intrinsic factors of the food play an important role in liking, whereas those consumers who are unfamiliar with them, may rely more on the extrinsic factors, such as information displayed on the product packaging [38]. We chose to present samples "as such" without any additional ingredients where possible. The added ingredients, such as berry flavors in oat yoghurts or added side dishes on top of breads or biscuits, would have affected the perceived familiarity and liking. This would have directed the perspective away from the features of the oat itself towards the added ingredients. Already the sample of granola with added oat milk in study 3 resulted in the highest hedonic ratings among all oat products. This shows the importance of a real life situation and context in conducting the acceptance tests, and calls for further studies in more narrow product categories served in meal type settings. Potentially, oat could be incorporated as an ingredient in local foods or products instead of introducing it in more unfamiliar product forms and concepts. On the other hand, the added ingredients should not undermine or compromise the healthiness of the oat. While it is globally true that the sensory properties of food are the primary driver in food choices of consumers, there are variations between people to the extent of readiness to trade the enjoyment from food for potential health benefits, which was also noted in the results of this study.

The number of new oat-based products has recently skyrocketed in Finland, and separate, sensory tests could be organized using many of the product categories alone. Many of the product categories (Table 3) or actual products (Table 2) are becoming more frequently used and thus more familiar to larger consumer groups. Some consumers may discard all animal-based or wheat-based foods from their diets, while many are ready to test new plant-based alternatives and use them in addition to the traditional options. At the same time, new innovative foods are introduced actively to the markets, some also with oat as the main, or additional yet emphasized ingredient. Especially, the meat or dairy replacers have increased their popularity in recent years. In these product categories, oats have provided a more local option to various other plant-based options, such as soy.

In addition to familiarity, consumer's food neophobia status has an important role in determining the acceptance of foods, especially with novel and unfamiliar foods [36,39–41]. Food neophobia has also been shown to be partially genetically determined [42]. Clusters 1–3 consisted mainly or mostly of less food neophobic participants, whereas clusters 4 and 5 had more neophobic participants (Table 3). In the comparison of clusters 1 and 4, where other factors such as gender, interest in healthiness of food and origin of the participants were considered, the more novel oat drinks were more liked by cluster 1. However, the food neophobia status of the participants in the online questionnaire affected the liking ratings notably less in comparison to participants' origin. Food neophobia may also affect the stronger flavored foods more than the mild ones [43]. Here, the attribute "mild" was linked to oat porridges (Figures 1 and 2) along with "natural" and "oat-like" attribute indicating the natural "oat" flavor being mild [13,14]. In Figure 2, various food items in the third group, which were more novel for the participants, correlated with the "strong" attribute, and thus, they may have been partially disliked due to food neophobia.

Interestingly, the oat porridges were associated with healthiness in both Finland and China (studies 3 and 4). At the same time, in both locations or in the online questionnaire, the porridges were not among the most or least liked samples or food categories. Potentially, this is due to simplicity and less processing which in some cases may be more liked and often associated with being healthier [44]. However, the liking and the interest to purchase the porridge (Table 4) were rated higher by the Finnish panel. The oat porridge has already been for a long time perceived as a healthy and nutritious option for a breakfast. The healthiness of oats has become increasingly known among the consumers, potentially also in China, especially among the higher educated participants in study 4, i.e., students and staff at the Jiangnan University. As the students may be more open to new food choices and perhaps more aware of the health benefits of oats, they are potential consumers of oat products. Interest to purchase oat products may be more dependent on the consumer's education level and the nutritive value of the product rather than the product price or the income level of the consumer [45]. Similar to study 4, also in Finland in studies 2 and 3, the participants were mainly highly educated university students and staff, and thus, extrapolation to the whole population should be done with caution [22]. More studies are needed also with more random sampling and participants with a wider socioeconomic background and food preference (deriving from regional differences). Concerning the healthiness of oat, the GHI status of the participants affected mainly the porridge category in the online questionnaire as the more health interested in cluster 1 gave higher ratings compared to the corresponding cluster 2 with lower health interest. As the healthiness of oat may be a potential marketing strategy for various product categories, familiarity and sensory properties likely have stronger impacts on the purchasing decision by consumers who do not emphasize the healthiness of foods.

In addition to being associated with healthiness, porridges were associated with the "natural" attribute in Figures 1 and 2. The naturalness of foods is a very important factor in food choices for most of the consumers and it can be classified by its origin, technologies and ingredients used or by the characteristics of the final product [46]. Here, the porridges in studies 3 and 4, as well as the plain muesli in study 2, were potentially perceived as less processed and with lesser ingredients in comparison to the other samples. Moreover, the oat-like odor and flavor perceived from the porridges were the most intense among the samples (Table 4). Some of the products included in this study, and also many oat products currently in market, may contain relatively little oat in comparison to the other ingredients, thus resulting in the lower intensity of oat-like odors and flavors.

In conclusion, the acceptance of oat products and product categories was studied in Finland and China. The ethnicity of the participants had an important role affecting the rated pleasantness and familiarity of oat product categories, whereas both food neophobia and health interest status also had an influence. The effects of these were observed in different oat product categories. For example, the ethnicity affected the acceptance of the traditional and generally liked foods, such as breads and food neophobia status, and the acceptance of novel and familiar foods, such as meat substitutes. The health interest status affected the acceptance of oat porridges, which were also universally described as "healthy". The sensory tests using the CATA method showed several similarities between the Finnish and Chinese panel in the selection of sensory descriptors for product. In addition to porridge being generally described as "healthy"', oat granolas were described as "sweet"' and "roasted"' in both locations. The generally liked sweet oat biscuits were described differently in two tests.

Based on this study, different strategies are needed for different oat products in order to introduce them successfully to consumers in Finland, where the oat is a significant part of consumers' daily diets, or China, where oats are only regionally known or only becoming more popular. However, the healthiness of the oat is a crucial factor affecting the choices of consumers in both countries, thus strongly affecting the acceptance of "oat-like" products, such as porridges. Sweetness is also a generally liked attribute among the products in this study, but too much added sugar results in reducing the potential healthiness of oats. However, more studies are needed focusing on the consumer acceptance and usage of oats and oat products, especially in different varying consumer segments, in order enhance their consumption. The study provides assistance to food companies in bringing products to new markets or to new consumer groups. This applies both to marketing oat products in Western countries, where the oat is traditionally consumed, and to exporting oat products to markets where the oat is not yet equally well utilized and accepted. The study also benefits researchers studying multicultural differences especially in relation to preference for grains or foods in general.

Supplementary Materials: The following are available online at http://www.mdpi.com/2304-8158/9/9/1234/s1, Table S1: List of Check-All-That-Apply (CATA) attributes for describing the samples as presented to panelists in Finland (English and Finnish) and in China (Chinese). Table S2: The most important reasons [a] for selecting oat products presented as percentages of the participants. Table S3: Mean hedonic ratings and standard deviations (on a scale 1–9) for samples included in the sensory tests.

Author Contributions: Conceptualization, K.M.L., B.Y. and O.L.; methodology, O.L.; software, O.L., E.P.; validation, K.M.L., O.L., E.P. and X.M.; formal analysis, O.L., E.P. and X.M.; investigation, O.L., E.P. and K.M.L.; resources, B.Y., P.Z. and K.M.L.; data curation, O.L. and K.M.L.; writing—original draft preparation, O.L. and K.M.L.; writing—review and editing, O.L., X.M., E.P., B.Y., P.Z., and K.M.L,; visualization, O.L. and K.M.L.; supervision and project administration, K.M.L.; funding acquisition, K.M.L. and B.Y. All authors have read and agreed to the published version of the manuscript.

Funding: This work was funded by Business Finland as part of the OATyourGUT "Oat-derived Wellbeing by Clinical and Perceived Assessment" project (grant number 5469/31/2016) cofunded by Finnish food companies and the University of Turku.

Acknowledgments: Volunteers who participated in the sensory studies or answered the questionnaire are warmly acknowledged. Companies providing oat products or product concepts without flavorings are acknowledged. Lina Zhang from Jiangnan University is thanked for valuable aid with shipments of the product concepts. Yulong Bao is thanked for his help in practical matters outside the research.

Conflicts of Interest: The authors declare no conflict of interest.

References

1. Ho, H.V.T.; Sievenpiper, J.L.; Zurbau, A.; Mejia, S.B.; Jovanovski, E.; Au-Yeung, F.; Jenkins, A.L.; Vuksan, V. The effect of oat β-glucan on LDL-cholesterol, non-HDL-cholesterol and apoB for CVD risk reduction: A systematic review and meta-analysis of randomised-controlled trials. *Br. J. Nutr.* **2016**, *116*, 1369–1382. [CrossRef] [PubMed]
2. Menon, R.; Gonzalez, T.; Ferruzzi, M.; Jackson, E.; Winderl, D.; Watson, J. Chapter One—Oats—From Farm to Fork. In *Advances in Food and Nutrition Research*; Henry, J., Ed.; Academic Press: Cambridge, MA, USA, 2016; Volume 77, pp. 1–55.
3. Shen, X.L.; Zhao, T.; Zhou, Y.; Shi, X.; Zou, Y.; Zhao, G. Effect of Oat β-Glucan Intake on Glycaemic Control and Insulin Sensitivity of Diabetic Patients: A Meta-Analysis of Randomized Controlled Trials. *Nutrients* **2016**, *8*, 39. [CrossRef]
4. Du, B.; Meenu, M.; Liu, H.; Xu, B. A Concise Review on the Molecular Structure and Function Relationship of β-Glucan. *Int. J. Mol. Sci.* **2019**, *20*, 4032. [CrossRef] [PubMed]
5. Rasane, P.; Jha, A.; Sabikhi, L.; Kumar, A.; Unnikrishnan, V.S. Nutritional advantages of oats and opportunities for its processing as value added foods—A review. *J. Food Sci. Technol.* **2015**, *52*, 662–675. [CrossRef] [PubMed]
6. EFSA Panel on Dietetic Products; Nutrition and Allergies (NDA). Scientific Opinion on the substantiation of health claims related to beta glucans and maintenance of normal blood cholesterol concentrations (ID 754, 755, 757, 801, 1465, 2934) and maintenance or achievement of a normal body weight (ID 820, 823) pursuant to Article 13(1) of Regulation (EC) No 1924/2006. *EFSA J.* **2009**, *7*, 1254. [CrossRef]
7. EFSA Panel on Dietetic Products; Nutrition and Allergies (NDA). Scientific Opinion on the substantiation of health claims related to beta-glucans from oats and barley and maintenance of normal blood LDL-cholesterol concentrations (ID 1236, 1299), increase in satiety leading to a reduction in energy intake (ID 851, 852), reduction of post-prandial glycaemic responses (ID 821, 824), and "digestive function" (ID 850) pursuant to Article 13(1) of Regulation (EC) No 1924/2006. *EFSA J.* **2011**, *9*, 2207. [CrossRef]
8. EFSA Panel on Dietetic Products; Nutrition and Allergies (NDA). Scientific Opinion on the substantiation of health claims related to oat and barley grain fibre and increase in faecal bulk (ID 819, 822) pursuant to Article 13(1) of Regulation (EC) No 1924/2006. *EFSA J.* **2011**, *9*, 2249. [CrossRef]
9. EFSA Panel on Dietetic Products; Nutrition and Allergies (NDA). Scientific Opinion on the substantiation of a health claim related to oat beta glucan and lowering blood cholesterol and reduced risk of (coronary) heart disease pursuant to Article 14 of Regulation (EC) No 1924/2006. *EFSA J.* **2010**, *8*, 1885. [CrossRef]
10. Natural Resources Institute Finland. *Balance Sheet for Food Commodities 2018, Preliminary and 2017 Final Figures 2019*; Natural Resources Institute Finland: Helsinki, Finland, 2019.
11. Welch, R.W. Cereal Grains. In *The Encyclopedia of Human Nutrition*; Academic Press: New York, NY, USA, 2006; pp. 346–357.
12. Heiniö, R.L.; Noort, M.W.J.; Katina, K.; Alam, S.A.; Sozer, N.; de Kock, H.L.; Hersleth, M.; Poutanen, K. Sensory characteristics of wholegrain and bran-rich cereal foods—A review. *Trends Food Sci. Technol.* **2016**, *47*, 25–38. [CrossRef]
13. Klensporf, D.; Jeleń, H.H. Effect of heat treatment on the flavor of oat flakes. *J. Cereal Sci.* **2008**, *48*, 656–661. [CrossRef]
14. Heiniö, R.-L.; Kaukovirta-Norja, A.; Poutanen, K. Flavor in Processing New Oat Foods. *Cereal Foods World* **2011**. [CrossRef]
15. Günther-Jordanland, K.; Dawid, C.; Dietz, M.; Hofmann, T. Key Phytochemicals Contributing to the Bitter Off-Taste of Oat (*Avena sativa* L.). *J. Agric. Food Chem.* **2016**, *64*, 9639–9652. [CrossRef]
16. Molteberg, E.L.; Solheim, R.; Dimberg, L.H.; Frølich, W. Variation in Oat Groats Due to Variety, Storage and Heat Treatment. II: Sensory Quality. *J. Cereal Sci.* **1996**, *24*, 273–282. [CrossRef]
17. Hu, X.-Z.; Zheng, J.-M.; Li, X.; Xu, C.; Zhao, Q. Chemical composition and sensory characteristics of oat flakes: A comparative study of naked oat flakes from China and hulled oat flakes from western countries. *J. Cereal Sci.* **2014**, *60*, 297–301. [CrossRef]

18. Lyly, M.; Salmenkallio-Marttila, M.; Suortti, T.; Autio, K.; Poutanen, K.; Lähteenmäki, L. Influence of Oat β-Glucan Preparations on the Perception of Mouthfeel and on Rheological Properties in Beverage Prototypes. *Cereal Chem. J.* **2003**, *80*, 536–541. [CrossRef]
19. Bekker, G.A.; Tobi, H.; Fischer, A.R.H. Meet meat: An explorative study on meat and cultured meat as seen by Chinese, Ethiopians and Dutch. *Appetite* **2017**, *114*, 82–92. [CrossRef]
20. Price, E.J.; Tang, R.; Kadri, H.E.; Gkatzionis, K. Sensory analysis of honey using Flash profile: A cross-cultural comparison of Greek and Chinese panels. *J. Sens. Stud.* **2019**, *34*, e12494. [CrossRef]
21. Jamir, S.M.R.; Stelick, A.; Dando, R. Cross-cultural examination of a product of differing familiarity (Hard Cider) by American and Chinese panelists using rapid profiling techniques. *Food Qual. Prefer.* **2020**, *79*, 103783. [CrossRef]
22. Ares, G. Methodological issues in cross-cultural sensory and consumer research. *Food Qual. Prefer.* **2018**, *64*, 253–263. [CrossRef]
23. Ares, G.; Giménez, A.; Vidal, L.; Zhou, Y.; Krystallis, A.; Tsalis, G.; Symoneaux, R.; Cunha, L.M.; de Moura, A.P.; Claret, A.; et al. Do we all perceive food-related wellbeing in the same way? Results from an exploratory cross-cultural study. *Food Qual. Prefer.* **2016**, *52*, 62–73. [CrossRef]
24. Brückner-Gühmann, M.; Banovic, M.; Drusch, S. Towards an increased plant protein intake: Rheological properties, sensory perception and consumer acceptability of lactic acid fermented, oat-based gels. *Food Hydrocoll.* **2019**, *96*, 201–208. [CrossRef]
25. Lyly, M.; Roininen, K.; Honkapää, K.; Poutanen, K.; Lähteenmäki, L. Factors influencing consumers' willingness to use beverages and ready-to-eat frozen soups containing oat β-glucan in Finland, France and Sweden. *Food Qual. Prefer.* **2007**, *18*, 242–255. [CrossRef]
26. Yen, D.A.; Cappellini, B.; Wang, C.L.; Nguyen, B. Food consumption when travelling abroad: Young Chinese sojourners' food consumption in the UK. *Appetite* **2018**, *121*, 198–206. [CrossRef] [PubMed]
27. Liu, R.; Hoefkens, C.; Verbeke, W. Chinese consumers' understanding and use of a food nutrition label and their determinants. *Food Qual. Prefer.* **2015**, *41*, 103–111. [CrossRef]
28. Liu, R.; Pieniak, Z.; Verbeke, W. Consumers' attitudes and behaviour towards safe food in China: A review. *Food Control* **2013**, *33*, 93–104. [CrossRef]
29. Yin, S.; Wu, L.; Du, L.; Chen, M. Consumers' purchase intention of organic food in China. *J. Sci. Food Agric.* **2010**, *90*, 1361–1367. [CrossRef]
30. Lu, Z.; Zhang, H.; Luoto, S.; Ren, X. Gluten-free living in China: The characteristics, food choices and difficulties in following a gluten-free diet—An online survey. *Appetite* **2018**, *127*, 242–248. [CrossRef]
31. Pliner, P.; Hobden, K. Development of a scale to measure the trait of food neophobia in humans. *Appetite* **1992**, *19*, 105–120. [CrossRef]
32. Roininen, K.; Lähteenmäki, L.; Tuorila, H. Quantification of consumer attitudes to health and hedonic characteristics of foods. *Appetite* **1999**, *33*, 71–88. [CrossRef]
33. Varela, P.; Ares, G. Sensory profiling, the blurred line between sensory and consumer science. A review of novel methods for product characterization. *Food Res. Int.* **2012**, *48*, 893–908. [CrossRef]
34. Ares, G.; Varela, P. Trained vs. consumer panels for analytical testing: Fueling a long lasting debate in the field. *Food Qual. Prefer.* **2017**, *61*, 79–86. [CrossRef]
35. Mielmann, A.; Brunner, T.A. Consumers' snack choices: Current factors contributing to obesity. *Br. Food J.* **2019**, *121*, 347–358. [CrossRef]
36. Tuorila, H.; Hartmann, C. Consumer responses to novel and unfamiliar foods. *Curr. Opin. Food Sci.* **2020**, *33*, 1–8. [CrossRef]
37. Wang, S. Fodder oats in China. In *Fodder Oats: A World Overview*; Suttie, J.M., Reynolds, S.G., Eds.; Food and Agriculture Organization of the United Nations: Rome, Italy, 2004; pp. 123–144.
38. Nacef, M.; Lelièvre-Desmas, M.; Symoneaux, R.; Jombart, L.; Flahaut, C.; Chollet, S. Consumers' expectation and liking for cheese: Can familiarity effects resulting from regional differences be highlighted within a country? *Food Qual. Prefer.* **2019**, *72*, 188–197. [CrossRef]
39. Tuorila, H.; Lähteenmäki, L.; Pohjalainen, L.; Lotti, L. Food neophobia among the Finns and related responses to familiar and unfamiliar foods. *Food Qual. Prefer.* **2001**, *12*, 29–37. [CrossRef]
40. Damsbo-Svendsen, M.; Frøst, M.B.; Olsen, A. A review of instruments developed to measure food neophobia. *Appetite* **2017**, *113*, 358–367. [CrossRef] [PubMed]

41. Dovey, T.M.; Staples, P.A.; Gibson, E.L.; Halford, J.C.G. Food neophobia and 'picky/fussy' eating in children: A review. *Appetite* **2008**, *50*, 181–193. [CrossRef]
42. Knaapila, A.; Tuorila, H.; Silventoinen, K.; Keskitalo, K.; Kallela, M.; Wessman, M.; Peltonen, L.; Cherkas, L.F.; Spector, T.D.; Perola, M. Food neophobia shows heritable variation in humans. *Physiol. Behav.* **2007**, *91*, 573–578. [CrossRef]
43. Laureati, M.; Spinelli, S.; Monteleone, E.; Dinnella, C.; Prescott, J.; Cattaneo, C.; Proserpio, C.; De Toffoli, A.; Gasperi, F.; Endrizzi, I.; et al. Associations between food neophobia and responsiveness to "warning" chemosensory sensations in food products in a large population sample. *Food Qual. Prefer.* **2018**, *68*, 113–124. [CrossRef]
44. Martins, I.B.A.; Oliveira, D.; Rosenthal, A.; Ares, G.; Deliza, R. Brazilian consumer's perception of food processing technologies: A case study with fruit juice. *Food Res. Int.* **2019**, *125*, 108555. [CrossRef]
45. Su, R.N.; Qiao, G.H. Consumers' willingness to buy oat products and the influencing factors. *Guizhou Agric. Sci.* **2013**, *6*, 212–215. (In Chinese)
46. Román, S.; Sánchez-Siles, L.M.; Siegrist, M. The importance of food naturalness for consumers: Results of a systematic review. *Trends Food Sci. Technol.* **2017**, *67*, 44–57. [CrossRef]

Article

Post-Ingestive Sensations Driving Post-Ingestive Food Pleasure: A Cross-Cultural Consumer Study Comparing Denmark and China

Mette Duerlund [1,*], Barbara Vad Andersen [1], Kui Wang [2,3], Raymond C. K. Chan [2,3] and Derek Victor Byrne [1]

[1] Food Quality Perception and Society, Department of Food Science, Aarhus University, Agro Food Park 48, 8200 Aarhus N, Denmark; barbarav.andersen@food.au.dk (B.V.A.); derekv.byrne@food.au.dk (D.V.B.)
[2] Neuropsychology and Applied Cognitive Neuroscience Laboratory, CAS Key Laboratory of Mental Health, Institute of Psychology, Chinese Academy of Sciences, Beijing 100049, China; wangk@psych.ac.cn (K.W.); rckchan@psych.ac.cn (R.C.K.C.)
[3] Department of Psychology, University of Chinese Academy of Sciences, Beijing 100049, China
* Correspondence: mette.duerlund@food.au.dk; Tel.: +45-871-560-00

Received: 13 March 2020; Accepted: 7 May 2020; Published: 11 May 2020

Abstract: Culture is one of the main factors that influence food assessment. This cross-cultural research aimed to compare Chinese and Danish consumers in their post-ingestive drivers of Post-Ingestive Food Pleasure (PIFP). We define PIFP as a "subjective conscious sensation of pleasure and joy experienced after eating". We conducted two in-country consumer studies in Denmark (n = 48) and in China (n = 53), measuring post-ingestive sensations and PIFP using visual analogue scale, for three hours following consumption of a breakfast meal. Key results revealed perceived Satisfaction, Mental, Overall and Physical wellbeing to be highly influential on PIFP in both countries. Moreover, Danish consumers perceived appetite-related sensations such as Satiety, Hunger, Desire-to-eat and In-need-of-food to be influential on PIFP, which was not the case in China. In China, more vitality-related sensations such as Energized, Relaxation and Concentration were found to be drivers of PIFP. These results suggest similarities but also distinct subtleties in the cultural constructs of PIFP in Denmark and in China. Focusing on Food Pleasure as a post-ingestive measure provides valuable output, deeper insights into what drives Food Pleasure, and, importantly, takes us beyond the processes only active during the actual eating event.

Keywords: cross-cultural; post-ingestive food pleasure; food reward; post-ingestive sensation; satisfaction; china; Denmark

---

1. Introduction

Culture is one of the main underlying factors that influence how we assess food, our attitudes and beliefs about food and our food choices [1]. Considering the globalization and emergent food markets, where foods are exported beyond national borders, we need to take into consideration cultural aspects when seeking to understand human eating behaviors on the respective markets [2]. Consequently, cross-cultural research has become increasingly more pertinent within Sensory and Consumer Science [3,4], and several studies suggest that cultural differences exist in the way we perceive food, in our associations with specific foods, and also within food-related concepts [5–11]. The contribution and importance of cross-cultural research studies thus bring new perspectives in the domain of Sensory and Consumer Science, and it contributes to the development and understanding of various food concepts.

Emphasis on cultural differences in food perceptions might be imperative in our association with Food Reward. Food Reward comes in many disguises and the concept can be defined and explained in many different ways. Reward is not a unitary construct, but comprises multiple psychological components [12], and different disciplines include different measurements of Food Reward. Investigating reward initially originates from disciplines such as psychology and neuroscience, where activation of brain circuits and neural pathways bring important knowledge. Reward research has been used to provide insights into several psychological and cognitive conditions including drug addictions, depression, eating disorders, gambling, obsession, sex addiction etc. [12,13]. Neuroscientists disclose three major components of reward namely Motivation, Learning and Affect. Each category comprises both explicit and implicit psychological constituents [12,14]. Motivation includes wanting, either as a conscious desire for incentives or as an underlying implicit motivation for reward. Learning represents association and prediction of future rewards based on experience, which includes both explicit cognitive expectancy and implicit associative conditioning. Affect includes liking and pleasantness, either as an implicit affect response or as a conscious pleasure in the ordinary sense of the term [13]. The present research situates and considers Post-Ingestive Food Pleasure (PIFP) as part of the Affect category with focus on conscious pleasure, inspired and described by [12,14].

Food Reward functions as an important conception for research within disciplines such as Health, Nutrition and Food Sciences [15]. For instance, several researchers within the Sensory and Consumer Science field seek to better understand Food Reward's role in appetite and hence our eating behaviors [16]. The most common Food Reward measures are self-reported liking [16,17], self-reported desire-to-eat [13] or wanting for a specific food [13,16]. Many Food Reward measures and tasks have been developed and are in active use [18]. Rogers and Hardman (2015) define Food Reward as the "momentary value of a food to the individual at the time of ingestion", measured directly with a rating of "desire to eat the entire portion right now" [19]. This approach and definition have also been used by, e.g., Ruddock et al. (2017) [19,20]. Explicit liking and desire to eat are most frequently evaluated using rating scales such as visual analogue scales [16]. The Leeds Food Preference Questionnaire (LFPQ) is a computer based measurement tool for Food Reward first proposed by Finlayson et al. (2007) [21]. LFPQ utilizes food image stimuli to evaluate explicit liking and wanting as well as implicit wanting in a choice task. The LFPQ has been applied and/or adapted by several researchers [16,20,22–24]. Additional Food Reward measures include grip force operant tasks [18,20,25], willingness to pay [20,26], and Emotional attentional blink [18,27]. An operant task could include tapping a space bar for 60 seconds, being told, the more you tap, the more of a given food you will receive [19,20], or by squeezing a handheld dynamometer as a response to specific food images [25].

The majority of the studies investigating Food Reward within the appetite space, measure and apply liking, wanting and other reward-associated tasks prior to or during consumption, so as part of the momentary eating event [17]. However, Food Reward derived from eating might also depend on sensations experienced in the time after eating. Møller (2015) argue that reward from eating depends also on mental and bodily wellbeing experienced after a meal, and that this focus is practically untouched in the scientific exploration of Food Reward [28]. There is a need to extend the concept of affective reward in human reward processes [17], to move further and beyond the actual eating event in seeking to quantify reward, wellbeing, pleasure, food joy and satisfaction derived from eating [28]. This gives rise to incorporating Food Pleasure to our post-ingestive experiences, as well as to explore how Post-Ingestive Food Pleasure associates to other post-ingestive sensations. Food Pleasure can thus also be a measure after intake and in the time following intake. Food Pleasure as a longer-lasting measure than just at the actual eating event can provide deeper insights into what drives Food Pleasure post intake.

In the present paper, we define Post-Ingestive Food Pleasure (PIFP) as a "subjective conscious sensation of pleasure and joy experienced after eating", measured directly by asking: "Please rate your sense of joy when thinking of the meal you ate today". Therefore, PIFP in this study represents explicit enjoyment experienced by the individual consumer. Studying post-ingestive drivers of Food Pleasure, whether it be appetite-related sensations or more bodily or mental sensations, requires measurements of our interoceptive states [28,29]. Interoception defines as the subjective experience of internal signals related to, e.g., satiety, hunger, heat, pain, energy, visceral and muscular sensations [29–32]. Interoceptive states also function as a basis for self-awareness and subjective feelings, representing a conscious evaluation of "how I feel" [28,29]. Including different interoceptive states appeals to consumers' ability to introspect at a conscious level and contributes to an in-depth picture of the extended appetite experience [28,33].

The overall purpose of this research study was to study post-ingestive sensations as drivers of PIFP in a cross-cultural comparative study between Denmark and China. Investigating the post-ingestive consumer experience including PIFP provides valuable novel insights into the more extended eating experience going beyond the momentary eating event. It enables us to study the dynamics and the interrelationships between sensations in the time after eating and provides a clearer picture as to which sensations that are important for PIFP for Chinese and Danish consumers. PIFP and other post-ingestive sensations can act as desired outcome goals for consumers, and these desired goals could guide the individual to regulate one's eating behaviors in order to obtain the desired goal. This, indeed, serves important knowledge for both academia, food industry and food policy. Product developers, marketing professionals and health advisers can benefit from this knowledge in order to facilitate and promote healthier eating behaviors and thus help people to better navigate in a challenging food environment. Therefore, investigating PIFP offers new ways to understand and predict food choice and intake behaviors. The research specifically aims to:

(1) Study post-ingestive drivers of self-reported Post-Ingestive Food Pleasure (PIFP), and study the development of drivers of PIFP over three hours post intake;
(2) Compare Danish and Chinese consumers in their post-ingestive drivers of self-reported PIFP.

The research included conducting of two in-country consumer studies in Denmark and in China, respectively. We investigated similarities and differences between Danish and Chinese consumers, and it was hypothesized that differences exist between the two cultures with respect to variation in drivers of PIFP. The constructs of PIFP were therefore expected to vary with consumers' cultural background. Note, data from the Danish in-country study have been used and published elsewhere, but with different aims and analyses [34]. In the present paper, we include Chinese in-country data, partake different aims and analyses and focus on a cross-cultural perspective comparing dimensions of PIFP between Denmark and China.

2. Materials and Methods

2.1. Participants and Recruitment

In China, participants (n_{total} = 53) were recruited from the University of Chinese Academy of Sciences (UCAS), north of Beijing. In Denmark, participants (n_{total} = 48) were recruited from Ollerup Sports Academy, south of Odense. In both studies, participants voluntarily signed up for the research study after advertisement via internal written communication channels. Inclusion criteria in both countries were: being between 18 and 25 years old, being a liker of yoghurt for breakfast, not suffer from any food allergies and being Chinese or Danish in nationality, respectively. Table 1 displays the characteristics of participants from Denmark and China. The study was approved by the Ethics Committee of Institute of Psychology, Chinese Academy of Sciences with the approval number: H18011. In Denmark, ethical approval is not required for this type of study according to the National Committee on Health Research Ethics in Denmark (Section 14 (2) in the Committee Act) [35]. Prior to participation, all participants gave their written consent.

Table 1. Participant characteristics from the Danish and Chinese central-location studies.

Characteristics	China	Denmark
n_{total}	53	48
Males/females	24/29	31/17
Age (years)	22 ± 1.1 (20–25) *	20.4 ± 1.1 (18–25) *
Weight (kg)	60.3 ± 10.8 (41–95) *	71.5 ± 12.0 (51–108) *
Height (cm)	168.6 ± 7.2 (154–183) *	175.7 ± 8.5 (162–192) *
BMI [1] (kg/m^2)	21.1 ± 2.7 (17.3–33.6) *	23.0 ± 2.4 (19–30.5) *

* Mean ± standard deviation (range); [1] BMI= body mass index

2.2. Procedure and Study Design

Two comparative central-location consumer studies were carried out in Denmark and in China, both with a randomized controlled crossover design. Data collection procedures were executed in identical manners. Data were collected within a reasonable timeframe, in this case within seven months, avoiding any disproportionate period of time between data collection which can hinder comparability [36]. Data were collected in March 2018 in Denmark and in October 2018 in China, with similar weather conditions, avoiding any bias related to seasonal heat or cold weathers during summer or winter. Participants came in for two breakfast sessions on two separate days. Breakfast meals were served in random order across participants. The study began at 7:30 a.m. in the morning, with participants coming in fasting state since 22:00 the night before, and ran for 3 hours until 10:30 a.m. To make sure, all participants consumed the same amounts, the breakfast meals were mandatory intake. Response variables were collected pre intake, immediately after intake and for three hours in 30 minute intervals post intake. Post-Ingestive Food Pleasure and Satisfaction were evaluated post intake, since they refer to sensations after eating.

2.3. Questionnaire

The questionnaire focused on post-ingestive sensations and PIFP, and the specific chosen response variables were developed from existing scientific literature linked to consumer food perception and investigated post-ingestive sensation variables [6,32,37–39]. PIFP in the present study was evaluated with focus and emphasis on food joy post intake using the question phrasing: "Please rate your sense of joy when thinking of the meal you ate today". Response variables were evaluated in randomized order using a visual analogue scale (VAS) via CompuSense® Cloud software (CompuSense Inc., Guelph, Ontario, Canada). Data were thus collected on a continuous scale ranging from 0, anchored "not at all" to 10, anchored "extremely". In addition, participants answered demographic question including gender, age, weight and height.

Emphasis was made on ensuring linguistic equivalence between the two languages, as well as developing the questionnaire in participants' native languages. It has been shown that non-native language questionnaires can lead to a preference for neutral answers due to lack of understanding of the question or lack of comfortability with the task in general, whereas native language questionnaires support the respondent to use the full range and to find the concepts more refined [40]. The questionnaires in each country included identical response variables in linguistic equivalent phrasings. For specific phrasings used in the Danish and Chinese consumer studies, see Table 2. To ensure and validate proper translations for cultural equivalence, question phrasings were translated using back-translation [41]. Chinese native speakers translated the questionnaire from English to Chinese, followed by a back-translation to English by another native speaking Chinese researcher. All translators were researchers within food science to ensure that the meaning of the questions were correctly communicated after translations into another language.

Table 2. Response variables with Danish, English, and Chinese phrasings as used in the questionnaires. Data were collected on a continuous visual analogue scale (VAS) ranging from 0 to 10.

Variable	English Phrasing	Chinese Phrasing	Danish Phrasing
Hunger	"How hungry are you right now?"	"您现在有多饿?"	"Hvor sulten er du lige nu?"
Satiety	"How full are you right now?"	"您现在有多饱?"	"Hvor mæt er du lige nu?"
Energized	"How energetic are you right now?"	"您现在精力有多充沛?"	"Hvor energisk er du lige nu?"
Relaxation	"How relaxed are you right now?"	"您现在有多放松?"	"Hvor afslappet er du lige nu?"
Concentration	"How is your concentration right now?"	"您现在精神有多集中?"	"Hvor koncentreret er du lige nu?"
Sleepiness	"How sleepy are you right now?"	"您现在有多累?"	"Hvor træt er du lige nu?"
Satisfaction	"How satisfied are you with the breakfast meal you ate today?"	"您对今天的早餐多满意?"	"Hvor tilfreds er du med det måltid du har spist i dag?"
Overall wellbeing	"Please rate your overall wellbeing as you feel it right now"	"请立即给您现在的总体舒适状况打分"	"I hvor høj grad fornemmer du en generel velvære lige nu?"
Physical wellbeing	"Please rate your physical wellbeing as you feel it right now"	"请您凭第一感觉对自己现在的身体舒适状况打分"	"Hvor fysisk veltilpas er du lige nu?"
Mental wellbeing	"Please rate your mental wellbeing as you feel it right now"	"请您凭第一感觉给自己现在的心情舒畅状况打分"	"Hvor mentalt veltilpas er du lige nu?"
Desire to eat	"How much do you desire to eat something right now?"	"您现在有多想吃点东西呢?"	"Hvor stor er din lyst til noget at spise lige nu?"
Sweet desire	"How much do you desire to eat something sweet right now?"	"您现在有多想吃些甜的食物?"	"I hvor høj grad har du lyst til noget sødt lige nu?"
Salty desire	"How much do you desire to eat something salty right now?"	"您现在有多想吃些咸的食物?"	"I hvor høj grad har du lyst til noget salt lige nu?"
Fatty desire	"How much do you desire to eat something fatty right now?"	"您现在有多想吃些油腻的食物?"	"I hvor høj grad har du lyst til noget fedt lige nu?"
In need of food	"How much do you need food right now?"	"您现在有多需要食物?"	"I hvor høj grad mangler du noget mad lige nu?"
Post-Ingestive Food Pleasure	"Please rate your sense of joy when thinking of the meal you ate today".	"请给您在回想今日所进食食物的时候的愉悦程度打分"	"I hvor høj grad fornemmer du en glæde ved den mad du har spist i dag?"

2.4. Breakfast Meals

The breakfast meals consisted of a non-flavored yoghurt added toppings of plain muesli, natural almonds, raisins and fresh blueberries. The breakfast meals were developed to resemble each other in each country. All ingredients were identical except for the yoghurt and fresh blueberries, which were purchased commercially in China and in Denmark, respectively. Thus, yoghurt and blueberries were locally bought in each country aiding familiarity and fresh produce. In both countries, we selected a non-flavored yoghurt with no added sugar, which was similar in content and calories for both countries. The yoghurt from China contained six more calories per 100g than the Danish yoghurt. To facilitate a span in evaluated response variables, we added whey protein isolate and glucose syrup to the breakfast meals stirred into the yoghurt. The added whey protein isolate (35g/131.5Cal) and added glucose syrup (42.35g/131.5Cal) were the exact same amounts in both countries. For meal content and specific ingredient details, see Table 3. Note, for this paper, the meals served as a tool to span post-ingestive sensations and PIFP, and to facilitate a comparison between Chinese and Danish consumers in post-ingestive drivers of PIFP. Results accounting product/meal differences have been published elsewhere [34]. The test meals were made following standardized procedures to ensure validity and standardization. The visual look of the breakfast meals were identical in China and in Denmark, and can be seen from the picture in Figure 1.

Table 3. Contents of the breakfast meals.

Food Ingredient	Breakfast A	Breakfast B
	Amount (g)/Calories (Cal)	Amount (g)/Calories (Cal)
Yoghurt in Denmark [1]	300/123	300/123
Yoghurt in China [2]	300/141	300/141
Lacprodan [3]	35/131.5	-
Glucose syrup [4]	-	42.35/131.5
Muesli [5]	30/99	30/99
Almonds [6]	6/30	6/30
Raisins [7]	2.5/5	2.5/5
Fresh blueberries [8]	2.5/5	2.5/5
Total Denmark	376/393	383.35/393
Total China	376/411	383.35/411

[1] Arla®lactose-free yoghurt natural (Arla Foods, Viby, Denmark); [2] JinShi Dai 今时代 low-fat, sugar-free yoghurt containing xylitol (Odward Dairy, Beijing, China); [3] Lacprodan®SP-9225 Instant (whey protein isolate) (Arla Foods, Viby, Denmark); [4] Dansukker®glucose syrup (Dansukker, Copenhagen, Denmark); [5] Kornkammeret muesli (Lantmännen Cerealia A/S, Vejle, Denmark); [6] Coop almonds natural (Coop Danmark A/S, Albertslund, Denmark); [7] Urtekram Sultana raisins (Urtekram International A/S, Mariager, Denmark); [8] Bought from a local supermarket the day before test day—origin unknown.

Figure 1. Picture of the breakfast meal (non-flavored yoghurt, muesli, almonds, raisins, fresh blueberries).

2.5. Data Analysis

Subjective reports on weight and height were used to calculate BMI: weight (kg)/height (m)^2. Repeated Measures Analysis of Variance (ANOVA) was performed separately for each country to analyze dynamics in post-ingestive sensations over time. p-values ≤ 0.05 were considered statistically significant, and Tukey's Honest Significant Differences (HSD) test was applied for pairwise comparisons between time points. Effect sizes were examined using Cohen's d values [42]. Partial Least Squares Regression (PLSR) models were applied to study drivers of PIFP in Denmark and in China, separately. This approach has also been employed by other researchers for Sensory and Consumer Science data [39,43]. PLSR deals with multi-collinearity between explanatory variables as well as takes into account the latent co-variance structure between PIFP (Y-variable) and explanatory post intake variables (X-variables). Data were centered and reduced, and all models were full cross-validated using Jackknife leave-one-out (LOO) validation. A cumulative Q^2 index above 0.5 was considered a good predictive quality of the models. Variable Importance in Projection (VIP-score) was analyzed to determine the influential variables on PIFP. Only VIP-scores above 0.8 were considered influential variables as defined by, e.g., [44–46]. As described by [47], a strategy to select subsets of variables is discarding of variables with small VIP values [45,47]. Therefore, and for visualization, the plots depicted in the manuscript only displays variables that the models found influential on PIFP, discarding small VIP values. PLSR models were applied across products (across breakfast A and B, thus not accounting meal differences) and separately for China and Denmark. PLSR models and ANOVA models were applied at the time points 'immediately post intake' as well as 'three hours post intake', to reflect the longest interval and variation in data and to represent the post-ingestive effects for the whole study period. All PLSR models were carried out using XLSTAT by Addinsoft, version 2019.2. (XLSTAT, Long Island, NY, USA) [48].

3. Results

3.1. Dynamics in Post-Ingestive Sensations

Significant main effects of time between the time points 'immediately post intake' and 'three hours post intake' was seen for 15 post-ingestive variables in Denmark and for 10 post-ingestive variables in China, see Table 4. In general, for both countries, Hunger, Desire-to-eat, Sweet desire, Salty desire, Fatty desire and In need of food significantly increased over three hours, whereas Satiety significantly decreased. Only in China, Overall, Mental and Physical wellbeing significantly increased over three hours, whereas the same variables significantly decreased in Denmark over three hours. In China, no time effects were seen for Energized, Relaxation, Concentration, Sleepiness, Satisfaction and PIFP, whereas only Sleepiness in Denmark did not show a significant time effect. In both countries, Hunger, Satiety, Desire to eat and In need of food yielded the largest effect sizes for time differences. Table 4 shows the least squares (LS) means across products at the two time points for each country. Level of significance, F values and Cohen's d values for effect sizes are included for each country separately.

Table 4. Least squares means ± standard deviations (China: $n = 53$, Denmark: $n = 48$) across products immediately post intake and three hours post intake for all post-ingestive response variables.

	China					Denmark				
	Immediately Post Intake	Three Hours Post Intake	p	F	d [1]	Immediately Post Intake	Three Hours Post Intake	p	F	d [1]
Hunger	$1.81^a \pm 1.8$	$4.52^b \pm 2.4$	***	121.8	1.3	$4.36^a \pm 2.6$	$7.90^b \pm 1.5$	***	205.3	1.7
Satiety	$7.61^a \pm 2.1$	$4.33^b \pm 2.4$	***	195.2	1.5	$5.66^a \pm 2.0$	$1.99^b \pm 1.6$	***	267.6	2.0
Energized	6.52 ± 2.0	6.37 ± 2.0	ns	-	-	$4.31^a \pm 1.6$	$3.72^b \pm 1.6$	**	10.9	0.4
Relaxation	6.70 ± 2.2	7.06 ± 1.7	ns	-	-	$5.70^a \pm 1.7$	$4.66^b \pm 1.8$	***	20.2	0.6
Concentration	6.96 ± 1.9	6.67 ± 1.8	ns	-	-	$4.73^a \pm 1.4$	$3.92^b \pm 1.5$	***	17.7	0.6
Sleepiness	3.17 ± 2.0	3.75 ± 2.1	ns	-	-	5.05 ± 1.9	4.86 ± 2.0	ns	-	-
Satisfaction	5.72 ± 2.4	6.05 ± 2.2	ns	-	-	$4.48^a \pm 2.1$	$3.28^b \pm 2.0$	***	52.5	0.6
Overall wellbeing	$6.19^a \pm 2.2$	$6.79^b \pm 1.8$	**	9.1	0.3	$5.37^a \pm 1.5$	$4.39^b \pm 1.9$	***	23.8	0.6
Physical wellbeing	$6.00^a \pm 2.2$	$6.65^b \pm 1.9$	**	10.5	0.3	$5.24^a \pm 1.5$	$4.16^b \pm 1.9$	***	29.8	0.6
Mental wellbeing	$6.26^a \pm 2.1$	$6.93^b \pm 1.8$	***	14.8	0.3	$5.36^a \pm 1.6$	$4.44^b \pm 1.7$	***	26.0	0.6
Desire to eat	$2.38^a \pm 2.3$	$4.83^b \pm 2.5$	***	89.3	1.0	$4.73^a \pm 2.5$	$8.15^b \pm 1.4$	***	202.2	1.7
Sweet desire	$2.57^a \pm 2.3$	$3.51^b \pm 2.5$	***	23.2	0.4	$4.02^a \pm 2.3$	$5.12^b \pm 2.4$	***	19.7	0.5
Salty desire	$3.14^a \pm 2.8$	$3.94^b \pm 2.5$	**	11.9	0.3	$2.95^a \pm 2.0$	$3.92^b \pm 2.5$	***	19.1	0.4
Fatty desire	$2.28^a \pm 2.4$	$2.96^b \pm 2.4$	**	10.9	0.3	$2.50^a \pm 1.9$	$3.74^b \pm 2.6$	***	37.5	0.5
In need of food	$2.28^a \pm 2.1$	$4.79^b \pm 2.5$	***	100.8	1.1	$4.61^a \pm 2.5$	$8.01^b \pm 1.4$	***	237.0	1.7
Post-Ingestive Food Pleasure	5.93 ± 2.4	6.16 ± 2.1	ns	-	-	$4.19^a \pm 2.2$	$3.08^b \pm 1.9$	***	35.0	0.5

Means with different superscript (^{a,b}) within each country differ significantly (Tukey $p < 0.05$); *** $p < 0.0001$, ** $p < 0.01$, ns = no significant time difference; [1] Effect size (Cohen's d); data were collected on a continuous visual analogue scale (VAS) ranging from 0 to 10.

3.2. Post-Ingestive Sensations Driving Post-Ingestive Food Pleasure in China and in Denmark

Partial Least Squares Regression models were employed to determine the main drivers of PIFP amongst the included fifteen post-ingestive sensations in each country at two different time points; immediately post intake and three hours post intake. All models showed good predictive quality, defined by how well the model fits the observed values (goodness of fit). Accordingly, from the Chinese data, the model quality presented a cumulative Q^2 index of 0.7 and 0.6; at time immediately post intake and three hours post intake, respectively. From the Danish data, the model quality presented a cumulative Q^2 of 0.5 and 0.9 immediately post intake and three hours post intake, respectively.

Analysis of Variables Importance in Projection (VIP-scores) revealed specific post-ingestive sensations to be drivers of PIFP. In China, the following post-ingestive sensations were main drivers of PIFP: Satisfaction, Mental wellbeing, Overall wellbeing, Physical wellbeing, Energized, Concentration and Relaxation. These post-ingestive sensations remained influential drivers from immediately post intake and until three hours post intake. In Denmark, immediately post intake, the following post-ingestive sensations were main drivers of PIFP: Satisfaction, Mental wellbeing, Overall wellbeing, Physical wellbeing, In need of food, Desire to eat, Satiety and Hunger. Three hours post-intake, the analysis revealed two additional post-ingestive sensations to become influential drivers (VIP-scores >0.8), namely Energized and Concentration. Table 5 shows the VIP-scores from the PLSR models for each country at the two time points.

Table 5. Variable Importance in Projection (VIP-scores) of Post-Ingestive Food Pleasure for all fifteen post-ingestive sensations immediately post intake and three hours post intake. Variables are computed in descending order according to VIP-scores immediately post intake.

China			Denmark		
Post-Ingestive Sensation	**Immediately Post Intake**	**Three Hours Post Intake**	**Post-Ingestive Sensation**	**Immediately Post Intake**	**Three Hours Post Intake**
Satisfaction	**1.7**	**2.4**	Satisfaction	**2.6**	**2.6**
Mental wellbeing	**1.6**	**1.4**	Mental wellbeing	**1.3**	**1.0**
Overall wellbeing	**1.5**	**1.1**	Overall wellbeing	**1.3**	**1.1**
Physical wellbeing	**1.5**	**1.1**	Physical wellbeing	**1.0**	**1.0**
Energized	**1.2**	**1.3**	In need of food	**1.0**	**0.9**
Concentration	**1.2**	**1.1**	Desire to eat	**0.9**	**1.0**
Relaxation	**1.1**	**1.3**	Satiety	**0.9**	**0.9**
Sleepiness	0.6	0.7	Hunger	**0.8**	**1.1**
In need of food	0.6	0.5	Relaxation	0.6	0.6
Satiety	0.6	0.6	Energized	0.6	**1.2**
Desire to eat	0.5	0.6	Salty desire	0.6	0.2
Salty desire	0.6	0.6	Sleepiness	0.5	0.1
Hunger	0.5	0.5	Sweet desire	0.5	0.2
Sweet desire	0.5	0.5	Concentration	0.5	**1.1**
Fatty desire	0.3	0.6	Fatty desire	0.5	0.2

VIP-scores are calculated based on Partial Least Squares Regression (PLSR) models with Post-Ingestive Food Pleasure as dependent variable (Y) and post-ingestive variables as explanatory variables (X). A VIP-score > 0.8 is considered significantly influential on Post-Ingestive Food Pleasure and highlighted in **bold** in this table.

The relationships between variables 'immediately post intake' are visually illustrated in Figures 2 and 3 for China and Denmark, respectively. The relationships between variables at 'three hours post intake' show the same overall relations (plots not shown). For easing visualization, only influential variables (VIP-scores >0.8) are included in the figures. From the Chinese results (Figure 2), the PLSR model explained in total 80% for the dependent variable (Y) and 84% for the explanatory variables (X). The first component explained 68.4% and 75.6% of Y- and X-data, respectively, and the second component explained 11.6% and 8.4% of Y- and X-data, respectively. Interpreting the plot in Figure 2, we see that component 1 explained the majority of variance in data displaying all the influential post-ingestive sensations (VIP-scores > 0.8) positively correlated with PIFP. Component 2 was somewhat explained by Satisfaction and PIFP, but with low explained variance compared to component 1. From the Danish results (Figure 3), the PLSR model explained in total 60% for the dependent variable (Y) and 65.4% for the explanatory variables (X). The first component explained 40.4% and 45.3% of Y- and X-data, respectively, and the second component explained 19.6% and 20.1% of Y- and X-data, respectively. Overall wellbeing, Mental wellbeing and Physical wellbeing contributed the most in explaining the first component, whereas Desire to eat, Hunger, In need of food and Satiety contributed the most in explaining component 2 dividing the appetite sensations into two, with Satiety opposite the other three appetite sensations and negatively correlated with them. Satisfaction and PIFP were shown to contribute to explaining both components, however, mainly the first component.

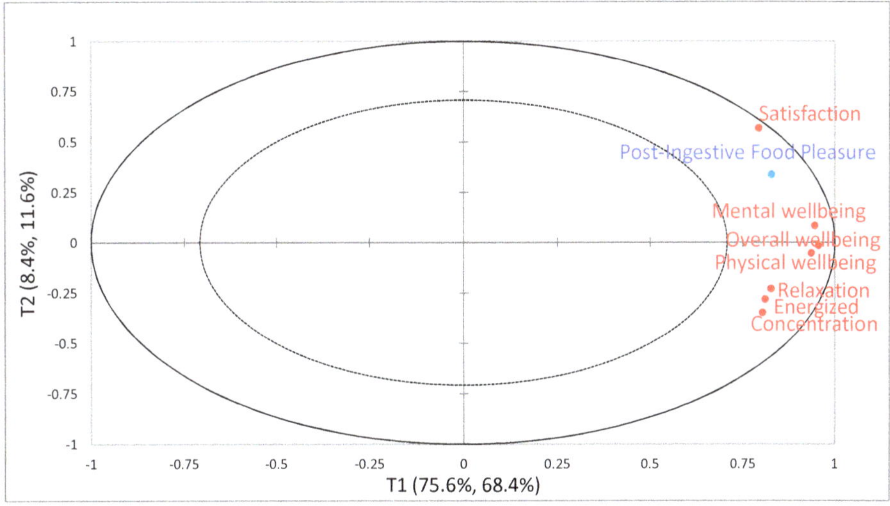

Figure 2. Partial Least Squares Regression (PLSR) correlation loadings plot from the Chinese consumer study immediately post intake, with Post-Ingestive Food Pleasure as dependent variable (Y) and Post-ingestive sensations as explanatory variables (X). Only explanatory variables with VIP-scores >0.8 are included. The plot displays component 1 (X explained variance: 75.6%, Y: 68.4%) vs component 2 (X explained variance: 8.4%, Y: 11.6%).

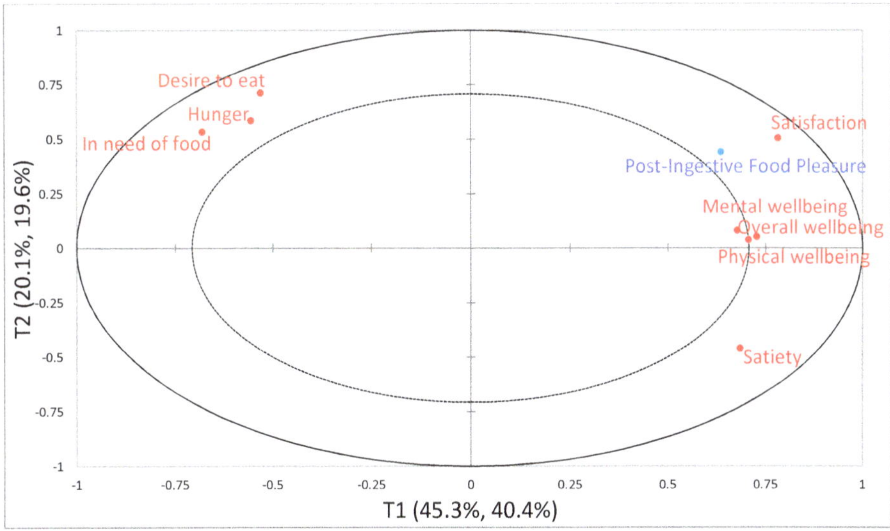

Figure 3. Partial Least Squares Regression (PLSR) correlation loadings plot from the Danish consumer study immediately post intake, with Post-Ingestive Food Pleasure as dependent variable (Y) and Post-ingestive sensations as explanatory variables (X). Only explanatory variables with VIP-scores >0.8 are included. The plot displays component 1 (X explained variance: 45.3%, Y: 40.4%) vs component 2 (X explained variance: 20.1%, Y: 19.6%).

4. Discussion

4.1. Cross-Cultural Differences in Drivers of Post-Ingestive Food Pleasure

This comparative research aimed to study and compare post-ingestive sensations as drivers of PIFP in Denmark and in China. Key results revealed both cross-cultural similarities as well as cross-cultural difference in post-ingestive drivers of PIFP in Denmark and in China. For both Danish and Chinese consumers, the post-ingestive variables Satisfaction, Mental wellbeing, Overall wellbeing and Physical wellbeing were highly influential on PIFP with VIP-scores > 1. This suggests a somewhat similar structure in the cultural constructs of Food Pleasure post intake and to some degree, a common conceptualization of PIFP in both countries (further elaborated in Section 4.2.). However, results also showed that other and different dimensions of post-ingestive sensations explained consumers' subjective PIFP in each country. Danish consumers perceived appetite-related sensations such as Satiety, Hunger, Desire-to-eat and In-need-of-food to be influential on PIFP, and this was not the case for Chinese consumers. On the contrary, in China, the more vitality- and energy-related post-ingestive sensations such as Relaxation, Energized and Concentration were found to be drivers of PIFP.

Importantly, these results thus indicate that Chinese and Danish consumers seem to experience and associate PIFP via different dimensions. Other studies, conducted in European countries, report appetite sensations to influence Satisfaction. For instance, Boelsma et al. (2010) reported that Satiety relate particularly to feelings of Satisfaction [49]. Andersen et al. (2017, 2015) found that main positive drivers of Food Satisfaction included Hunger, Fullness and Energy for Danish consumers [39,43]. Researching Food Reward, Rogers and Hardman (2015) demonstrated Hunger to independently have an effect, this with UK consumers [19]. They, however, measured Food Reward as Desire to eat a portion, defined as the momentary value of a food at the time of ingestion, and not Food Pleasure as a post intake measure such as in this study. The present results from the Danish study agree with above mentioned studies, that appetite sensations such as Hunger and Satiety demonstrate influential drivers in our perception of food. However, the mentioned research studies focused mainly on Food Satisfaction and not PIFP as evaluated in this study.

The question arises whether Danish people are more driven by the homeostatic aspects of appetite such as Hunger and Satiety than the Chinese people are. Furthermore, one might interpret that Chinese people have a different perspective in relation to PIFP with more focus on vitality terms, including Energy and Relaxation. It seems that the Danish consumers experience PIFP as more related to the body's physical needs, whereas the Chinese consumers experience PIFP as more connected to the body's 'mental' needs. Previous research has also reported differences in the values associated to food and eating between Western countries and China [5,6], e.g., results have pointed towards cultural differences in the way we associate for instance wellbeing and 'feeling good'. Sulmont-Rossé et al. (2019) report cross-cultural perspectives on 'feeling good' in the context of food using qualitative approaches. Their results demonstrated that 'feeling good' was associated to emotional, physical (health-related) and social dimensions across countries. However, when specifically looking at China, they found that Chinese people expressed fewer words related to specific food items than other countries in a word association task, but instead provided more mentions related to happiness, joy and enthusiasm. Oppositely, mentions associated to nutrition and healthy diet tended to be more frequent in Western countries [5]. Furthermore, exploring cross-cultural associations with food-related wellbeing, Ares et al. (2016) found the biggest country difference for emotional and spiritual aspects of wellbeing, with especially Chinese people integrating an aspect of nature in association to food and wellbeing [6].

In order to try to understand such cultural differences, it is essential to look at cultural values in broader perspectives. Ma (2015) describes food in China to represent social status and symbolic meaning [50]. In Denmark, food is often more fixated around nutrient content, and whether the food is good or bad for your physical health [51]. Hofstede (1980, 2001) has proposed a model where cultures are compared based on values. The model includes six dimensions: Power distance, Uncertainty Avoidance, Individualism/Collectivism, Masculinity, Long term orientation and Indulgence with index scores ranging from 0-100 [52,53]. Large difference have been reported amongst many countries, and it is evident that, within Hofstede's framework, cultural differences also exist between Denmark and China. Denmark and China especially differ for the dimensions Individualism (China 20, Denmark 74) and Long term orientation (China 87, Denmark 35)—data retrieved from [54]. China is characterized as a collectivistic culture and Denmark as an individualistic culture. This implies greater focus on harmony and sense of belonging to larger environments in China rather than the individual environment in Denmark. Furthermore, China scores high in the Long term orientation index, referring to a greater importance put on future events rather than the present events [52]. These large value differences between China and Denmark might help us to understand differences in eating behaviors as well. This type of approach can contribute to our understanding of the cultural determinants of differences in food-related behaviors [55]. Nevertheless, care should always be taken in the generalizing of results when comparing and interpreting cross-cultural differences.

4.2. Post-Ingestive Food Pleasure, Satisfaction, and Wellbeing Associations

PIFP demonstrated to have the same four main drivers in both China and in Denmark. Especially Satisfaction revealed to be the biggest driver of PIFP in both countries, regardless of time point, this with VIP-scores as high as 2.6. Moreover, the three wellbeing variables (Overall, Mental and Physical wellbeing), also proved to drive PIFP with big impact in both countries. This indicates that variables such as Satisfaction and subjective Wellbeing can be seen as holistic responses comparable with PIFP (as defined in this study). Supporting this argument, Ares et al. (2015) studied consumer's association with wellbeing in a food-related context, and found that wellbeing was mainly associated with calmness, happiness and satisfaction [7]. In a qualitative study on consumer reflections on post-ingestive sensations, Duerlund et al. (2019) found pleasure and feeling good to be part of consumers' perceptions of post-ingestive wellbeing [32]. Furthermore, post-ingestive Psychological Wellbeing was found to be part of Food Satisfaction, in a study from Andersen et al. (2017) [43]. Furthermore, in a study from Sulmont-Rossé et al. (2019), the effects of consuming food included emotional aspects when conceptualizing feeling good. Particularly, consumers refereed to positive emotions such as happy, enthusiastic and satisfied [5]. The present results together with above-mentioned research, supports the link and associations between PIFP, Satisfaction and Wellbeing.

Interestingly, all these holistic variables; Satisfaction, Overall wellbeing, Mental wellbeing, Physical wellbeing and PIFP, demonstrated to either significantly increase or stay the same over three hours in China, but conversely significantly decreased in Denmark. Hence, we here see a cultural difference in the development of these holistic sensations over time, specifically with a longer-lasting effect in China compared to Denmark. This could indicate a somewhat disconnect between food and the holistic concepts in China compared to Denmark. As demonstrated by Sulmont-Rossé et al. (2019), China associated 'feeling good' with emotional and hedonic dimensions rather than to specific food or beverages items, whereas Western countries more often associated it with specific food items [5]. This might help explain the present differences in the development of holistic sensations in China compared to Denmark.

PIFP was in this study considered and defined as a "subjective conscious sensation of pleasure and joy experienced after eating", with emphasis on a rewarding sensation of joy lasting after eating rather than reward (liking or desire to eat) whilst eating. PIFP thus included aspects of joy and pleasure lasting after eating with a memory of the food eaten. Memory for recent eating and Food Pleasure are linked. We can for instance recall from memory how enjoyable a food was, which then can also influence the additional after effects from eating [56]. Hence, we cannot be entirely sure how consumers rated PIFP, whether they evaluated joy right NOW relative to the food they ate, or joy with the food WHEN they ate it. Seen in a Denmark-China perspective, and interpreting the fact that PIFP in Denmark decreased over three hours, something could indicate that Danish consumers focus a lot on the sensation they have in their body right now, i.e., that they sense more joy with the food right after it is eaten than after three hours. In Denmark, it seems that before consumers experience PIFP, the food must meet some physical bodily needs. These are often related to the food's ability to keep one sated, but it must also ensure to keep one energized for daily chores and to maintain one's concentration. As time passes, the food is 'put to the test' by the consumer in order to keep his/her energy level up and to facilitate concentration. This could explain why Energized and Concentration become explanatory for PIFP in Denmark three hours after eating. In China, on the other hand, the ratings of PIFP were constant over time, which indicates that they do not have the same requirements for food to satisfy some physical bodily needs, but that food has a more mental effect longer term. This indicates that the Chinese consumes rated PIFP as the joy with the food when they ate it, and that this joy is not vulnerable to change over time, maybe because it is not tied to the same requirements for food to meet certain needs as in Denmark.

The fact that different sensations influenced PIFP in China and in Denmark in this study, suggest some cross-cultural differences in the way we perceive food and the association we have with food. Understanding the differences and elucidating 'why' is multi-faceted. One thing is for sure, cross-cultural research is becoming more important as a result of rapid globalization, also within Sensory and Consumer Science [3,55,57]. As we know, culture is one of the main factors underlying how we assess and choose food [1], and the present results can illuminate what is important when exposing consumers to new products in new markets. It is clear that more research is warranted to elucidate cross-cultural differences in our eating behavior and perception of food.

As mentioned by Møller (2015), Food Reward comes in many and different disguises, both in terms of immediate liking and motivational wanting of a food whilst eating, but also as a longer lasting feeling of for instance wellbeing after a meal. Both aspects are included and suggested to be valid components of Reward in Berridge and Robinson's powerful model of Reward components [12]. In this study, we situated and considered PIFP as part of the Affect category within reward with focus on conscious pleasure derived from eating. Quantifying joy and pleasure obtained from eating food as a Food Pleasure measure provides pertinent and relevant output, and, importantly, the results take us beyond the processes only active during the actual eating event [12,28].

4.3. Limitations

An important aspect to consider when interpreting the results from this cross-cultural comparative study between China and Denmark is the composition and nature of the participants in each country. Recruitment resulted in unequal gender distribution with particularly more males than females in Denmark. In China, gender was more equally distributed, but with a slight majority of female participants. Quota sampling, rather than convenience sampling, is recommended for comparing consumer perceptions across cultures [55]. Participant characteristics, therefore, also differed, with higher average weight and height in Denmark than in China. However, this, in general, reflects the average size of the people living in the two countries as well. Furthermore, participants in Denmark were recruited from a Sports Academy, possibly placing, even more than normal, emphasis on physical performance, which could be reflected in the results around PIFP.

Standardization between the served breakfasts meals in the two countries were attained with the same amounts served in Denmark and in China. With this said, some of the Danish participants commented on too small amounts, whereas some of the Chinese participants commented on too large amounts. This could, certainly, also be explained by the before-mentioned differences in weight and height for the participants, with larger males naturally needing greater amounts of food. Different yoghurts were used in each country in order to aid fresh produce and familiarity. However, a perspective to consider here is the familiarity in general for yoghurt as a breakfast meal. Familiarity could have affected PIFP or other ratings in unknown ways in this study. Participants, though, were recruited being likers of yoghurt for breakfast. That being said, commercially available yoghurts in China consists mainly of flavored yoghurts, whereas Danish people perhaps are more used to non-flavored yoghurts. Additionally, our results, whilst representative for young consumers, may differ when considered in relation to the general population in both countries. Furthermore, whilst our results apply to eating breakfast, results may vary if applied in other contexts such as lunch or dinner. Therefore, more confirmatory research studies are needed with differing population groups, as well as application in different contexts, in order to establish the generalizability of the results from the present study.

In cross-cultural research, it is important to acknowledge possible cultural differences in response style, because culture is known to have a strong influence on how people use scales to answer questions [55,58]. Especially extreme response style, middle response style, and acquiescence response style are common response styles to acknowledge and be aware of [40,55,59,60]. An extreme response style can be characterized by the tendency to use the end-points more frequently than for instance acquiescence response style characterized by the tendency to respond with agreement or affirmation. Especially Likert scales and semantic differential scales (i.e., good to bad, dirty to clean) are vulnerable to response style tendencies. Furthermore, scale anchors indicating the degree of agreement or importance are more vulnerable to acquiescence response style in cultures with high Power distance and Collectivism [40]. In this study, we did not use Likert scales indicating the degree of agreement or importance. We used intensity scales with semantic similar anchor points (i.e., "not at all energetic" to "extremely energetic"). Different approaches are used to investigate response style tendencies, where the most commonly used are frequency or percentage of answers in particular categories of a scale [40,59,60], or a multisample confirmatory factor analysis if scales are to be directly compared [55,61,62]. Calculating percentages of answers in the top 25% and bottom 25% of the intensity scales in both countries, showed equal distribution of answers in both low, middle, and high end of the intensity scale for both countries. No formal multisample confirmatory factor analysis was conducted to test direct scale comparability, since we did not directly compare the scales. All analyses were conducted separately for each country with no direct comparisons in raw data. Actions can be taken to account for difference in response style, e.g., standardization and/or centering of data. However, it should be taken into account that standardization can potentially remove some of the true differences among cultures [40]. It is, nevertheless, important to acknowledge possible hidden differences in response styles for the two countries, and in general consider potential cultural bias in methodology when doing cross-cultural research [36,55]. Unknown differences in response style and understanding of terms might have influenced the results in these studies [60], but could very well also be a true difference in cultures for the post-ingestive experiences.

4.4. Research Contribution and Future Perspectives

The present research findings contribute new knowledge about the cultural differences between China and Denmark. Specifically, we contribute to a better understanding of the differences in consumers' post-ingestive experiences including PIFP and its drivers in each country. This knowledge helps to unravel and recognize some aspects of cultural differences when addressing food culture and eating behaviors. A key contribution to the cross-cultural research field, and a major advantage of this study, is the actual serving of food and evaluation of consumers' perceptive responses to actual intake. The two studies thus provide centrally collected data in each country with native consumers and in-country behavior. Other cross-cultural research often contributes with knowledge collected online with no serving or eating of food, and/or with expatriates as consumers who no longer reside in their original culture [2].

The present results and knowledge can support researchers as well as the food industry to a better understanding of the cultural differences between Denmark and China. Particularly, food companies could use this knowledge when applying and introducing new products to new markets and cultures. Furthermore, continuing conducting cross-cultural research is important and relevant. The contribution to the scientific field and to fundamental research is highly important, since we still have limited knowledge as to how we perceive food and associate food between cultures in today's rising globalization.

Future perspectives for this particular research focus within PIFP and in the Sensory and Consumer Science area should include alternative ways to measure and collect data in cross-cultural research. Examples could include the application of more implicit measures, such as behavioral, indirect approaches to scaling, and biometric measures, which would provide usable data. Targeting Food Pleasure and other aspects of Food Reward from different angles including both explicit and implicit measures could add to the validation and application of the concept. This would require multidisciplinary approaches combining Sensory and Consumer science with, e.g., neuroscience, biology, psychology and human nutrition utilizing a multimodal approach.

5. Conclusions

This cross-cultural research study aimed to compare Danish and Chinese consumers in their post-ingestive drivers of Post-Ingestive Food Pleasure (PIFP). The work involved conducting two in-country consumer studies in Denmark and in China, respectively, measuring self-reported PIFP after eating together with other post-ingestive sensations. Key results revealed that for both Danish and Chinese consumers, the post-ingestive variables Satisfaction, Mental wellbeing, Overall wellbeing and Physical wellbeing were highly influential on PIFP. This suggests a somewhat similar structure in the cultural constructs of PIFP in both countries. However, the results also revealed that disjointed and different dimensions of post-ingestive variables drove consumers' PIFP in each country. Danish consumers perceived appetite-related sensations such as Satiety, Hunger, Desire-to-eat and In-need-of-food to be influential on PIFP, which was not the case for the Chinese consumers. On the contrary, in China, the more vitality- and energy-related post-ingestive variables such as Relaxation, Energized and Concentration were found to be drivers of PIFP post intake. These results resonate with our research hypothesis showing distinct subtleties in the cultural constructs of PIFP in Denmark and in China.

These findings serve relevance to various areas. The contribution to science is highly important, since we still have limited knowledge as to how we perceive and associate food between cultures. Moreover, the food industry can indeed benefit from taking into account consumers' both physical and mental sensations when designing products for different markets, including knowing that cultural differences exist as to which sensations drive Food Pleasure after eating. Investigating the post-ingestive consumer experience thus matters, because it provides valuable novel insights into the elaborated eating experience, which goes beyond the momentary eating event.

Cross-cultural differences exist in the way we perceive food and the association we have with food, and the constructs of PIFP also vary with consumers' cultural background. Understanding and explaining 'why' these differences exist is multi-faceted and they do not have a single answer. One thing is certain, cross-cultural research becomes extra relevant in the context of rapid globalization, especially within sensory and consumer science. Undoubtedly, more research is needed to elucidate cross-cultural differences in our eating behavior and perception of food.

Author Contributions: Conceptualization, M.D., B.V.A., R.C.K.C., D.V.B.; methodology, M.D., B.V.A.; software, M.D.; validation, M.D., B.V.A.; formal analysis, M.D., B.V.A.; investigation, M.D.; obtaining ethics, M.D., K.W., R.C.K.C.; resources, M.D., K.W., R.C.K.C.; data curation, M.D.; writing—original draft preparation, M.D.; writing—review and editing, M.D., B.V.A., K.W., R.C.K.C., D.V.B.; visualization, M.D.; supervision, B.V.A., R.C.K.C., D.V.B.; project administration, M.D., B.V.A, D.V.B.; funding acquisition, B.V.A., D.V.B. All authors have read and agreed to the published version of the manuscript.

Funding: This work was supported by the university partnership Denmark-China, Sino Danish Centre (SDC), within the 'Food and Health Research Theme', Aarhus, Denmark and the Aarhus University Graduate School of Science and Technology (GSST), Aarhus, Denmark.

Acknowledgments: The authors would like to thank Yu Xinyang for helping with recruitment of Chinese participants, as well as Ye Junyan, Wang Lingling, and Zheng Hong for practical help with conducting of the study at UCAS, China. We would also like to thank Thomas Lykke Pedersen, event and relation coordinator in the Chinese office at the Sino-Danish Centre, China, for help with internal coordination and communication. Furthermore, we would like to thank Aimei Wang, Joe Hoefler Tyler, and Qian Janice Wang for help with translations and back-translations of the questionnaire. For practical help with conducting the study in Denmark, we would like to thank Britta Hansen, Birgitte Foged, and Kamilla Hall Kragelund, as well as Hanne Gyntzel for helping with coordination, recruitment, and internal communication at the Ollerup Sports Academy. Without all these wonderful people, we could not have done the research.

Conflicts of Interest: The authors declare no conflict of interest.

References

1. Rozin, P. Cultural Approaches to Human Food Preferences. In *Nutritional Modulation of Neural Function*; Morley, J.E., Sterman, M.B., Walsh, J.H., Eds.; Academic Press: San Diego, CA, USA, 1988; pp. 137–153.
2. Goldman, A. Important Considerations in Cross-Cultural Sensory and Consumer Research — Introductory Overview. *Food Qual. Prefer.* **2006**, *17*, 646–649. [CrossRef]
3. Meiselman, H.L. The Future in Sensory/Consumer Research: Evolving to a Better Science. *Food Qual. Prefer.* **2013**, *27*, 208–214. [CrossRef]
4. Rodrigues, H.; Otterbring, T.; Piqueras-Fiszman, B.; Gómez-Corona, C. Introduction to Special Issue on Global Perspectives on Sensory and Consumer Sciences: A Cross-Cultural Approach. *Food Res. Int.* **2019**, *116*, 135–136. [CrossRef] [PubMed]
5. Sulmont-Rossé, C.; Drabek, R.; Almli, V.L.; Van Zyl, H.; Patricia, A.; Kern, M.; Mcewan, J.A.; Ares, G. A Cross-Cultural Perspective on Feeling Good in the Context of Foods and Beverages. *Food Res. Int.* **2019**, *115*, 292–301. [CrossRef]
6. Ares, G.; Giménez, A.; Vidal, L.; Zhou, Y.; Krystallis, A.; Tsalis, G.; Symoneaux, R.; Cunha, L.M.; de Moura, A.P.; Claret, A.; et al. Do We All Perceive Food-Related Wellbeing in the Same Way? Results from an Exploratory Cross-Cultural Study. *Food Qual. Prefer.* **2016**, *52*, 62–73. [CrossRef]
7. Ares, G.; de Saldamando, L.; Giménez, A.; Claret, A.; Cunha, L.M.; Guerrero, L.; de Moura, A.P.; Oliveira, D.C.R.; Symoneaux, R.; Deliza, R. Consumers' Associations with Wellbeing in a Food-Related Context: A Cross-Cultural Study. *Food Qual. Prefer.* **2015**, *40*, 304–315. [CrossRef]
8. Kim, H.J.; Chung, S.J.; Kim, K.O.; Nielsen, B.; Ishii, R.; O'Mahony, M. A Cross-Cultural Study of Acceptability and Food Pairing for Hot Sauces. *Appetite* **2018**, *123*, 306–316. [CrossRef]
9. Kim, S.H.; Petard, N.; Hong, J.H. What Is Lost in Translation: A Cross-Cultural Study to Compare the Concept of Nuttiness and Its Perception in Soymilk among Korean, Chinese, and Western Groups. *Food Res. Int.* **2018**, *105*, 970–981. [CrossRef]
10. Lee, H.S.; Lopetcharat, K. Effect of Culture on Sensory and Consumer Research: Asian Perspectives. *Curr. Opin. Food Sci.* **2017**, *15*, 22–29. [CrossRef]
11. Torrico, D.D.; Fuentes, S.; Gonzalez Viejo, C.; Ashman, H.; Dunshea, F.R. Cross-Cultural Effects of Food Product Familiarity on Sensory Acceptability and Non-Invasive Physiological Responses of Consumers. *Food Res. Int.* **2019**, *115*, 439–450. [CrossRef]
12. Berridge, K.C.; Robinson, T.E. Parsing Reward. *Trends Neurosci.* **2003**, *26*, 507–513. [CrossRef]
13. Berridge, K.C. "Liking" and "Wanting" Food Rewards: Brain Substrates and Roles in Eating Disorders. *Physiol. Behav.* **2009**, *97*, 537–550. [CrossRef] [PubMed]
14. Berridge, K.C.; Kringelbach, M.L. Affective Neuroscience of Pleasure: Reward in Humans and Animals. *Psychopharmacology* **2008**, *199*, 457–480. [CrossRef] [PubMed]
15. Mela, D.J. Eating for Pleasure or Just Wanting to Eat? Reconsidering Sensory Hedonic Responses as a Driver of Obesity. *Appetite* **2006**, *47*, 10–17. [CrossRef]
16. Oustric, P.; Thivel, D.; Dalton, M.; Beaulieu, K.; Gibbons, C.; Hopkins, M.; Blundell, J.; Finlayson, G. Measuring Food Preference and Reward: Application and Cross-Cultural Adaptation of the Leeds Food Preference Questionnaire in Human Experimental Research. *Food Qual. Prefer.* **2020**, *80*, 103824. [CrossRef]
17. Pool, E.; Sennwald, V.; Delplanque, S.; Brosch, T.; Sander, D. Measuring Wanting and Liking from Animals to Humans: A Systematic Review. *Neurosci. Biobehav. Rev.* **2016**, *63*, 124–142. [CrossRef]

18. Arumäe, K.; Kreegipuu, K.; Vainik, U. Assessing the Overlap between Three Measures of Food Reward. *Front. Psychol.* **2019**, *10*, 1–11. [CrossRef]
19. Rogers, P.J.; Hardman, C.A. Food Reward. What It Is and How to Measure It. *Appetite* **2015**, *90*, 1–15. [CrossRef]
20. Ruddock, H.K.; Field, M.; Hardman, C.A. Exploring Food Reward and Calorie Intake in Self-Perceived Food Addicts. *Appetite* **2017**, *115*, 36–44. [CrossRef]
21. Finlayson, G.; King, N.; Blundell, J. Is It Possible to Dissociate "liking" and "Wanting" for Foods in Humans? A Novel Experimental Procedure. *Physiol. Behav.* **2007**, *90*, 36–42. [CrossRef]
22. Miguet, M.; Fillon, A.; Khammassi, M.; Masurier, J.; Julian, V.; Pereira, B.; Lambert, C.; Boirie, Y.; Duclos, M.; Blundell, J.; et al. Appetite, Energy Intake and Food Reward Responses to an Acute High Intensity Interval Exercise in Adolescents with Obesity. *Physiol. Behav.* **2018**, *195*, 90–97. [CrossRef] [PubMed]
23. Dalton, M.; Finlayson, G. Psychobiological Examination of Liking and Wanting for Fat and Sweet Taste in Trait Binge Eating Females. *Physiol. Behav.* **2014**, *136*, 128–134. [CrossRef] [PubMed]
24. Charlot, K.; Malgoyre, A.; Bourrilhon, C. Proposition for a Shortened Version of the Leeds Food Preference Questionnaire (LFPQ). *Physiol. Behav.* **2019**, *199*, 244–251. [CrossRef] [PubMed]
25. Ziauddeen, H.; Subramaniam, N.; Cambridge, V.C.; Medic, N.; Farooqi, I.S.; Fletcher, P.C. Studying Food Reward and Motivation in Humans. *J. Vis. Exp.* **2014**, *85*, 1–10. [CrossRef] [PubMed]
26. Hardman, C.A.; Herbert, V.M.B.; Brunstrom, J.M.; Munafò, M.R.; Rogers, P.J. Dopamine and Food Reward: Effects of Acute Tyrosine/Phenylalanine Depletion on Appetite. *Physiol. Behav.* **2012**, *105*, 1202–1207. [CrossRef]
27. Piech, R.M.; Pastorino, M.T.; Zald, D.H. All I Saw Was the Cake. Hunger Effects on Attentional Capture by Visual Food Cues. *Appetite* **2010**, *54*, 579–582. [CrossRef]
28. Møller, P. Satisfaction, Satiation and Food Behaviour. *Curr. Opin. Food Sci.* **2015**, *3*, 59–64. [CrossRef]
29. Craig, A.D. How Do You Feel? Interoception: The Sense of the Physiological Condition of the Body. *Nat. Rev. Neurosci.* **2002**, *3*, 655–666. [CrossRef]
30. Simmons, W.K.; DeVille, D.C. Interoceptive Contributions to Healthy Eating and Obesity. *Psychology* **2017**, *17*, 106–112. [CrossRef]
31. Stevenson, R.J.; Mahmut, M.; Rooney, K. Individual Differences in the Interoceptive States of Hunger, Fullness and Thirst. *Appetite* **2015**, *95*, 44–57. [CrossRef]
32. Duerlund, M.; Andersen, B.V.; Grønbeck, M.S.; Byrne, D.V. Consumer Reflections on Post-Ingestive Sensations. A Qualitative Approach by Means of Focus Group Interviews. *Appetite* **2019**, *142*. [CrossRef] [PubMed]
33. Yeomans, M.R. Measuring Appetite and Food Intake. In *Methods in Consumer Research, Volume 2: Alternative Approaches and Special Applications*; Ares, G., Varela, P., Eds.; Woodhead Publishing: Sawston, UK, 2018; pp. 119–149.
34. Duerlund, M.; Andersen, B.V.; Byrne, D.V. Dynamic Changes in Post-Ingestive Sensations after Consumption of a Breakfast Meal High in Protein or Carbohydrate. *Foods* **2019**, *8*. [CrossRef] [PubMed]
35. Act on Research Ethics Review of Health Research Projects. Available online: https://www.retsinformation.dk/Forms/r0710.aspx?id=192671 (accessed on 8 May 2020).
36. Buil, I.; De Chernatony, L.; Martínez, E. Methodological Issues in Cross-Cultural Research: An Overview and Recommendations. *J. Target. Meas. Anal. Mark.* **2012**, *20*, 223–234. [CrossRef]
37. Murray, M.; Vickers, Z. Consumer Views of Hunger and Fullness. A Qualitative Approach. *Appetite* **2009**, *53*, 174–182. [CrossRef]
38. Andersen, B.V.; Hyldig, G. Consumers' View on Determinants to Food Satisfaction: A Qualitative Approach. *Appetite* **2015**, *95*, 9–16. [CrossRef]
39. Andersen, B.V.; Hyldig, G. Food Satisfaction: Integrating Feelings before, during and after Food Intake. *Food Qual. Prefer.* **2015**, *43*, 126–134. [CrossRef]
40. Harzing, A. Response Styles in Cross-National Survey Research: A 26-Country Study. *Int. J. Crosscultural Manag.* **2006**, *6*, 243–266. [CrossRef]
41. Arnold, B.R.; Smith, J.L. Methodologies for Test Translation and Cultural Equivalence. In *Handbook of Multicultural Mental Health: Assessment and Treatment of Diverse Populations: Second Edition*; Yamada, A.-M., Paniagua, F.A., Eds.; Elsevier Inc.: Amsterdam, The Netherlands, 2013; pp. 243–263. [CrossRef]
42. Cohen, J. *Statistical Power Analysis for the Behavioral Sciences*, 2nd ed.; Lawrence Erlbaum Associates: New York, NY, USA, 1988.

43. Andersen, B.V.; Mielby, L.H.; Viemose, I.; Bredie, W.L.P.; Hyldig, G. Integration of the Sensory Experience and Post-Ingestive Measures for Understanding Food Satisfaction. A Case Study on Sucrose Replacement by Stevia Rebaudiana and Addition of Beta Glucan in Fruit Drinks. *Food Qual. Prefer.* **2017**, *58*, 76–84. [CrossRef]
44. Wold, S.; Johansson, E.; Cocchi, M. PLS—Partial Least Squares Analysis. In *3D QSAR in Drug Design. Theory Methods and Applications*; Kubinyi, H., Ed.; KLUWER; ESCOM: Ludwigshafen, Germany, 2000; pp. 523–550.
45. Rossini, K.; Verdun, S.; Cariou, V.; Qannari, E.M.; Fogliatto, F.S. PLS Discriminant Analysis Applied to Conventional Sensory Profiling Data. *Food Qual. Prefer.* **2012**, *23*, 18–24. [CrossRef]
46. Eriksson, L.; Byrne, T.; Johansson, E.; Trygg, J.; Vikström, C. *Multi- and Megavariate Data Analysis Basic Principles and Applications*, 3rd ed.; Umetrics Academy: Umeå, Sweden, 2013.
47. Chong, I.G.; Jun, C.H. Performance of Some Variable Selection Methods When Multicollinearity Is Present. *Chemom. Intell. Lab. Syst.* **2005**, *78*, 103–112. [CrossRef]
48. Addinsoft. *XLSTAT Statistical and Data Analysis Solution*; Addinsoft: Long Island, NY, USA, 2019.
49. Boelsma, E.; Brink, E.J.; Stafleu, A.; Hendriks, H.F.J. Measures of Postprandial Wellness after Single Intake of Two Protein-Carbohydrate Meals. *Appetite* **2010**, *54*, 456–464. [CrossRef] [PubMed]
50. Ma, G. Food, Eating Behavior, and Culture in Chinese Society. *J. Ethn. Foods* **2015**, *2*, 195–199. [CrossRef]
51. Sørensen, S.; Markedal, K.E.; Sørensen, J.C. Food, Nutrition, and Health in Denmark (Including Greenland and Faroe Islands). In *Nutritional and Health Aspects of Traditional and Ethnic Foods of Nordic Countries*; Andersen, V., Bar, E., Wirtanen, G., Eds.; Elsevier Inc.: Amsterdam, The Netherlands, 2018; pp. 99–125. [CrossRef]
52. Hofstede, G. Dimensionalizing Cultures: The Hofstede Model in Context. *Online Readings Psychol. Cult.* **2011**, *2*, 1–26. [CrossRef]
53. Hofstede, G. *Culture's Consequences: International Differences in Work-Related Values*; Sage: Beverly Hills, CA, USA, 1980.
54. Hofstede Insights. Available online: https://www.hofstede-insights.com/product/compare-countries/ (accessed on 8 May 2020).
55. Ares, G. Methodological Issues in Cross-Cultural Sensory and Consumer Research. *Food Qual. Prefer.* **2018**, *64*, 253–263. [CrossRef]
56. Higgs, S. Cognitive Processing of Food Rewards. *Appetite* **2016**, *104*, 10–17. [CrossRef]
57. Ares, G.; Giménez, A.; Deliza, R. Methodological Approaches for Measuring Consumer-Perceived Well-Being in a Food-Related Context. In *Methods in Consumer Research, Volume 2: Alternative Approaches and Special Applications*; Ares, G., Varela, P., Eds.; Woodhead Publishing: Sawston, UK, 2018; pp. 183–200.
58. Johnson, T.; Kulesa, P.; Cho, Y.I.; Shavitt, S. The Relation between Culture and Response Styles: Evidence from 19 Countries. *J. Cross. Cult. Psychol.* **2005**, *36*, 264–277. [CrossRef]
59. Batchelor, J.H.; Miao, C. Extreme Response Style: A Meta-Analysis. *J. Organ. Psychol.* **2016**, *16*. [CrossRef]
60. Rammstedt, B.; Danner, D.; Bosnjak, M. Acquiescence Response Styles: A Multilevel Model Explaining Individual-Level and Country-Level Differences. *Pers. Individ. Dif.* **2017**, *107*, 190–194. [CrossRef]
61. Cheung, G.W.; Rensvold, R.B. Assessing Extreme and Acquiescence Response Sets in Cross-Cultural Research Using Structural Equations Modeling. *J. Cross. Cult. Psychol.* **2000**, *31*, 187–212. [CrossRef]
62. O'Sullivan, C.; Scholderer, J.; Cowan, C. Measurement Equivalence of the Food Related Lifestyle Instrument (FRL) in Ireland and Great Britain. *Food Qual. Prefer.* **2005**, *16*, 1–12. [CrossRef]

Article

Consumer Perception of Food Quality and Safety in Western Balkan Countries: Evidence from Albania and Kosovo

Rainer Haas [1], Drini Imami [2,3], Iliriana Miftari [4], Prespa Ymeri [4], Klaus Grunert [5] and Oliver Meixner [1,*]

1 Department of Economics and Social Sciences, Institute of Marketing & Innovation, University of Natural Resources and Life Sciences, 1180 Vienna, Austria; rainer.haas@boku.ac.at
2 Faculty of Economics and Agribusiness, Agricultural University of Tirana, 1025 Tirana, Albania; dimami@ubt.edu.al
3 Faculty of Tropical Agri Sciences, Czech University of Life Sciences Prague and CERGE EI, 16500 Prague, Czech Republic
4 Department of Agricultural Economics, Faculty of Agriculture and Veterinary, University of Prishtina, 10000 Prishtina, Kosovo; Iliriana.Miftari@uni-pr.edu (I.M.); prespaymerii@gmail.com (P.Y.)
5 Centre for Research on Customer Relations in the Food Sector MAPP, Aarhus University, 8210 Aarhus, Denmark; klg@mgmt.au.dk
* Correspondence: oliver.meixner@boku.ac.at; Tel.: +43-1-47654-73515

Abstract: Domestic food markets are of significant importance to Kosovar and Albanian companies because access to export markets is under-developed, partly as a result of the gaps in food safety and quality standards. Kosovar and Albanian consumers' use of food safety attributes and their evaluation of the quality of domestic food versus imported food are the research objectives of this study. The paper is based on a structured consumer survey of 300 Kosovars and 349 Albanians analyzing their perceptions of issues related to food safety and quality, measured through two respective batteries of items using a 5-point Likert scale. We used the *t*-test to identify differences between populations, correlation analysis and the bootstrapping method. Despite the prevalent problems with food safety, consumers in both countries consider domestic food to be safer as well as of higher quality than imported products. Kosovars are more likely than Albanians to perceive domestic food products to be significantly better than imported products. Female and better educated consumers use information related to food safety more often. Expiry date, domestic and local origin, and brand reputation are the most frequently used safety and quality cues for both samples. International food standards such as ISO or HACCP are less frequently used as quality cues by these consumer groups. It is important to strengthen the institutional framework related to food safety and quality following best practices from EU countries.

Keywords: food safety; food quality; Kosovar consumers; Albanian consumers; Western Balkan countries; bootstrapping

Citation: Haas, R.; Imami, D.; Miftari, I.; Ymeri, P.; Grunert, K.; Meixner, O. Consumer Perception of Food Quality and Safety in Western Balkan Countries: Evidence from Albania and Kosovo. *Foods* **2021**, *10*, 160. https://doi.org/10.3390/foods10010160

Received: 4 December 2020
Accepted: 11 January 2021
Published: 14 January 2021

1. Introduction

Food quality can be defined as "fitness for consumption" and as "the requirements necessary to satisfy the needs and expectations of the consumer" [1] (p. 4). Food quality consists of an objective and a subjective dimension [2], and food producers are only successful if they are able to combine these two dimensions by meeting consumers' expectations (i.e., subjective quality) and transforming these into specific physical attributes (i.e., objective quality). Subjective food quality often refers to process attributes such as organic production or attention to animal welfare, but it also covers attributes such as taste or price [2]. Objective quality covers the chemical, microbiological, and physical attributes of a food product. For example, the origin of food influences a subjective quality perception by consumers and has an objective quality difference based on the "terroir", the unique constellation of micro climate, soils, precipitation leading to a unique composition of nutrients in foods of different origin [3].

"Quality cues are defined as information stimuli that are related to the quality of the product and can be ascertained by the consumer through the senses prior to consumption" [4] (p. 312). To form a quality judgement, consumers can draw from single quality cues or a combination of cues [5]. When a desirable product attribute is unknown, consumers infer it from quality cues, which can be any evidence or information that consumers believe may be predictive of the desired attribute [6]. The physical characteristics of the product can also be used as cues: for example, the color or other aspects of the product's appearance may be used as an indicator of the product quality [7]. The fact that some quality cues can be ascertained prior to consumption means that these quality cues are either search (e.g., brand name, food safety certificates) or credence attributes (e.g., food certificates, organic production); this differentiation between search, experience and credence attributes goes back to the early 1970s [8–10]. Search attributes are attributes that can be inspected by the consumer in the store or the supermarket and are used as quality cues to predict the quality of the food product [11].

Food safety, from a consumer's perspective, is part of the perceived (subjective) quality of a food product and is considered to be an inherent part of food quality [1]. Food safety is a credence attribute, as it is unknown before the purchase whether the food is in fact safe; safety cannot be ascertained by experience after purchase either [2], unless, of course, the consumer becomes sick and the sickness can be attributed to the food with some certainty.

West Balkan Countries (WBCs) belong to the countries with higher corruption and lowest GDP (Gross Domestic Product) in Europe. Insofar it is not surprising that their national food safety control systems are characterized by incoherent legislation and a lack of human and capital resources, as well as reliable data [12], resulting in weak law enforcement. Even though legislation improved food safety regulations during the last years (also due to the growing support and pressure emerging from the EU integration process as well as the request of international food companies requiring certifications of food safety standards) [13], these problems still exist and create real and perceived safety risks for consumers.

Developing and transition countries face serious challenges related to food safety due to weak animal disease controls that result in a higher prevalence of endemic infectious animal diseases [14]. Problems in the agricultural health and food safety systems have been identified by several studies, especially in the meat [15,16] and dairy sector [17,18]. Brucellosis has been a major health concern in Albania and Kosovo [17,19], and aflatoxin in maize for feed was reported in other WBCs [20,21]. Reportedly, aflatoxin also represents a serious problem for small dairy farmers in Albania and Kosovo [22,23], and there are similar concerns in Serbia [18]. In general, mycotoxins and aflatoxins are reported to be of high significance concerning food safety in these transition economies [24]. Other studies report high levels of contamination of plants and soil with cadmium, lead, and other heavy metals in these countries [25].

Public agencies are responsible for ensuring food safety enforcement; however, their capacity in the context of developing or transition countries (like Albania or Kosovo) is limited, conditioned by the weak institutional framework and corruption [26]. This finding is especially relevant because consumers tend to derive food safety based on their trust into regulators and food manufacturers [27]. Due to the lack of capacity of public agencies Albanian consumers trust more in retailers than in governmental regulators for guaranteeing food safety [15]. Due to this context Albanian consumers also prefer to buy food directly from producers [28]. In Albania and Kosovo, most farmers across agri-food sectors lack information or awareness related to food safety standards. More specifically, in the case of the livestock sector (which is also the focus of this paper and also which among most sensitive sectors related to food safety), most farmers lack information about which institutions are in charge of basic standards control related to food safety or animal welfare—lack of awareness about standards results in standards non-compliance [17,29].

Considering these food safety issues, consumers in WBCs unsurprisingly perceive food safety as critical. However, consumers are in general unfamiliar with international food safety standards [15,30]. Zaric et al. [31] found that quality is by far the most important factor influencing consumer purchasing behavior in Serbia. A previous study found that Kosovar consumers perceive origin, food safety certificates, and brands as important means to identify food safety and quality [32]. Brand reputation is a widely used cue to reduce the perceived risks associated with food purchases. Strong brand reputation leads to higher perceived quality [5,33]. Consumers also use provenance of a food product—may it be regional, domestic, or international—as an indicator for quality [34–36] but "indication of origin may only become a signal of enhanced quality if the source of origin is associated with higher food safety or quality" [37]. A study by Miftari [38] showed that a majority of Kosovar consumers had a positive bias towards domestic (versus foreign) dairy products, which points towards a prevalence of consumer patriotism concerning food. A qualitative means-end chain analysis of Croatian consumers concerning their motives for buying traditional food products found that domestic origin was an important quality cue that consumers connected with the absence of a risk to personal health. The Croatian consumers even associated a domestic origin with better flavor/taste, and connected the attribute "domestic origin" with the use of traditional ingredients and with trustworthy producers [39]. Food labels and certificates can act as quality cues if they are available, are understood by consumers, and are regarded as trustworthy [2]. Verbeke and Ward [36] found that there is high consumer interest in quality seals and expiration dates as quality cues. Consumers also use organic production (a specific form of a food certificate) as a quality cue. The extent to which they use organic food certificates as a quality cue is influenced by their environmental concerns, their trust in local producers in the country and their ethnocentrism [40]. According to a previous consumer study in Albania, most consumers consider factors surrounding health to be the most important dimension of the organic products, while the impact of organic food production on the environment does not appear to be important at all [41]. Comprehensive information about ingredients and the origin of raw materials is an important cue to create consumer trust in brands, based on a broad study of German consumers [42]. In the absence of trustworthy third-party certification of food products, personal trust between customer and seller can compensate as a quality cue. The reputation or personal knowledge of the seller or manufacturer of a food product is used to judge the quality and safety of food products [43,44]. For example, the majority of respondents of a study analyzing consumer behavior in Albania (purchasing lamb meat) answered that their main source of trust is knowing the butcher or seller [16].

Rising consumer concern about food safety has led to an increased demand for standards related to quality assurance such as ISO 9000 or HACCP, which focus on objective food quality [45,46]. Despite the fact that HACCP and ISO 9000 have been implemented in the market for decades, most consumers do not know about them [47]. This is not surprising, because they have been established as business-to-business standards, often treated by retailers as an insurance policy to protect them against food scandals [48]. WBCs are known for weak governmental institutions. This is the reason why consumers in Albania trust more the retailer concerning food safety [15] and if the option is available, consumers tend to buy directly from the farmer, because they believe the food to be of higher quality and safer [28]. For example, the biggest knowledge gaps among food handlers in Serbia are related to temperature control and sources of food contamination [49]. In response to a possible EU accession in the future, WBCs are in the process of implementing and harmonizing their food quality and safety standards, and it is also necessary for them to fulfill these standards for trade with the EU.

Perception of food safety might also be affected by socio-demographic factors such as gender and education. In particular, female and better educated persons are assumed to pay more attention to food safety and related issues [16,50]. Gkana and Nychas [51], in a study of Greek consumers, found no differences in gender and age in respect to perceptions of food safety, but a higher education level resulted in a higher awareness and knowledge of food safety issues. A study of Turkish consumers reported that households with higher income and higher education levels were more interested in food safety, and so were female and older consumers [52]. Another study of Turkish consumers found a positive influence of education on knowledge of, attitudes towards, and practices related to food safety [53].

In a nutshell, previous studies looked at singular aspects, how consumers in WBCs evaluate food quality and safety. To our knowledge, none of the previous studies collected data in more than one country. This study is the first one to deliver comparative results for Albania and Kosovo. The consumer survey is based on the evaluation of milk and cheese products, for which food safety is very sensitive. Accordingly, milk and cheese products serve as a representative food product category to measure consumers' perception of quality and safety standards.

2. Materials and Methods

A quantitative, structured survey with urban consumers from Albania and Kosovo was conducted in 2019, with the data being collected in Prishtina and Tirana. These are the capital and largest cities of the respective countries (home to almost 1/3 of the population), where purchasing power is concentrated, and as such, the findings are important for the industry. Furthermore, both cities have high diversity of cultures and origins (most residents have migrated from all parts of the respective countries, including rural areas), and as such, the findings can be considered indicative also for the rest of the country. In April and May 2019 students, previously trained in taking face to face interviews, collected the data in Prishtina (Kosovo) and Tirana (Albania). The interviews were done outdoor on market places and public squares by using a convenience sample technique without quotas.

Within the questionnaire, two important batteries of items measured the perception of food safety and quality; the first had a 5-point Likert scale for the perception of food safety and quality for domestic milk and cheese, taken from the literature [54], while the second battery of items presented a list of quality and safety attributes. These attributes were taken from the literature review above and included brand reputation, expiry date, list of ingredients, organic production, food safety certificates, HACCP and ISO 9000, origin (local, domestic, foreign, or EU), and, finally, knowledge about the seller and the producer. We asked the consumers how often they bought cheese and how often they used these attributes related to quality and food safety. The answers measured the frequency, from 1 = never to 5 = always. Pre-test interviews to check the questionnaire were conducted in Pristina and Tirana with randomly selected consumers. The intended sample size was 300 interviews for each country. Based on the literature review, we formulated three hypotheses for further testing:

1. The quality and safety of imported food is rated higher than the quality and safety of domestic food. In both countries, the import of cheese is dominated by EU countries where the food safety standards tend to be much higher (imports from non-EU countries are negligible).
2. Kosovo (Prishtina) and Albania (Tirana) differ with respect to the frequency of use of cues related to food safety and quality.
3. Socio-demographic variables influence the perception of food safety and quality cues.

We used the *t*-test to identify differences between populations, and correlation analysis. In order to improve the reliability of the test results, we implemented the bootstrapping method [55]. Bootstrapping is usually used when a standard normal distribution cannot be guaranteed. It is "a computationally intensive method that involves repeatedly sampling from the data set and estimating the indirect effect in each resampled data set" [55] (p. 80). Developed in the 1970s [56], the bootstrapping method supports the reliability of analytical interpretations in quantitative research [57] and delivers valuable information about a possible distribution of analytical findings. There are a number of comparable publications available that incorporate bootstrapping into quantitative analysis to improve reliability to test, for example, the significance of differences in consumer evaluations between populations [58]. In accordance with recently published research about consumer perception and food safety, we used 5000 bootstrap samples and the 95% bootstrap confidence interval [59].

To test the hypotheses, we collected data in Prishtina (n = 300) and Tirana (n = 349). The overall sample size therefore amounts to 647 valid responses. The samples had a similar structure with respect to age and household size to statistical data for the populations of Kosovo and Albania. The samples were, however, biased with respect to the variables "gender" (more females than males in the Albanian sample), "education level" (the samples were more highly educated) and, to some extent, "income"—the structure of the income across the sample was not completely comparable to the general statistics, since the distribution of the income within the sample seems to deviate towards a higher income compared to the overall population (Table 1). Therefore, the following analysis will probably not deliver perfectly transferable results. Although this is not the main intention of this study, we have to take this into account when interpreting the following results.

Table 1. Socio-demographic variables of the sample and population.

		Sample % n = 300	Kosovo % [1]	Sample % n = 349	Albania % [2]
Age	15–24	9.4	15.0	11.5	11.0
	25–40	34.2	36.8	34.9	27.9
	41–54	35.9	26.3	25.4	24.6
	55–64	14.4	11.0	15.3	17.9
	65+	6.0	10.8	13.0	18.7
Gender	Male	42.0	50.3	35.5	49.9
	Female	58.0	49.7	64.5	50.1
Education	Basic to Middle School	24.0	66.5	10.5	58.8
	High School	30.3	20.6	35.2	29.9
	Higher Education, University	45.7	12.9	54.4	11.3
Household size	1–2 persons	4.7	9.3	22.1	28.2
	3 persons	7.7	8.3	16.9	19.5
	4 persons	22.3	15.8	29.4	26.8
	5 persons or more	65.3	66.6	31.7	25.5
Monthly income	Up to 500 EUR	14.0		42.7	
	501 to 800 EUR	32.7		41.9	
	801 to 1200 EUR	35.7		12.2	
	More than 1200 EUR	17.7		3.2	
	Mean (approx.) per capita		370		330

[1] Kosovo Agency of Statistics, 2017 (http://askdata.rks-gov.net); [2] Institute of Statistics for Albania (http://www.instat.gov.al/): Age, gender (total population): 2019; education: Census 2011, household budget survey 2018.

3. Results

Confirming the literature review, there are obviously certain problems with food standards in Kosovo and Albania. Therefore, we assumed that the safety and quality of imported food would be rated higher than the safety and quality of domestic food (tested on the example of cheese and fresh milk). To assess the perception of domestic food in comparison to imported food, the respondents were asked to specify their agreement with four statements concerning the safety and quality of domestic and imported foods, by means of a 5-point Likert scale (1 = total disagreement with the statement; 5 = total agreement with the statement). In addition, the respondents were asked whether they considered imported food to be of high quality.

The descriptive analysis clearly shows that in Kosovo (Prishtina), consumers considered domestic dairy food to be much safer and of much higher quality than imported dairy food (column M "Prishtina" in Table 2). Also, in Albania (column M "Tirana" in Table 2), the four comparison statements (domestic vs. imported dairy food) on average received an agreement score of more than three (i.e., a greater preference for domestic dairy food). However, the values are much lower than for the Kosovo sample and much closer to the mid-value of three (3 = "neither agree, nor disagree with statement", which should be interpreted as a perception of no difference between domestic and imported dairy food). For both samples, the average for the last "imported cheese" statement was around three: no clear tendency is visible and we assume that, in general, imported cheese is not considered to be of a high quality.

Table 2. Differences in the assessment of domestic in comparison to imported dairy food in Kosovo (Prishtina) and Albania (Tirana) ($n = 649$); unpaired two sample t-test including bootstrap sampling.

Statements [1]	*M*		*MD*	*SE*	*t*	*p*	Bootstrap [3]		
	Prishtina	Tirana					*MD* 95% Confidence Interval		*p*
Domestic cheese is safer than imported [2]	4.14	3.33	+0.814	0.095	8.636	≤0.001	+0.628	+1.001	≤0.001
Domestic milk is safer than imported [2]	4.19	3.29	+0.894	0.096	9.474	≤0.001	+0.704	+1.077	≤0.001
Domestic cheese is of higher quality than imported [2]	4.36	3.28	+1.080	0.086	12.842	≤0.001	+0.913	+1.245	≤0.001
Domestic milk is of higher quality than imported [2]	4.13	3.23	+0.901	0.090	10.151	≤0.001	+0.723	+1.072	≤0.001
Imported cheese is of high quality	2.93	2.96	−0.024	0.082	−0.288	0.773	−0.182	+0.144	0.772

M = Mean; *MD* = Mean Difference; *SE* = Standard Error; *p* = Significance; [1] Likert scale: 1 = "disagree" to 5 = "agree"; [2] Levene test on homogeneity of variances: variances are not equal, Welch test used; [3] 5000 bootstrap samples.

To test whether the mean for the perception of domestic vs. imported food (M) differs significantly between the Prishtina and the Tirana samples, we used an unpaired two sample t-test including bootstrap sampling with 5000 random bootstrap samples (Table 2). The differences for the four comparison statements were significant ($p \leq 0.001$), and most of them were considerable ($MD = +0.814$ to $+1.080$; $t = 8.6$ to 12.8). The 95% confidence interval based on bootstrap sampling confirms this interpretation: the conclusion is that Prishtina consumers have significantly more trust in domestic products. The value for M is considerably higher, which means that these consumers agreed more strongly with the four comparison statements and rated domestic products as safer and of higher quality than imported products.

The following test of H1 shows whether the deviations from three are significant. Based on the descriptive results and the results of the t-test presented above, we assume that the Kosovo sample has, in general, a much higher preference for domestic food.

3.1. Test of Hypothesis 1: Perception of Food Quality and Safety of Imported and Domestic Food

As a result of certain food safety problems in the WBCs, we formulated H1: The quality and safety of imported food is rated higher than the quality and safety of domestic food. However, the descriptive results clearly suggest that H1 should be rejected and that, in general, domestic food is rated as safer and of higher quality. Obviously, consumer patriotism is relevant for Kosovar, and also Albanian, consumers. The following analysis tests whether the responses to the five statements in Table 3 are significantly higher or lower than three ("neither agree, nor disagree with statement"). If the differences are significant, we interpret the deviations in view of their metric size (positive or negative). If they are not significant, there is no empirical evidence that imported and domestic food were rated differently, and, instead, domestic and imported food were considered to be equal in view of safety and quality. To analyze the deviations, we performed a t-test with a test value of three including a bootstrapping method with 5000 random bootstrap samples.

The means M of the Prishtina and Tirana samples are found to be significantly higher/lower than three. The t-test with bootstrapping shows that the deviations within the Prishtina sample ($n = 300$) are highly significant for the comparison statements ($M = 4.13$ to 4.36; $t = 19.5$ to 26.1; $p \leq 0.001$; Table 3). All the deviations are significantly positive ($MD = +1.133$ to $+1.363$), meaning that the respondents on average clearly agreed with the statements that domestic food is safer and of higher quality than imported food. The deviations from three for the "imported cheese" statement are not significant ($M = 2.93$; $MD = -0.067$; $t = -1.095$; $p = 0.275$; Table 3). Within the Tirana sample ($n = 349$), the deviations are also significant (except for the "imported cheese" statement), but they are much lower ($M = 3.23$ to 3.33; $t = 3.448$ to 4.772; $p \leq 0.001$; Table 3). Again, with respect to the "imported cheese" statement, the deviations from three are not significant ($M = 2.96$; $MD = -0.043$; $t = -0.776$; $p = 0.438$). The 95% confidence intervals of the bootstrapping show the principal reliability of these results and that it is advisable to reject H1: The quality and safety of imported food is not rated higher than the quality and safety of domestic food. In Prishtina, consumers generally evaluated domestic food to be safer and of higher quality. In Tirana, too, the consumers tended to evaluate domestic food more highly (there are significant, positive deviations from three); however, although the deviations are significant, they are, on average, much lower, and the results are therefore much less clear than for the Prishtina sample.

Table 3. Perception of domestic in comparison to imported dairy food in Kosovo (Prishtina) and Albania (Tirana); *t*-test including bootstrap sampling with test value = 3 ("neither agree nor disagree with statement").

Statement [1]	*M*	*MD*	*SE*	*t*	*p*	Bootstrap [2]		
						MD 95% Confidence Interval		*p*
Kosovo, Prishtina (n = 300)								
Domestic cheese is safer than imported	4.14	+1.143	0.065	17.830	≤0.001	+1.014	+1.272	≤0.001
Domestic milk is safer than imported	4.19	+1.187	0.064	18.640	≤0.001	+1.062	+1.314	≤0.001
Domestic cheese is of higher quality than imported	4.36	+1.363	0.053	26.172	≤0.001	+1.260	+1.466	≤0.001
Domestic milk is of higher quality than imported	4.13	+1.133	0.058	19.575	≤0.001	+1.021	+1.244	≤0.001
Imported cheese is of high quality	2.93	−0.067	0.061	−1.095	0.275	−0.184	+0.055	0.275
Albania, Tirana (n = 349)								
Domestic cheese is safer than imported	3.33	+0.330	0.069	4.772	≤0.001	+0.194	+0.463	≤0.001
Domestic milk is safer than imported	3.29	+0.292	0.069	4.193	≤0.001	+0.156	+0.426	≤0.001
Domestic cheese is of higher quality than imported	3.28	+0.284	0.066	4.299	≤0.001	+0.156	+0.411	≤0.001
Domestic milk is of higher quality than imported	3.23	+0.232	0.067	3.448	0.001	+0.099	+0.362	0.001
Imported cheese is of high quality	2.96	−0.043	0.055	−0.776	0.438	−0.152	+0.064	0.433

M = Mean; *MD* = Mean Difference; *SE* = Standard Error; *p* = Significance level; [1] Likert scale: 1 = "disagree" to 5 = "agree"; n = 300; 299 degrees of freedom (*df*) (Prishtina Sample); n = 349; 348 *df*; (Tirana Sample) [2] 5000 bootstrap samples.

3.2. Test of Hypothesis 2: Perception of Food Safety and Quality Cues

To analyze perceptions of food safety and quality cues, the respondents were asked how often they checked specific characteristics that are connected to food safety and quality (tested on the example of cheese). The relevant question (Q11) was: "If you want to know about the safety of cheese you are going to buy, which of the following characteristics do you check?" The respondents used a 5-point semantic scale, with a minimum value of 1 (never) to a maximum of 5 (always). The semantic meaning of the in-between values are: 2 = occasionally (about 1 to 2 times per week), 3 = frequently (about half the time or 3 to 4 times every week); 4 = often (about 5 times per week).

As we can see from Table 4 the consumers in Prishtina checked the food safety and quality cues much less frequently than those in Tirana—with one exception, the expiration date. Overall, the expiration date was the most important food safety and quality cue for consumers in both samples. Other important cues were the list of ingredients, organic production, origin, and knowledge of the producer and seller (particularly for consumers in Tirana). The food safety and quality cues with the lowest importance (within both samples) were the international food standards HACCP and ISO. To see whether these differences in the perception of food safety and quality cues are significant, we tested Hypothesis 2.

According to Hypothesis 2 consumers from Kosovo (Prishtina) and Albania (Tirana) should differ in respect to the frequency at which they used food safety and quality related cues. This hypothesis was proposed because the average income in Albania is much higher than the average in Kosovo, and the level of education is also higher in Albania (see sample description in Table 1). As we can see from Table 4, most quality cues were rated to be more important (that is, the consumers checked them more regularly) by consumers in Tirana. The mean differences are quite large. Neither of the samples identified safety and quality cues that were of the highest importance, with average values greater than four, and it was only "expiration date" that reached a mean close to four. In particular, in Prishtina only a few food safety and quality cues seem to have high importance. We tested the differences between the samples by means of an unpaired two sample *t*-test including bootstrap sampling with 5000 random bootstrap samples (Table 4). All of the differences were significant, and all, except for the differences for expiration date, were considerable.

MD ranges from −0.148 (expiration date) to −1.236 (organic production). This means that consumers from Tirana seem to use food safety and quality cues much more frequently than consumers from Prishtina. Bootstrapping further emphasizes the accuracy of this interpretation, with the 95% confidence intervals shown in Table 4.

Table 4. Differences in the rating of food safety and quality cues between consumers from Kosovo (Prishtina) and consumers from Albania (Tirana) (n = 649); unpaired two sample t-test including bootstrap sampling.

Characteristics [1]	*M*		*MD*	*SE*	*t*	*p*	Bootstrap [3]		
	Prishtina	Tirana					*MD* 95% Confidence Interval		*p*
Brand reputation	2.41	3.13	−0.723	0.076	−9.499	≤0.001	−0.869	−0.576	≤0.001
Expiration date [2]	3.63	3.78	−0.148	0.051	−2.900	0.004	−0.249	−0.045	0.003
List of ingredients [2]	2.54	2.86	−0.322	0.116	−2.787	0.005	−0.550	−0.097	0.004
Organic production [2]	1.94	3.17	−1.236	0.104	−11.909	≤0.001	−1.440	−1.027	≤0.001
Food safety certificate	1.86	3.06	−1.201	0.092	−13.009	≤0.001	−1.377	−1.019	≤0.001
HACCP [2]	1.35	2.11	−0.756	0.077	−9.832	≤0.001	−0.907	−0.611	≤0.001
ISO [2]	1.36	2.14	−0.782	0.078	−10.038	≤0.001	−0.935	−0.633	≤0.001
Local origin [2]	2.25	3.24	−0.983	0.085	−11.515	≤0.001	−1.149	−0.813	≤0.001
Domestic origin	2.87	3.24	−0.360	0.078	−4.639	≤0.001	−0.511	−0.214	≤0.001
Foreign origin	2.16	2.93	−0.777	0.085	−9.183	≤0.001	−0.943	−0.616	≤0.001
EU origin	1.88	2.86	−0.985	0.084	−11.785	≤0.001	−1.143	−0.826	≤0.001
Knowing the seller [2]	2.10	2.84	−0.742	0.091	−8.147	≤0.001	−0.923	−0.561	≤0.001
Knowing the producer	2.63	3.05	−0.419	0.091	−4.581	≤0.001	−0.594	−0.244	≤0.001

M = Mean; *MD* = Mean Difference; *SE* = Standard Error; *p* = Significance; [1] Likert scale: 1 = "never check characteristic to assess food safety and quality", 5 = "always check characteristic to assess food safety and quality"; [2] Levene test on homogeneity of variances: variances are not equal, Welch test used; [3] 5000 bootstrap samples.

3.3. Test of Hypothesis 3: The Influence of Socio-Demographic Variables on the Perception of Food Safety and Quality Cues

As pointed out in the literature, socio-demographic factors might influence the perception of food and quality cues, which leads to H3: Socio-demographic variables influence the perception of food safety and quality cues. In particular, we tested the variables "gender", "age", "household size", "education", and "income" as predictors for the rating of domestic in comparison to imported food and the perception of food safety and quality cues. To evaluate the significance of the socio-demographic variables we (1) compared means by the use of appropriate statistical methods (independent t-test, correlation analysis); and (2) if significant differences were found, analyzed the effects of the independent variables as predictors in order to ensure the reliability of the analysis, including the use of bootstrapping (5000 bootstrap samples; valid data without missing values in Prishtina n = 299 and in Tirana n = 344).

As we can see from Table 5 (which shows only the significant correlation coefficients), the dependency of the perception of domestic in comparison to imported dairy food on socio-demographic variables is rather low or almost non-existent. Significant correlation coefficients were only found for education and household size (in Prishtina), and age and, to some extent, education in Tirana. The variables "gender" and "income" had no influence in either sample. Concerning statements one to four (domestic vs. imported foods), in Prishtina, there was a weak negative correlation to education and a weak positive correlation to household size. This could be interpreted as showing that well educated people had (slightly) more trust in imported food than less well-educated people, and that larger households perceived domestic dairy food as more trustworthy. In contrast to these findings for Prishtina, the Tirana respondents showed different patterns, where age seems to have an influence, but a very small one, on the perception of domestic versus imported dairy food, and education seems to have a positive influence on the statement that imported cheese is of higher quality. However, even the significant correlations are quite low, with the minimum and maximum r being −0.222 and +0.299, respectively. Given the bootstrap 95% confidence interval, r would not undercut −0.325 or exceed +0.391. Even the minimum/maximum correlation coefficients do not reach a moderate *r*-level of beyond or below ±0.5.

Table 5. Correlation between socio-demographic variables and perception of domestic in comparison to imported dairy food for Kosovo and Albania; correlation coefficients including bootstrap sampling.

Statements [1]	*r*	*SE*	*p*	Bootstrap [2] *r* 95% Confidence Interval	
Prishtina: Education	Spearman-Rho				
Domestic cheese is safer than imported	−0.222	0.053	≤0.001	−0.325	−0.118
Domestic milk is safer than imported	−0.174	0.054	0.003	−0.281	−0.065
Domestic cheese is of higher quality than imported	−0.147	0.056	0.011	−0.256	−0.038
Domestic milk is of higher quality than imported	−0.143	0.056	0.013	−0.254	−0.033
Prishtina: Household size	Pearson's *r*				
Domestic cheese is safer than imported	0.287	0.049	≤0.001	0.190	0.383
Domestic milk is safer than imported	0.299	0.048	≤0.001	0.205	0.391
Domestic cheese is of higher quality than imported	0.226	0.054	≤0.001	0.119	0.330
Domestic milk is of higher quality than imported	0.194	0.056	0.001	0.084	0.301
Tirana: Age	Pearson's *r*				
Domestic cheese is safer than imported	0.185	0.053	0.001	0.080	0.286
Domestic milk is safer than imported	0.139	0.052	0.010	0.035	0.239
Domestic cheese is of higher quality than imported	0.135	0.055	0.012	0.026	0.241
Domestic milk is of higher quality than imported	0.108	0.056	0.045	−0.003	0.216
Imported cheese is of high quality	−0.192	0.054	≤0.001	−0.299	−0.081
Tirana: Education	Spearman–Rho				
Imported cheese is of high quality	0.210	0.053	≤0.001	0.104	0.311

r = Correlation coefficient Spearman-Rho (ordinal data) or Pearson's *r* (metric data); *SE* = Standard Error; *p* = Significance; [1] Likert scale: 1 = "disagree" to 5 = "agree"; [2] 5000 bootstrap samples.

Concerning the perception of food safety and quality cues, the influence of socio-demographic variables seems to be more important than in the previous findings. In particular, the gender of the respondents seems to have a significant influence on the perception of food safety and quality cues (Table A1 in the Appendix A). In general, women considered food safety and quality cues, such as lists of ingredients or seals of quality, more often than the men (negative *MD* means that the average Likert value for the men is lower than that for the women; the men less often checked food safety and quality cues than the women). Table A1 in the Appendix A contains only significant differences for both samples. As we can see from A1, the findings are quite similar, although some differences can be observed: origin seems to be more relevant for gender differences in the Kosovar sample, while brand reputation and certificates (HACCP, ISO 9000) seem to be slightly more important in the Albanian sample. After bootstrap sampling, the 95% confidence interval shows, however, that the "true" difference in the perception between men and women might—although it is significant—be much lower or higher. It appears that for some variables the differences might also be almost non-existent (the lower boundary of the bootstrap 95% confidence interval is close to 0; e.g., for local origin). For other variables (e.g., organic production), these differences are well supported by the bootstrapping method.

Concerning the other socio-demographic variables, education, in particular, seems to influence the perception of food safety and quality cues significantly in both samples. In the Kosovar sample, income slightly influences the perception of food safety and quality cues (in particular, brand reputation and knowledge of producer). Further, the negative *r* for the variable "household size" might be negligible (Table A2 in Appendix A). In the Albanian sample, age also seems to have a negative, but minor, influence on the perception of the food safety and quality cues (the older people less frequently checked the cues). However, here too, the influence approximated through correlation analysis is rather low for almost all the socio-demographic variables (see r in Table A2 for the Kosovar sample and in Table A3 for the Albanian sample). These findings are further supported by the bootstrapping method; no upper/lower boundaries in the 95% confidence interval exceed/undercut ±0.5—a level at which one could suppose at least a moderate effect of socio-demographic variables on the perception of food safety and quality cues.

In a nutshell, one might consider socio-demographic variables as relevant predictor variables. Even though their influence seems to be rather low, H3 is clearly supported: Socio-demographic variables do have an influence on the perception of food safety and quality in Kosovo and Albania.

4. Discussion

Based on the literature review [15–18,32], which indicated a lower level of food safety in Albania and Kosovo [22,23] (and other WBCs [20,25]), we expected that respondents from both countries would evaluate imported better than nationally produced food products (H1). However, we did not observe this in our study, so H1 could not be confirmed and had to be rejected. The descriptive analysis and the test for significant differences showed that in Kosovo (Prishtina) consumers evaluated domestic dairy food to be safer and of higher quality than imported dairy food, which is in accordance with [38]. Albanian (Tirana) consumers also showed a greater preference for domestic over imported cheese and milk products. However, the values are lower than in the Kosovar sample. Both samples were indifferent about the quality of imported cheese. Consumer patriotism also seems to be of relevance for Kosovar and Albanian consumers.

For H2, we assumed that Kosovar consumers would differ from Albanians in respect to the frequency at which they used food safety and quality related cues. We used a list of quality and food safety cues derived from the literature to test this assumption. Important quality and food safety cues mentioned in the literature are brand reputation [5,33], information on labels (expiry date, list of ingredients) [36], food certificates [36], organic production [2,40], quality related standards (ISO 9000, HACCP) [45,46], and country of origin [32,34–36,39]. In the absence of reliable quality standards or third party food certificates, knowing the producer or seller of a product can act as another quality cue [44]. H2 was confirmed, with the differences between Kosovar and Albanian consumers being significant for all the food safety and quality cues tested.

In our study, the Kosovar consumers used food safety and quality cues less frequently than Albanians. Expiration date was the most commonly used quality cue for both consumer groups. The Albanians used food certificates more often than the Kosovars. HACCP or ISO related information was the least frequently used quality cue for both consumer groups, which is not surprising because HACCP and ISO standards are primarily used for business-to-business communication and are normally not communicated to consumers. The frequency of use (see Table 4) showed that information about expiry date, domestic origin/local origin, organic production, knowing the producer, and brand reputation were the most frequently used food safety and quality cues for the Albanians. The Kosovar consumers used expiry date, domestic origin, knowing the producer, brand reputation and local origin, in descending frequency. The Kosovar respondents used information about organic production and food certificates much less often than the Albanian respondents (see Table 4, biggest mean difference of 1.236 (organic) and −1.201 (certificates)). This may be an indicator of a lower availability of organic food supplies in Kosovar or also can reflect the strategy of Albanian consumers to pay more attention to organic certification to tackle their concern for food safety situation [41]. As suggested by the study of Thøgersen et al. (2019) it could also be an indicator of an absence of environmental concerns among consumers or insufficient trust in the countries of origin of the organic food [40]. We found that food certificates were less frequently used by Kosovars than by Albanians. According to Grunert (2005), this could be an indicator either that the Kosovar consumers did not consider the food certificates on dairy products to be trustworthy, because of the lack of trust in formal institutions [26] or that they did not understand them [2]. In our sample, Kosovar consumers used the quality cue "knowing the producer" in third place after expiry date and domestic origin, which is in accordance with several international studies and studies from the WBCs [15,28,43,44]. In the absence of trustworthy third-party certification of food products, the personal trust between customer and seller can compensate as a quality cue.

Previous studies have reported that female consumers and better educated consumers pay more attention to information related to food safety [16,50]. Our study found similar, statistically significant results (H3: Socio-demographic variables influence the perception of food safety and quality cues). Women and better educated consumers checked food safety and quality related information more often than men or those with lower education. Women obviously considered food safety and quality cues, such as list of ingredients or food certificates, more often than men. Origin seemed to be more relevant with respect to gender differences in the Kosovar sample, while brand reputation and quality assurance standards (HACCP, ISO 9000) seemed to be slightly more important in the Albanian sample. The other socio-demographic variables (age, income, household size) were either not statistically significant or, if they were significant, showed only small correlation coefficients. Older Albanian consumer groups, for instance, perceived the quality and safety of domestic dairy products to be higher than did younger consumer groups, which makes sense as older consumer groups tend to show more ethnocentrism than younger consumer groups (see [53]).

5. Limitations and Conclusions

The major limitation of this study, besides the convenient nature of the samples, is the fact that the respondents were selected from the countries' two main cities, and thus the samples do not represent all consumers but rather urban consumers. As such, the study findings cannot be considered representative of the whole population. However, as highlighted earlier in the paper, Tirana and Pristina are the capital and largest cities of the respective countries, with high diversity of cultures and origins, and as such, the findings can be considered indicative also for the rest of the country. Furthermore, these two cities also represent the largest markets where purchasing power is concentrated, and as such, the findings are important for the industry. Future studies could broaden the context of this study by including social and cultural conditions that are perhaps able to further explain differences in consumer perspectives between Kosovars and Albanians on food safety and quality.

Confirming literature [12], in Albania and Kosovo there are serious gaps in the food safety system (in particular in the meat [15,16] and dairy sector [17,18]), which is reflected in real and perceived food safety risk. However, according to the study findings, overall, consumers have higher preference for local products, the safety and quality of which is perceived higher compared to imports, although cheese imports are dominated by EU countries, which are characterized by high food safety standards. Kosovar consumers' perception of food safety is higher when compared to Albanians—one factor that may contribute to this difference is the fact that Kosovo has been under the supervision and heavily supported by the international community in its efforts to build up the institutional system after the war. Another reason could be a high prevalence of ethnocentrism of Kosovar consumers, as identified in a previous study [38].

As expected, women and higher educated consumers used food quality and safety related information more often than other consumer groups in both countries. In accordance with this result of our study, it is advisable for governmental institutions who want to inform the public or food companies releasing advertising campaigns to focus on these consumer groups to improve the effectiveness of their communication efforts. However, stakeholders could also focus especially on less well-educated consumers to reduce their information deficits concerning food safety and quality standards in their home countries.

We found that Kosovar consumers used food certificates less often than Albanians. This could be either because they do not trust them or because they do not understand them. In the first case, this would emphasize the need to establish trustworthy third-party certification, while the latter case would point towards a promotion of existing food certificates by governmental or privately funded educational campaigns (see [53]). In our opinion, these findings could be used by policymakers and food companies to improve the level of trust in public/private institutions to guarantee food safety, and to increase consumers' awareness and knowledge in this area. They underline the importance of food labeling information such as expiry date/best before date, of trustworthy food brands and of clear and transparent communication of the country of origin or the local origin.

Although not directly addressed within this study, we have to take into account that there is still a lack of capacity due to weak institutional frameworks and corruption [26]—despite the improved legislation in Kosovo and Albania in recent years due to EU integration processes [13]. An important institutional measure could be the strengthening of the existing or establishment of new national food safety and quality organizations, following best practices from EU countries, such as AMA Marketing in Austria (AMA = Agricultural Market Austria). AMA Marketing is a governmental organization independent from existing ministries, comparable to SOPEXA in France, with a focus on establishing and controlling food quality and safety standards, and implementing communication strategies to reach consumers. The advantage of establishing a new organization independent from existing agricultural or health ministries could be an important sign to consumers that the new organization also stands for an improved, trustworthy regulatory capacity. This approach could help to gain more trust in the national food safety system and help to further improve the food safety perception of consumers, which is advisable, according to our empirical findings.

Author Contributions: Conceptualization, R.H., D.I., I.M., and K.G.; methodology, R.H., D.I., I.M., P.Y., O.M. and K.G.; software, O.M.; validation, R.H. and O.M.; formal analysis, O.M.; investigation, R.H., D.I., I.M. and P.Y.; resources, R.H., D.I., I.M., P.Y., O.M. and K.G.; data curation, R.H., D.I., I.M. and P.Y.; writing—original draft preparation, R.H., D.I., I.M., P.Y., O.M. and K.G.; writing—review and editing, R.H., O.M.; visualization, O.M.; supervision, R.H. and O.M.; project administration, R.H.; funding acquisition, R.H., D.I. and I.M. All authors have read and agreed to the published version of the manuscript.

Funding: The study was supported by financed funds of the Republic of Kosovo and the Austrian Development Cooperation (HERAS Grant: Higher Education, Research and Applied Science) (K-09-2018).

Institutional Review Board Statement: Ethical review and approval were waived for this study, due to full anonymity of interviewees and as no sensitive data were collected. Personal data (e.g., age) were collected in public spaces and cannot be traced back to individuals.

Informed Consent Statement: Informed consent was obtained from all subjects involved in the study.

Data Availability Statement: The data presented in this study are available on request from the corresponding author. The data are not publicly available due to privacy.

Acknowledgments: We would like to thank the four anonymous reviewers for their valuable comments and the respondents who participated in our survey.

Conflicts of Interest: The authors declare that they have no conflicts of interest. The funders had no role in the design of the study; in the collection, analysis, or interpretation of the data; in the writing of the manuscript; or in the decision to publish the results.

Appendix A

Table A1. Significant differences in the rating of food safety and quality cues with respect to the variable "gender"; samples: consumers from Kosovo (Prishtina; $n = 299$) and Albania (Tirana; $n = 346$); t-test including bootstrap sampling.

Characteristics [1]	*MD*	*SE*	*DF*	*t*	*p*	Bootstrap [3]		
						MD 95% Confidence Interval		*p*
Prishtina								
List of ingredients	−0.405	0.177	298.0	−2.296	0.022	−0.732	−0.061	0.017
HACCP [2]	−0.294	0.082	296.7	−3.600	≤0.001	−0.452	−0.131	≤0.001

Table A1. *Cont.*

Characteristics [1]	*MD*	*SE*	*DF*	*t*	*p*	Bootstrap [3]		
						MD 95% Confidence Interval		*p*
ISO [2]	−0.287	0.082	297.0	−3.493	≤0.001	−0.446	−0.122	0.001
Local origin	−0.267	0.131	298.0	−2.044	0.042	−0.518	−0.007	0.047
Domestic origin [2]	−0.288	0.116	256.1	−2.492	0.013	−0.514	−0.067	0.011
Foreign origin	−0.462	0.124	298.0	−3.728	≤0.001	−0.706	−0.228	0.001
EU origin	−0.423	0.116	298.0	−3.658	≤0.001	−0.643	−0.206	0.001
Tirana								
Brand reputation [2]	−0.255	0.117	214.5	−2.181	0.030	−0.488	−0.020	0.034
Expiry date [2]	−0.265	0.075	178.6	−3.534	0.001	−0.423	−0.121	0.001
List of ingredients	−0.452	0.156	345.0	−2.899	0.004	−0.758	−0.152	0.002
Organic production	−0.532	0.150	345.0	−3.550	≤0.001	−0.824	−0.222	≤0.001
HACCP [2]	−0.384	0.128	265.4	−2.992	0.003	−0.633	−0.130	0.004
ISO	−0.365	0.134	345.0	−2.728	0.007	−0.616	−0.102	0.007
Local origin [2]	−0.246	0.123	211.4	−2.006	0.046	−0.486	−0.010	0.044
Knowing the producer	−0.328	0.134	345.0	−2.442	0.015	−0.596	−0.062	0.016

MD = Mean Difference; *SE* = Standard Error; *DF* = Degrees of Freedom; *p* = Significance; [1] Likert scale: 1 = "never check characteristic to assess food safety and quality", 5 = "always check characteristic to assess food safety and quality"; [2] Levene test on homogeneity of variances: variances are not equal, Welch test used; [3] 5000 bootstrap samples.

Table A2. Significant correlation coefficients for socio-demographic variables and rating of food safety and quality cues in Albania (Prishtina; n = 299); correlation coefficients including bootstrap sampling.

Characteristics [1]	*r*	*SE*	*p*	Bootstrap [2] *r* 95% Confidence Interval	
Education	Spearman-Rho				
Brand reputation	0.260	0.057	≤0.001	0.147	0.369
Expiry date	−0.217	0.052	≤0.001	−0.318	−0.113
List of ingredients	0.210	0.058	≤0.001	0.100	0.326
Organic production	0.166	0.057	≤0.001	0.052	0.275
HACCP	0.335	0.046	≤0.001	0.242	0.423
ISO	0.341	0.046	≤0.001	0.249	0.428
Local origin	0.223	0.056	≤0.001	0.111	0.330
Foreign origin	0.178	0.055	0.002	0.069	0.284
Knowing the seller	0.127	0.058	0.028	0.014	0.243
Knowing the producer	0.150	0.055	0.009	0.044	0.258
Household size	Pearson's *r*				
Brand reputation	−0.120	0.057	0.038	−0.232	−0.009
Local origin	−0.161	0.056	0.005	−0.265	−0.044
Knowing the producer	−0.124	0.059	0.032	−0.237	−0.003
Income	Pearson's *r*				
Brand reputation	0.245	0.056	≤0.001	0.131	0.349
Expiry date	−0.165	0.060	0.004	−0.283	−0.050
Food safety certificate	0.198	0.055	0.001	0.089	0.304
HACCP	0.197	0.064	0.001	0.073	0.324
ISO	0.203	0.064	≤0.001	0.078	0.329
Local origin	0.189	0.053	0.001	0.084	0.292
Domestic origin	0.170	0.055	0.003	0.061	0.272
Knowing the seller	0.212	0.054	≤0.001	0.102	0.318
Knowing the producer	0.260	0.053	≤0.001	0.155	0.359

r = Correlation coefficient Spearman-Rho (ordinal data) or Pearson's *r* (metric data); *SE* = Standard Error; *p* = Significance; [1] Likert scale: 1 = "never check characteristic to assess food safety and quality" to 5 = "always check characteristic to assess food safety and quality"; [2] 5000 bootstrap samples.

Table A3. Significant correlation coefficients for socio-demographic variables and rating of food safety and quality cues in Albania (Tirana; $n = 343$); correlation coefficients including bootstrap sampling.

Characteristics [1]	*r*	*SE*	*p*	Bootstrap [2] *r* 95% Confidence Interval	
Age	Pearson's *r*				
List of ingredients	−0.196	0.053	≤0.001	−0.301	−0.092
Food safety certificate	−0.133	0.056	0.014	−0.243	−0.020
HACCP	−0.216	0.048	≤0.001	−0.309	−0.120
ISO	−0.216	0.048	≤0.001	−0.309	−0.119
Foreign origin	−0.140	0.058	0.009	−0.253	−0.024
EU origin	−0.172	0.056	0.001	−0.280	−0.063
Education	Spearman–Rho				
Brand reputation	0.259	0.055	≤0.001	0.153	0.365
Expiry date	0.190	0.057	≤0.001	0.078	0.299
List of ingredients	0.231	0.051	≤0.001	0.128	0.330
Organic production	0.210	0.052	≤0.001	0.105	0.314
Food safety certificate	0.321	0.054	≤0.001	0.210	0.424
HACCP	0.235	0.052	≤0.001	0.133	0.338
ISO	0.276	0.051	≤0.001	0.175	0.376
Local origin	0.335	0.052	≤0.001	0.232	0.437
Domestic origin	0.233	0.055	≤0.001	0.126	0.340
Foreign origin	0.314	0.053	≤0.001	0.209	0.416
origin	0.321	0.052	≤0.001	0.220	0.422
Knowing the producer	0.186	0.056	0.001	0.076	0.295
Income	Pearson's *r*				
Expiry date	0.109	0.050	0.043	0.008	0.207
HACCP	0.106	0.051	0.049	0.006	0.208
ISO	0.137	0.051	0.011	0.038	0.237

r = Correlation coefficient Spearman–Rho (ordinal data) or Pearson's *r* (metric data); *SE* = Standard Error; *p* = Significance; [1] Likert scale: 1 = "never check characteristic to assess food safety and quality" to 5 = "always check characteristic to assess food safety and quality"; [2] 5000 bootstrap samples. [54].

References

1. Peri, C. The universe of food quality. *Food Qual. Prefer.* **2006**, *17*, 3–8.
2. Grunert, K.G. Food quality and safety: Consumer perception and demand. *Eur. Rev. Agric. Econ.* **2005**, *32*, 369–391.
3. Marek, G.; Dobrzański, B.; Oniszczuk, T.; Combrzyński, M.; Ćwikła, D.; Rusinek, R. Detection and differentiation of volatile compound profiles in roasted coffee arabica beans from different countries using an electronic nose and GC-MS. *Sensors* **2020**, *20*, 2124.
4. Steenkamp, J.-B.E.M. Conceptual Model of the Quality Formation Process. *J. Bus. Res.* **1990**, *21*, 309–333.
5. Akdeniz, B.; Calantone, R.J.; Voorhees, C.M. Effectiveness of Marketing Cues on Consumer Perceptions of Quality: The Moderating Roles of Brand Reputation and Third-Party Information. *Psychol. Mark.* **2013**, *30*, 76–89.
6. Marchesini, S.; Huliyeti, H.; Regazzi, D. Literature review on the perception of agro-foods quality cues in the international environment. In Proceedings of the 105th EAAE Seminar on "International Marketing and International Trade of Quality Food Products", Bologna, Italy, 8–10 March 2007; pp. 729–738.
7. Grunert, K.G. What's in a steak? A cross-cultural study of the perception of beef. *Appetite* **1995**, *24*, 188.
8. Darby, M.R.; Karni, E. Free Competition and the Optimal Amount of Fraud. *J. Law Econ.* **1973**, *16*, 67–88.
9. Nelson, P. Information and Consumer Behavior. *J. Polit. Econ.* **1970**, *78*, 311–329.
10. Nelson, P. Advertising as Information. *J. Polit. Econ.* **1974**, *82*, 729–754.
11. Becker, T. The economics of food quality standards. In Proceedings of the Second Interdisciplinary Workshop on Standardization Research, University of the Federal Armed Forces Hamburg, Hamburg, Germany, 24–27 May 1999; Volume 24, p. 27.
12. Gurinović, M.; Milešević, J.; Novaković, R.; Kadvan, A.; Djekić-Ivanković, M.; Šatalić, Z.; Korošec, M.; Spiroski, I.; Ranić, M.; Dupouy, E.; et al. Improving nutrition surveillance and public health research in Central and Eastern Europe/Balkan Countries using the Balkan Food Platform and dietary tools. *Food Chem.* **2016**, *193*, 173–180.
13. Djekic, I.; Tomasevic, I.; Radovanovic, R. Quality and food safety issues revealed in certified food companies in three Western Balkans countries. *Food Control* **2011**, *22*, 1736–1741. [CrossRef]
14. Jaffee, S. Sanitary and Phytosanitary Regulation: Overcoming Constraints. In *Trade, Doha, and Development. A Window into the Issues*; Newfarmer, R., Ed.; The World Bank, Trade Department: Washington, DC, USA, 2005; pp. 357–374, ISBN 0-8213-6437-5.

15. Imami, D.; Chan-Halbrendt, C.; Zhang, Q.; Zhllima, E. Conjoint analysis of consumer preferences for lamb meat in central and southwest urban Albania. *Int. Food Agribus. Manag. Rev.* **2011**, *14*, 111–126.
16. Zhllima, E.; Imami, D.; Canavari, M. Consumer perceptions of food safety risk: Evidence from a segmentation study in Albania. *J. Integr. Agric.* **2015**, *14*, 1142–1152. [CrossRef]
17. Gjeci, G.; Bicoku, Y.; Imami, D. Awareness about food safety and animal health standards—The case of dairy cattle in Albania. *Bulg. J. Agric. Sci.* **2016**, *22*, 339–345.
18. Udovicki, B.; Djekic, I.; Kalogianni, E.P.; Rajkovic, A. Exposure assessment and risk characterization of aflatoxin m1 intake through consumption of milk and yoghurt by student population in Serbia and Greece. *Toxins* **2019**, *11*, 205. [CrossRef]
19. Shkodra, B. Brucellosis in Gjilan Region during 2000–2012. In Proceedings of the International Conference on Research and Education—Challenges towards the Future, Shkodra, Albania, 24–25 May 2013.
20. Schatzmayr, G.; Streit, E. Global occurrence of mycotoxins in the food and feed chain: Facts and figures. *World Mycotoxin J.* **2013**, *6*, 213–222. [CrossRef]
21. van Asselt, E.D.; van der Fels-Klerx, H.J.; Marvin, H.J.P.; van Bokhorst-van de Veen, H.; Groot, M.N. Overview of Food Safety Hazards in the European Dairy Supply Chain. *Compr. Rev. Food Sci. Food Saf.* **2017**, *16*, 59–75. [CrossRef]
22. Skreli, E.; Imami, D. *Agricultural Risk—A Rapid Assessment and Recommended Mitigation Mechanisms*; Technical Report Prepared for EBRD; Agricultural University of Tirana: Tirana, Albania, 2019.
23. Camaj, A.; Meyer, K.; Berisha, B.; Arbneshi, T.; Haziri, A. Aflatoxin M 1 contamination of raw cow's milk in five regions of Kosovo during 2016. *Mycotoxin Res.* **2018**, *34*, 205–209. [CrossRef]
24. Bhat, R.; Vasanthi, S. Food safety in food security and food trade. Mycotoxin food safety risk in developing countries. *Int. Food Policy Res. Inst. Focus* **2003**, *10*, 3–17.
25. Manojlović, M.; Singh, B.R. Trace elements in soils and food chains of the Balkan region. *Acta Agric. Scand. Sect. B Soil Plant Sci.* **2012**, *62*, 673–695. [CrossRef]
26. Imami, D.; Skreli, E.; Valentinov, V. Food safety and value chain coordination in the context of a transition economy: The role of agricultural cooperatives. *Int. J. Commons* **2021**. accepted.
27. De Jonge, J.; Van Trijp, H.; Jan Renes, R.; Frewer, L. Understanding consumer confidence in the safety of food: Its two-dimensional structure and determinants. *Risk Anal.* **2007**, *27*, 729–740. [CrossRef] [PubMed]
28. Imami, D.; Zhllima, E.; Canavari, M.; Merkaj, E. Segmenting Albanian consumers according to olive oil quality perception and purchasing habits. *Agric. Econ. Rev.* **2013**, *14*, 97–112.
29. Food and Agriculture Organization of the United Nations. *Smallholders and Family Farms in Albania—Country Study Report 2019*; Food and Agriculture Organization of the United Nations: Rome, Italy, 2019.
30. Imami, D.; Zhllima, E.; Merkaj, E.; Chan-Halbrendt, C.; Canavari, M. Albanian consumer preferences for the use of powder milk in cheese-making: A conjoint choice experiment. *Agric. Econ. Rev.* **2016**, *17*, 20–33.
31. Zaric, V.; Bodganovic, V.; Vasiljevic, Z.; Petkovic, D.; Mastilovic, J. Consumer attitudes to the animal food quality products in Serbia. In *Consumer Attitudes to Food Quality Products*; Klopčič, M., Kuipers, A., Hocquette, J.-F., Eds.; Wageningen Academic Publishers: Wageningen, The Netherlands, 2012; pp. 217–231.
32. Haas, R.; Canavari, M.; Imami, D.; Gjonbalaj, M.; Gjokaj, E.; Zvyagintsev, D. Attitudes and Preferences of Kosovar Consumer Segments Toward Quality Attributes of Milk and Dairy Products. *J. Int. Food Agribus. Mark.* **2016**, *28*, 407–426. [CrossRef]
33. Roselius, T. Consumer Rankings of Risk Reduction Methods. *J. Mark.* **1971**, *35*, 56. [CrossRef]
34. Ahmed, Z.U.; Johnson, J.P.; Yang, X.; Fatt, C.K.; Teng, H.S.; Boon, L.C. Does country of origin matter for low-involvement products? *Int. Mark. Rev.* **2004**, *21*, 102–120. [CrossRef]
35. Eroglu, S.A.; Machleit, K.A. Effects of Individual and Product-specific Variables on Utilising Country of Origin as a Product Quality Cue. *Int. Mark. Rev.* **1989**, *6*, 27–45. [CrossRef]
36. Verbeke, W.; Ward, R.W. Consumer interest in information cues denoting quality, traceability and origin: An application of ordered probit models to beef labels. *Food Qual. Prefer.* **2006**, *17*, 453–467. [CrossRef]
37. Loureiro, M.L.; Umberger, W.J. A choice experiment model for beef: What US consumer responses tell us about relative preferences for food safety, country-of-origin labeling and traceability. *Food Policy* **2007**, *32*, 496–514. [CrossRef]
38. Miftari, I. *Kosovo Consumer Buying Behavior Preferences and Demand for Milk and Dairy Products*; Norwegian University of Life Sciences: Ås, Norway, 2009.
39. Cerjak, M.; Haas, R.; Brunner, F.; Tomić, M. What motivates consumers to buy traditional food products? Evidence from Croatia and Austria using word association and laddering interviews. *Br. Food J.* **2014**, *116*, 1726–1747. [CrossRef]
40. Thøgersen, J.; Pedersen, S.; Aschemann-Witzel, J. The impact of organic certification and country of origin on consumer food choice in developed and emerging economies. *Food Qual. Prefer.* **2019**, *72*, 10–30. [CrossRef]
41. Imami, D.; Engjell, S.; Zhllima, E.; Chanb, C. Consumer attitudes towards organic food in the Western Balkans—The case of Albania. *Econ. Agro-Aliment.* **2017**, *19*, 245–260. [CrossRef]
42. Markenvertrauen: Starke Brands Legen Kräftig zu. Available online: http://www.lebensmittelzeitung.net/industrie/-Food-Brands-legen-kraeftig-zu-91347?crefresh=1 (accessed on 22 September 2020).
43. Ameseder, C.; Meixner, O.; Haas, R.; Fritz, M.; Schiefer, G. Measurement of the importance of trust elements in agrifood chains: An application of the analytic hierarchy process. *J. Chain Netw. Sci.* **2008**, *8*, 153–160. [CrossRef]

44. Purohit, D.; Srivastava, J. Effect of Manufacturer Reputation, Retailer Reputation, and Product Warranty on Consumer Judgments of Product. *J. Consum. Psychol.* **2001**, *10*, 123–134. [CrossRef]
45. Gellynck, X.; Kühne, B. Bottlenecks and Success Factors for the Introduction of Quality Assurance Schemes in the Agri-Food Sector. In Proceedings of the 104th (Joint) EAAE-IAAE Seminar Agricultural Economics and Transition: "What We Expected, What We Observed, the Lessons Learned", Budapest, Hungary, 6–8 September 2007.
46. Capmany, C.; Hooker, N.H.; Ozuna, T.; Van Tilburg, A. ISO 9000—A marketing tool for U.S. agribusiness. *Int. Food Agribus. Manag. Rev.* **2000**, *3*, 41–53. [CrossRef]
47. Trienekens, J.; Zuurbier, P. Quality and safety standards in the food industry, developments and challenges. *Int. J. Prod. Econ.* **2008**, *113*, 107–122. [CrossRef]
48. Poetz, K.; Haas, R.; Balzarova, M. CSR schemes in agribusiness: Opening the black box. *Br. Food J.* **2013**, *115*, 47–74. [CrossRef]
49. Smigic, N.; Djekic, I.; Martins, M.L.; Rocha, A.; Sidiropoulou, N.; Kalogianni, E.P. The level of food safety knowledge in food establishments in three European countries. *Food Control* **2016**, *63*, 187–194. [CrossRef]
50. Nganje, W.; Kaitibie, S. *Food Safety Risk Perception and Consumer Choice of Specialty Meats*; North Dakota State University: Fargo, ND, USA, 2003.
51. Gkana, E.N.; Nychas, G.J.E. Consumer food safety perceptions and self-reported practices in Greece. *Int. J. Consum. Stud.* **2018**, *42*, 27–34. [CrossRef]
52. Goktolga, Z.G.; Bal, S.G.; Karkacier, O. Factors effecting primary choice of consumers in food purchasing: The Turkey case. *Food Control* **2006**, *17*, 884–889. [CrossRef]
53. Ozilgen, S. Food safety education makes the difference: Food safety perceptions, knowledge, attitudes and practices among Turkish university students. *J. Verbrauch. Leb.* **2011**, *6*, 25–34. [CrossRef]
54. Bruner, G.C. *Marketing Scales Handbook. A Compilation of Multi-Item Measures for Consumer Behavior & Advertising Research*; GCBII Productions: Fort Worth, TX, USA, 2012; Volume 6, ISBN 978-0-615-63068-7.
55. Preacher, K.J.; Hayes, A.F. Asymptotic and resampling strategies for assessing and comparing indirect effects in multiple mediator models. *Behav. Res. Methods* **2008**, *40*, 879–891. [CrossRef] [PubMed]
56. Efron, B. The 1977 Rietz Lecture—Bootstrap methods: Another look at the jackknife. *Ann. Stat.* **1979**, *7*, 1–26. [CrossRef]
57. DiCiccio, T.J.; Efron, B. Bootstrap confidence intervals. *Stat. Sci.* **1996**, *11*, 189–228.
58. Haas, R.; Schnepps, A.; Pichler, A.; Meixner, O. Cow Milk versus Plant-Based Milk Substitutes: A Comparison of Product Image and Motivational Structure of Consumption. *Sustainability* **2019**, *11*, 5046. [CrossRef]
59. Vainio, A.; Kaskela, J.; Finell, E.; Ollila, S.; Lundén, J. Consumer perceptions raised by the food safety inspection report: Does the smiley communicate a food safety risk? *Food Control* **2020**, *110*, 106976. [CrossRef]

Article

Optimistic Bias, Food Safety Cognition, and Consumer Behavior of College Students in Taiwan and Mainland China

Guan-Yun Wang [1] and Hsiu-Ping Yueh [1,2,*]

[1] Department of Psychology, National Taiwan University, No.1, Sec 4, Roosevelt Rd., Taipei 10617, Taiwan; gywang.tp@gmail.com

[2] Department of Bio-Industry Communication and Development, National Taiwan University, No.1, Sec 4, Roosevelt Rd., Taipei 10617, Taiwan

* Correspondence: yueh@ntu.edu.tw

Received: 17 September 2020; Accepted: 27 October 2020; Published: 2 November 2020

Abstract: The purpose of this paper is to investigate how optimistic bias, consumption cognition, news attention, information credibility, and social trust affect the purchase intention of food consumption. Data used in this study came from a questionnaire survey conducted in college students in Taipei and Beijing. Respondents in the two cities returned 258 and 268 questionnaires, respectively. Samples were analyzed through structural equation modelling (SEM) to test the model. Results showed that Taiwanese college students did not have optimistic bias but Chinese students did. The models showed that both Taiwanese and Chinese students' consumption cognition significantly influenced their purchase intention, and news attention significantly influenced only Chinese students' purchase intention. Model comparison analysis suggested significant differences between the models for Taiwan and mainland China. The results revealed that optimistic bias can be reduced in different social contexts as that of the Taiwan model and the mainland Chinese model found in this study were indeed different. This study also confirmed that people had optimistic bias on food safety issues, based on which recommendations were made to increase public awareness of food safety as well as to improve government's certification system.

Keywords: optimistic bias; social trust; information behavior; food safety; certification mark; purchase intention

---

1. Introduction

FAO/WHO (Food and Agriculture Organization of the United Nations/World Health Organization) defined food safety risk management as conducting suitable undertakings, controlling the risk, and protecting public hygiene, and the primary goal of the management of risks associated with food is to protect public health by controlling such risks as effectively as possible through the selection and implementation of appropriate measures [1]. WHO further estimated of the global burden of foodborne diseases; the report suggested that, whether in developed or developing countries, about 10% of people worldwide will die from foodborne diseases. Therefore, effective risk communication to the public may be an important solution to improve people's health [2]. Food safety is a cross-cutting issue related to human behaviors such as psychology, consumer behavior, and information technology [2–4]. In particular, the messages conveyed in risk communication can help people decide whether they can adopt some strategies to protect them from risk [2].

In 2014, food safety in Taiwan almost collapsed due to food scandals related to cooking oil containing recycled waste oil and animal feed oil. In 2016, some restaurants in mainland China illegally added opium poppy shell powder to hot pot, causing consumers to develop opiate addictions.

As indicated by these major food safety incidents, although governments and food manufacturers should bear absolute responsibility, consumers must also have the wisdom to distinguish safe food.

Optimistic bias is one of the factors that affect people's purchase intention [5]. Despite the numerous food safety incidents in Taiwan and mainland China, people remain optimistic about food consumption. This optimistic bias may be due to people ignoring the fact that everyone is in the same situation and has the same possibility of eating unsafe food; as a result, they believe they are less likely than others to face risks. People in Taiwan and mainland China generally do not trust their governments or the certification and authentication systems [6–9]. They are dissatisfied with the current state of food safety and tend to control it themselves. In other words, people think and act differently from the government policy regarding food safety, and they tend to take individual control by themselves with optimistic biased thinking.

This study aimed to identify the factors that affect college student's consumption of certified food and agricultural products, and to understand their cognitive processes, including consumer awareness, information credibility, attention to news, and consumers' social trust in government. These variables were examined through a model and further explored for policy making. In addition, this study attempted to compare the differences in the cognitive processes of food consumption of Taiwanese and Chinese college students under the impact of similar dietary habits and food safety incidents in response to different lifestyle, civil culture, and social systems.

2. Literature Review

2.1. Optimistic Bias and Food Safety Risks

Consumers tend to take responsibility and have the ability to avoid risk in food risk management [10]. Individuals may be able to assert that their chances of negative risks are less than average. However, it is practically impossible to prove that one is right. If everyone thinks they are at less than average risk, it could lead to systematic errors [11]. This error is called optimistic bias, which means that people tend to believe that they are less likely than others to encounter negative consequences, and that they are more likely than others to experience positive events [12]. Optimistic bias occurs when individuals compare the likelihoods that a risk event will occur to themselves and to others [11]. If the difference in the probability of assessing individual and others risks is greater, higher optimistic bias will result. Optimistic bias is essentially unrealistic optimism, which means that people tend to underestimate the likelihood of risky events, so they may not respond appropriately to those events [13].

Many studies in the field of social psychology or health communication have shown that optimistic bias exists [14]. Researchers have found that young people have higher self-efficacy in food safety, and they believe that they have sufficient knowledge to avoid risks [15,16]. In addition, research in Taiwan and mainland China has indicated that people have optimistic biases on public health issues such as bird flu outbreaks [16], prostate cancer screening [17], and the Influenza A virus subtype (H1N1) Vaccine [18]. Food safety, as the abovementioned issues, was also a risky problem in daily life for individuals, especially with many food safety crises happened in recent years. Although related studies supported that risk perception affected individual food purchase intentions and actual behaviors, further examination of how optimistic bias occurred when people perceived risky event was needed [19]. On the assumption that people have optimistic bias about food safety, previous studies suggested that people with higher safety awareness actually possessed purchasing intentions on safer food [20]. In order to facilitate individuals to make appropriate risk assessment, reducing the optimistic bias could be helpful to increase purchasing intentions.

2.2. Consumer Cognition of Certified Food and Agricultural Products

On the other hand, although consumers may have a neutral response to food safety issues, most of them agree that other people need more food safety advice than they do [21]. Generally, consumers worried about food safety of pesticides, antibiotics, and heavy metal pollution problems within the production and also worried about additives or industrial hazardous substances in food processing [22]. Consumers are the most important stakeholders in overall food safety management. Consumer cognition indicates consumer value, and this cognition is a key factor in the success of agricultural products with traceability markers [23]. Effective interventions should improve consumers' risky food-handling behaviors via communication strategies or educational interventions [24,25], especially under consumers' weak awareness of active prevention of food safety. Previous study supported that consumers seldom complain about food safety despite of their roles as the stakeholders involved and impacted by the problem [22]. Other studies have found that the source and certification of food access are the reasons consumers decide to buy food [6,20,26]. Therefore, it can be assumed that the attributes of certified foods will influence consumers' purchase intentions because the certification mark provides some of the information that consumers expect.

2.3. Information Credibility and Attention to Food Safety

Information processing includes information attention, information access channel, and information source preference [9]. Researchers on risk communication of food safety have highlighted the importance of information accessibility, demand, and consumer-oriented risk perception [21,27,28]. The current study mainly focused on information attention and the credibility of access channels because certification marks may attract people's attention and convince consumers that "certified products are safe."

When people are aware of food safety risks, they need access to enough information to protect them, and the current food scandals have increased media attention on food safety. When people are exposed to this risk information, their concerns about food safety will be aroused, and they will temporarily change their behavior to mitigate risk [29,30]. If people have more demand for search information, they may take some steps to face the risk and try to solve the problem from the risk [31,32]. Therefore, this study assumed that if people pay more attention to food safety issues, they will be more willing to purchase certified food products.

Regarding the credibility of information, previous research mainly examined two aspects: information channels and information sources [9,31,33]. Wei et al. [32] pointed out that if the message is more credible, it can reduce optimistic bias and may change consumers' behavior. Hu and Yueh [34] further proposed that the use of government information sources enhanced the credibility of risk information and reduced the public's optimism bias. However, the public lacks incentives to report unsafe food to the government, and government regulations and policies need to be improved to further increase public awareness of food safety [35]. Therefore, consumer behavior and purchase intention depend on the credibility of the source of information, including regulations and government. The credibility of information is related to the legal effect and reputation [36–38]. Since the credibility of the information source reduces the uncertainty of the information seeker and increases the willingness to take risks, the credibility of the information may be an effective factor in consumer decision-making [17]. Therefore, there is reason to believe that if people think the information on food safety issues is more credible, they may be more willing to buy certified food.

2.4. Social Trust in Government in Food Safety Management

Risk management includes assessment, evaluation, and monitoring. The Taiwanese and mainland Chinese governments have both implemented certification systems to inform the public about food safety issues and to supervise food manufacturers. Most people tend to read food safety advice on food packaging [21]. In Taiwan and mainland China, when consumers want to understand the traceability of food, certification is the most important consideration because it can enhance consumer confidence and purchase willingness, and at the same time, it plays a role in educating consumers [8,39]. However, the reasons why consumers purchase certified food product rather than uncertified product are still called for further investigations. With reference to the compilation of cross-strait food certification marks by Liao et al. [33] and the food certification marks in mainland China introduced by Yin et al. [40], this study has compiled a comparison table for Taiwan and mainland China certification marks, as shown in Table 1.

Table 1. Common certification marks in Taiwan and mainland China.

Taiwan		Mainland China	
Good Agricultural Practice, GAP	This mark represents that the product contains allowable levels of pesticides and that farmers and the public are taught about proper pesticide usage.	QS mark	"QS" is the abbreviation of "Quality and Safety." In mainland China, all foods sold on the market should carry this mark.
Certified Agricultural Standards	This mark indicates that agricultural products meet Taiwan's agricultural standards.	Green Food	This symbol indicates that agricultural products are grown in an environment-friendly manner with less pollution and good quality.
Traceability Agricultural Product, TAP	This mark means that agricultural products can be traced back to their production processes and are environmentally friendly.	Harmless Agricultural Product	This mark indicates that the pollution content of agricultural products is within the safe range and meets standards.
CAS organic agricultural Product	This mark refers to organic products that meet standards during processing.	Organic Food	This mark refers to organic foods and confirms the absence of artificial materials, genetic engineering and environmental pollution during the production process.

Huang et al. [6] pointed out that although people lack sufficient background knowledge, they still like to see the government develop policies and encourage agricultural tourism. From the consumer perspective, they expect the government to provide a reliable certification mark to enhance consumer trust [41]. In addition, despite of the fact that consumers do not totally trust the government, they are still more willing to purchase government-certified products and trust the certification systems to a certain extent [20,42]. Previous study has found that the more consumers trust the food certification organization, the greater their willingness to purchase certified food [43]. Accordingly, the credibility and trust of government might be key points to consider when consumers buy food product.

Recent studies have also focused on the strategies and effects of regulators and governments in communicating with the public [38,44]. However, these studies have only discussed consumer or government issues separately, and a few have examined the relationship between consumers' intentions to buy food and their social trust in government. Social trust raises individuals' concerns about social issues and causes divisions in civil society [45]. Moreover, social trust also reflects people's assessment of the structure of society, which is based on the individual's knowledge that society consists of different groups having the same destiny.

Social trust can be measured by investigating government credibility. When the public lacks background information or is at a high potential risk, people will make decisions based on social trust [18]. If a food safety risk event occurs suddenly, people do not actually have enough information, which can also lead to a decline in government credibility. As suggested by Hu and Yueh [34], to improve risk communication, the government and media should value the interaction function of social media, reasonably encourage communication transparency, avoid providing incomplete information, and reduce the number of "unspoken secrets."

Food safety issues are very important for college students; in particular, research has shown that a large proportion of them tend to eat out and therefore pay more attention to food safety incidents [46].

Although they may lack a positive attitude towards food safety practices and food safety [47,48], however, such students are also more flexible in accepting new things. Besides, well-educated young adults have higher acceptance of certified foods and more demand for consumer choice [8].

3. Materials and Methods

3.1. Hypothesis and Research Model

Based on the review of the literature, this study selected five concepts as the independent variables: optimistic bias; consumption cognition; information process, comprising news attention and information credibility; and social trust. From the perspective of optimistic bias, when consumers prepare food, their confidence in food safety behaviors is related to their strong belief in risk prevention [21]. Consumption cognition is related to what consumers will consider when purchasing food. Information processing includes attention to food safety news and the credibility of information sources and messages per se. Social trust is a public assessment of government credibility. Purchase intention of certified food product was selected as a dependent variable since it reflected the efficiency of governmental policies on food safety.

In sum, this study first examined college students' optimistic bias (H1), and further considered a framework consisted of abovementioned variables. Figure 1 presents the model developed by this research, assuming the relationship between the above variables and purchase intention, and the interaction between these variables will be verified. The hypotheses for testing the model were as follows (H2–H6).

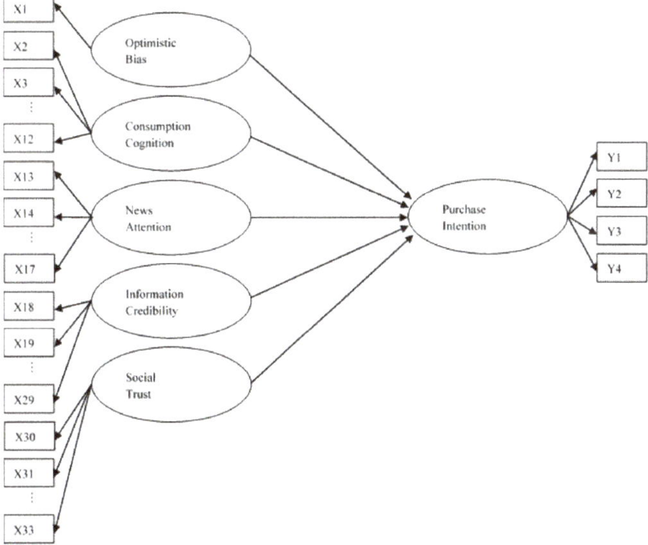

Figure 1. The conceptual model of food safety cognition and consumption attitude. Note: X1 to X33 and Y1 to Y4 refers to the sequential numbers of questionnaire items.

Hypothesis 1 (H1). *Students have optimistic bias on food safety incidents.*

Hypothesis 2 (H2). *Optimistic bias has a negative impact on purchase intention of certified food products.*

Hypothesis 3 (H3). *Consumption cognition has a positive impact on purchase intention of certified food products.*

Hypothesis 4 (H4). *News attention of food safety issues has a positive impact on purchase intention of certified food products.*

Hypothesis 5 (H5). *Information credibility of food safety issues has a positive impact on purchase intention of certified food products.*

Hypothesis 6 (H6). *Social trust has a positive impact on purchase intention of certified food products.*

Furthermore, this study compared models of food safety cognition and consumption attitudes among students in Taiwan and mainland China. Wilson et al. [38] conducted a cultural comparison study and found no significant differences in consumer perceptions of food safety regulators in Australia, New Zealand, and the United Kingdom. However, a study by Van Dijk et al. [44] found that communications with consumers, especially on food safety issues, may differ among cultures and countries.

Recent research on food safety issues in Taiwan and mainland China compares policies, government regulations, regulations, and certification marks [33,49], suggesting that social systems, cultures, and governments can cause different food safety issues and situations. Although Taiwan and mainland China have similar dietary habits (Chinese food), the two societies and civil cultures are different. To compare the two samples, this study tested the following hypothesis.

Hypothesis 7 (H7). *There are differences in the two models of certified food purchase intentions for Taiwanese and mainland Chinese students.*

3.2. Data Distribution and Reliability

R 3.5.3 in RStudio 1.1.414 and LISREL 8.8 were initially used for analysis. This study consisted of multi-item scales of optimistic bias, consumption cognition, news attention, information credibility, social trust, and purchase intention. The study first checked that the data distribution conformed to the normality distribution. Items with skewness higher than 3 and kurtosis higher than 10 would be deleted.

3.3. Structural Equation Modelling

Using structural equation modelling (SEM) provides researcher a comprehensive picture of the theoretical framework [50]. The parameters of SEM used maximum likelihood estimation. The models were built and tested with a two-step paradigm [50] wherein the measurement model was tested first and the structural model was tested next. The measurement model was developed by confirmatory factor analysis. For determining the goodness of fit of CFA model, limiting the standard factor load value to between 0.5 and 0.95 is suggested [51].

Several indices were used to determine the goodness of fit of structural model. According to previous recommendations, normed chi-square (NC) (χ^2/df) < 5 is acceptable, and <2 is a good model fit. As for root mean square error of approximation (RMSEA), <0.8 is good and <1 is an acceptable model fit [52]. In addition, confirmatory fit index (CFI) > 0.9 is a good model fit. If the analysis fails to meet the criteria for model fitting, the hypothetical model requires modification of alternatives.

3.4. Invariance Tests with Structural Equation Modelling

This study used SEM and the method proposed by Koufteros and Marcoulides [53] to perform two stages of invariance tests. For comparison, following Koufteros and Marcoulides [53], the invariance tests consisted of two phases, namely, the measurement model test and the structural model test.

Measurement equivalence can solve whether different groups of respondents interpret a given measure in a similar manner, so according to the suggestion of Vandenberg and Lance [54], the invariance of the measurement model was tested with 6 steps. After conducting the invariance of the measurement model, models can be ensured that they were measured in a similar way.

These 6 steps method will generate 6 models; in order to compare them, the χ^2 differences would be tested in these 6 invariance test models. In addition, this study also considers the comparative fit index (ΔCFI), which represented in the results if it will be less than −0.01 or not. According to Cheung and Rensvold [55], for detecting cross-group differences in measurement invariance, values of ΔCFI < −0.01 indicate no significant differences between different invariance models.

To test the structural model invariance, the current study also followed the research method of Teng and Lu [56] to set the path coefficients from consumption cognition to purchase intention to be equal and freely estimated other path coefficients for the two samples.

3.5. Participants and Data Collection

Considering that urban residents have higher levels of knowledge and consumption than rural areas [8], their awareness of food safety knowledge and certified foods may be higher. This study was conducted using a purposeful sampling strategy, which targeted university students in two metropolises, Taipei and Beijing.

The respondents of this study were fully informed about the research purpose, and their participation was voluntary and autonomous. No additional ethical approval was required in this study, in accordance with national and institutional requirements. The numbers of completed questionnaires collected in Taipei and Beijing were 258 and 268, respectively, which satisfied a reliable samples size of 200 for SEM [57]. The survey included male and female students with different majors from different school year in the National Taiwan University and Peking University. The majority of respondents (92.6% in Taipei and 84.7% in Beijing) were aged between 18 and 23.

3.6. Measurement Scales of Variables

This study used multi-item scales to measure the major constructs, including students' "optimistic bias," "consumption cognition," "news attention," "information credibility" and "social trust" on food safety issues, as well as their "purchase intention." Based on the theoretical framework shown in Figure 1, these scales were modified from previous studies, and all items of scales can be regarded as observed measures for the latent variables with 5 major constructs [50]. To test the "optimistic bias," two kinds of judgments, i.e., comparative judgment and absolute judgment, were measured. For absolute judgment, numerical evaluation was used to find the relationship between self and personal experience [13]. Absolute judgment was evaluated with two questions: one to first assess the possibility that the "self" would experience risk and the other to assess the possibility that the "others" would experience risk.

This study referred to and revised the tools used by Lu et al. [31], Lu et al. [18], and Weng et al. [5] to measure optimistic bias. Two items were used to assess optimism, namely, "I think I myself may eat unsafe food" and "I think others may eat unsafe food" on a 6-point Likert-type scale (ranging from 1 = "strongly disagree" to 6 = "strongly agree"). Furthermore, the degree of optimistic bias was calculated by subtracting the self-optimistic evaluation value from the optimistic evaluation value of others. A higher score indicated higher optimistic bias.

The variables "consumption cognition," "information credibility," "news attention," "information credibility," and "social trust" were measured using a 6-point Likert-type scale (ranging from 1 = "strongly disagree" to 6 = "strongly agree"). The "consumption perception scale" consisted of 11 items and was a tool modified from Huang et al. [6]. The "news attention scale" consisted of 5 items and was modified from the scales developed by Lu et al. [18] and Wei et al. [32]. The "credibility of information scale," modified from Lu et al. [46] and Wei et al. [32], was composed of 4 items related to message channels and 7 items related to message sources. The "social trust scale" contained four items revised with reference to the tools of Lu et al. [18]. Finally, the "purchase intention scale" contained 4 items, which were revised with reference to Lu et al. [18] and Weng et al. [5].

4. Results

4.1. Descriptive Statistics

The results of the analysis confirmed that all survey items were with skewness lower than 3 and kurtosis lower than 10. Table 2 provides the measurement items of the Taiwan and mainland China samples, the scale reliability (Cronbach's α coefficient), and the factor load value. In this study, the reliability values of the subscales of the two samples were calculated, and the Cronbach's α coefficients ranged from 0.7047 to 0.9481, indicating that the constructs included in the scale had high internal consistency.

Table 2. Descriptive statistics and confirmatory factor analysis of scale items in food safety (Taiwan, N = 258; mainland China, N = 268).

Statement	M [1] (T)	M (C)	SD (T)	SD (C)	Factor Loading (T)	Factor Loading (C)
Optimistic bias (α_T = 0.9021, α_C = 0.9281) [2]						
I think I myself may eat unsafe food.	4.91	4.83	0.92	1.42	–	–
I think others may eat unsafe food.	4.96	4.95	0.96	1.36	–	–
X1. "others" minus "self"	0.05	0.12	0.56	0.72	–	–
Consumption cognition (α_T = 0.8765, α_C = 0.9050)						
I value						
X2. food products having certification	3.90	4.28	1.11	1.62	0.71	0.71
X3. the package	4.60	4.40	1.17	1.53	0.65	0.70
X4. the expiration date	5.33	5.51	0.88	1.16	–	–
X5. food products having safe and hygienic marks	4.57	4.62	1.04	1.56	0.66	0.83
X6. food products made in my nation	3.52	2.88	1.18	1.46	0.64	–
X7. place of production	3.88	3.14	1.21	1.51	0.64	–
X8. freshness	5.15	5.24	0.92	1.25	0.57	0.67
X9. food products being contaminated	4.88	4.64	1.05	1.52	0.54	0.85
X10. food products having pesticide residue	4.48	4.45	1.14	1.58	0.66	0.82
X11. the record of production	3.74	3.61	1.13	1.60	0.74	0.69
X12. the manufacturers	3.97	3.75	1.19	1.66	0.63	0.56
News attention (α_T = 0.7353, α_C = 0.8358)						
I pay attention to food safety news on ...						
X13. newspaper	3.54	2.99	1.45	1.63	0.72	–
X14. television	4.09	3.74	1.30	1.63	0.80	0.80
X15. the Internet	4.76	2.68	1.07	1.55	0.58	0.60
X16. radio	2.59	4.56	1.38	1.49	0.53	0.77
X17. mobile phone (e.g., scanning the QR code)	3.10	3.59	1.53	1.63	–	0.60
Information credibility (α_T = 0.9155, α_C = 0.9481)						
Information about food safety on/from ... is credible.						
X18. newspapers	3.61	3.40	1.09	1.40	0.90	0.87
X19. television	3.48	3.44	1.05	1.41	0.94	0.89
X20. the Internet	3.38	3.21	1.01	1.33	0.76	0.87
X21. radio	3.38	2.87	1.11	1.27	0.84	0.79
X22. mobile apps	3.39	3.12	1.03	1.33	0.78	0.79
X23. media	3.28	3.13	1.10	1.33	0.81	0.83
X24. government	4.05	3.85	1.14	1.45	–	0.74
X25. research institutes	4.55	4.31	0.99	1.44	–	0.74
X26. consumer protection agency (NGO)	4.34	2.99	1.02	1.32	0.52	0.77
X27. manufacturers	3.00	2.76	1.09	1.30	–	0.68
X28. family members and friends	2.94	3.14	1.20	1.40	0.51	0.64
X29. Facebook pages/WeChat official accounts	2.65	2.64	1.11	1.24	0.57	0.68
Social trust (α_T = 0.7047, α_C = 0.8151)						
I think governments ... in/on food safety.						
X30. are correct to adopt policies	3.57	3.97	1.11	1.49	0.86	0.85
X31. are credible to adopt policies	3.40	3.59	1.14	1.41	0.96	0.90
X32. are able to solve problems	3.31	3.70	1.30	1.48	0.65	0.67
X33. should develop long-term plans	5.15	5.28	1.06	1.33	–	0.51
Purchase intention (α_T = 0.9150, α_C = 0.9444)						
I tend to ... food products which have certification marks.						
Y1. buy	4.74	4.68	1.09	1.58	0.97	0.96
Y2. eat	4.80	4.75	1.07	1.52	0.96	0.97
Y3. increase instances of buying	4.66	4.67	1.13	1.55	0.78	0.85
Y4. pay more money to buy	4.19	4.30	1.20	1.61	0.68	0.76

[1] The abbreviations used in this table and throughout this paper are as following: M = mean and SD = standard deviation. In the table, Taiwan also was abbreviated T, whereas mainland China was abbreviated C. [2] The value of α indicated Cronbach's alpha. Taiwan was abbreviated T, whereas mainland China was abbreviated C. In addition, X1 to X33 and Y1 to Y4 indicated the sequential numbers of questionnaire numbers. Measures of latent exogenous variables represented X, whereas measures of latent endogenous variables represented Y.

This study checked the standard factor load value to between 0.5 and 0.95. Based on the analysis results, 5 items in the Taiwan questionnaire and 6 items in the mainland China questionnaire were deleted because they did not meet the loading score criteria. The factor loading values of all questionnaire items are listed in Table 2. After modification of the hypothesized items, all exogenous variables had good fit.

4.2. Identified Optimistic Bias

To determine whether students are optimistic about food safety incidents, this study proposed the hypothesis that respondents believe they are less likely than others to face this risk. Results of paired *t*-test analysis showed that respondents in Beijing believed the risk of eating unsafe food "themselves" (M = 4.83) was lower than that of "others" (M = 4.95) (t = −2.6977, df = 267, $p < 0.01$). However, Taipei's respondents reported that there was no significant difference in the likelihood of "themselves" (M = 4.91) and "others" (M = 4.96) eating unsafe food (t = −1.3375, df = 257, $p = 0.0911$). Therefore, the H1 hypothesis was supported by the test results of mainland Chinese respondents but rejected by the test of Taiwanese respondents.

4.3. Hypothesis Testing of Taiwan Sample Structural Model

The hypothesized model included five latent exogenous variables (optimistic bias, consumption cognition, news attention, information credibility, and social trust) and one latent endogenous variable (purchase intention). This study used SEM to test the relationships between variables. In the Taiwan sample, the results of structural modelling analysis showed adequate fit to the data, as evidenced by χ^2/df (NC) = 3.20, CFI = 0.91, and RMSEA = 0.093. However, the correlations between "optimistic bias" and other variables did not reach significance. Therefore, the correlation between "optimistic bias" and other variables was rejected, and an alternative model was constructed accordingly.

SEM analysis of the alternative model in the Taiwan sample indicated acceptable data fit: χ^2/df (NC) = 3.17, CFI = 0.91 (≥0.9), and RMSEA = 0.092 (≤0.1). Table 3 presents the correlation coefficients between variables. As shown in Figure 2, the results of hypothesis testing demonstrated that only consumption cognition had a positive impact on purchase intention (β = 0.60, t = 8.41; $p < 0.001$); hence, H3 was supported. On the other hand, optimistic bias (β = 0.08, t = 1.55; $p > 0.05$), news attention (β = 0.09, t = 1.20; $p > 0.05$), information credibility (β = 0.02, t = 0.33; $p > 0.05$), and social trust (β = −0.01, t = −0.23; $p > 0.05$) did not predict purchase intention; therefore, H2, H4, H5, and H6 were all rejected.

Table 3. Correlation coefficients between variables in Taiwan sample model.

	Optimistic Bias [1]	Consumption Cognition	News Attention	Information Credibility	Social Trust
Optimistic bias	1.0000				
Consumption cognition	0.0338	1.0000			
News attention	−0.0914	0.3721 *** [2]	1.0000		
Information credibility	−0.0429	0.2809 ***	0.4468 ***	1.0000	
Social trust	0.0447	0.2554 ***	0.1744 **	0.3395 ***	1.0000

[1] In the alternative model, the correlation relationships between optimistic bias and other variables were rejected.
[2] *** $p < 0.001$, ** $p < 0.01$.

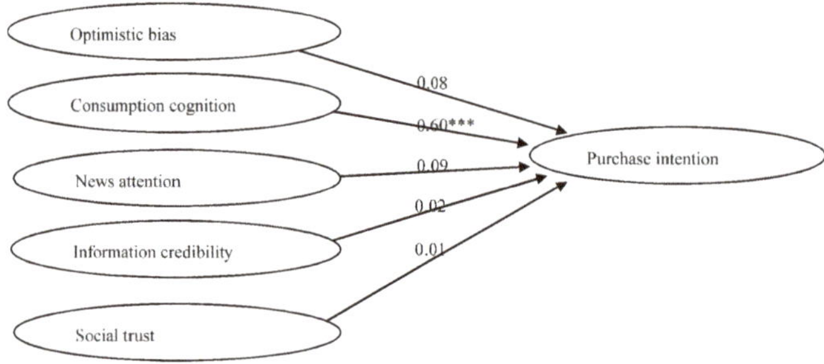

Figure 2. Path coefficient results of alternative model of the Taiwan sample. Note: *** $p < 0.001$; $\chi^2 = 1348.26$; df = 424; χ^2/df (NC) = 3.17; CFI = 0.91; RMSEA = 0.092.

4.4. Hypothesis Testing of Mainland China Sample Structural Model

In the mainland China sample, the results of structural modelling showed adequate fit to the data, as evidenced by χ^2/df (NC) = 3.56, CFI = 0.94, and RMSEA = 0.098. As in the Taiwan model, no significant correlations between "optimistic bias" and any other variables were found, and an alternative model was constructed for further analysis.

SEM analysis results of the alternative model in the mainland China sample indicated acceptable data fit: χ^2/df (NC) = 3.55, CFI = 0.94 (≥0.9), and RMSEA = 0.098 (≤0.1). Table 4 shows the correlation coefficients between variables. As shown in Figure 3, the results of hypothesis testing demonstrated that both consumption cognition (β = 0.38, t = 5.25; $p < 0.001$) and news attention (β = 0.24, t = 2.21; $p < 0.01$) had positive impacts on purchase intention; thus, H3 and H4 were supported. On the other hand, optimistic bias (β = −0.03, t = −0.60; $p > 0.05$), information credibility (β = 0.14, t = 1.59; $p > 0.05$), and social trust (β = 0.07, t = 1.08; $p > 0.05$) did not predict purchase intention; therefore, H2, H5, and H6 were all rejected.

Table 4. Correlation coefficients between variables in the mainland China model.

	Optimistic Bias [1]	Consumption Cognition	News Attention	Information Credibility	Social Trust
Optimistic bias	1.0000				
Consumption cognition	−0.0068	1.0000			
News attention	−0.0459	0.5580 *** [2]	1.0000		
Information credibility	−0.0878	0.4170 ***	0.6359 ***	1.0000	
Social trust	−0.0968	0.4575 ***	0.5278 ***	0.5740 ***	1.0000

[1] In the alternative model, the correlation relationships between optimistic bias and other variables were rejected.
[2] *** $p < 0.001$

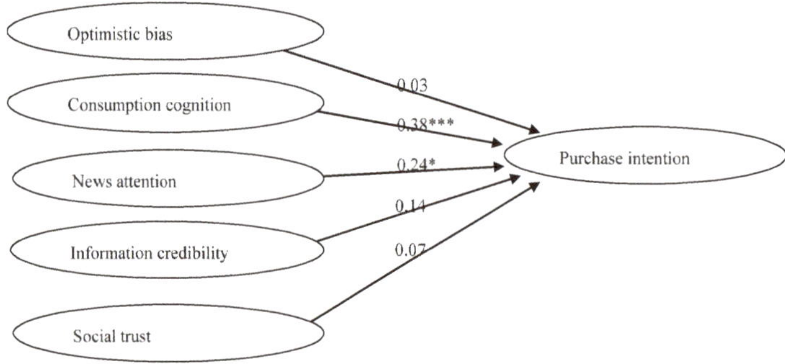

Figure 3. Path coefficient results of alternative model of the mainland China sample. Note: (1) *** $p < 0.001$, * $p < 0.5$; $\chi^2 = 1723.19$; $df = 485$; χ^2/df (NC) = 3.55; CFI = 0.94; RMSEA = 0.098.

4.5. Invariance Test for Comparison of Taiwan and Mainland China Models

For H7, it was hypothesized that differences in the two models of certified food purchase intentions would exist for the Taiwanese and mainland Chinese students. As shown in the previous model analysis, H3 was supported in both models; H4 was rejected in the Taiwan model but supported in the mainland China model; and H2, H5, and H6 were rejected. To verify the statistically significant differences, this study examined the degrees of invariance of the two models and then set up other fixed paths to calculate the difference in the causal coefficient of the H3 path.

Results of invariance of the measurement model analyses are shown in Table 5. To ensure that they were measured in a similar way, some items were removed because their variances or means were outliers.

Table 5. Six steps of measurement invariance test between the two models.

Model [1]	χ^2	$\Delta\chi^2$	df	Sign. Level	RMSEA	NNFI	CFI	ΔCFI
1. Configural Invariance	1564.0978	–	486		0.09820	0.9215	0.9309	–
1 versus 2	–	41.4246	–	0.000	–	–	–	−0.0014
2. Metric Invariance	1605.5224	–	505		0.09730	0.9229	0.9295	–
2 versus 3	–	144.0124	–	0.000	–	–	–	−0.0054
3. Scalar Invariance	1749.5348	–	528		0.9870	0.9207	0.9241	–
3 versus 4	–	48.2809	–	0.000	–	–	–	−0.0024
4. Factor Covariance Invariance	1797.8157	–	538		0.09930	0.9197	0.9217	–
4 versus 5	–	36.4852	–	0.000	–	–	–	−0.0015
5. Factor Variance Invariance	1834.3009	–	533		0.1008	0.8930	0.9202	–
5 versus 6	–	136.8797	–	0.000	–	–	–	−0.0094
6. Error Variance Invariance	1971.1806	–	552		0.1047	0.9108	0.9108	–

[1] χ^2 represents chi-square. df represents degree of freedom. Sign. Level represents significant level (α = 0.05). RMSEA represents root mean square error of approximation. NNFI represents Non-normed Fit Index. CFI represents Comparative Fit Index. Δ represents the differences.

Model 1 was used to test for configural invariance and freely estimate its factor loadings, factor variances, and covariance. Model 1 fits indices to the data RMSEA = 0.09820, NNFI = 0.92, and CFI = 0.93, and it was suggested to be plausible in all measurement situations. Next, model 2 tested metric invariance and suggested that the factor loading values of items were equal in the measurement scenario. Model 3 was a scalar invariance model; model 4 was used to test factor covariance invariance; model 5 was to test factor variance invariance; and model 6 was used for error variance invariance testing. After all analyses were performed in the LISREL program, the χ^2 difference and the change in CFI values were calculated.

Although for the overall items in the Taiwan model and the mainland China models, the χ^2 differences were significant in the invariance test models, the ΔCFI represented in the results was less than −0.01. According to the values of ΔCFI, therefore, the results showed that the invariance test of the measurement model was equal between the two groups and that the Taiwanese and mainland Chinese respondents in the study responded to the questionnaires in similar manners.

To test the structural model invariance, the current study also followed the research method of Teng and Lu [56]. As shown in Table 6, the causal coefficients of the baseline model of the two samples were freely estimated, and the path 2 model was constrained only by the causal coefficients of path 2, while other paths were freely estimated. According to the analysis results, the χ^2 difference indicated that the causal coefficient of the H3 path in the Taiwan model was significantly different from that of the mainland China model (Δx^2 = 4.4513, Δdf = 1, p = 0.03). Therefore, H7 was supported.

Table 6. Invariance test of the two-group structural model.

Model	χ	df	$\Delta\chi$	Δdf	Significance Level
Baseline model	1377.669	432			
Consumption cognition → Intention model	1382.12	433	4.4513	1	0.03 *

χ represents chi-square, df represents degree of freedom, and Δ represents the differences. * $p < 0.05$.

5. Discussion

5.1. Optimistic Bias of Taiwan and Mainland China

The reason why Taiwan college students did not have optimistic bias may be due to their personal life experience. Weinstein [11] pointed out that if people have experienced related risk events in past, they may increase the perceived likelihood that they will encounter risks, thus reducing the optimistic bias. In 2014, the news media in Taiwan reported food scandals related to edible oils, recycled waste oil, and animal feed oil. Cooking oil is involved in food preparation, and almost all foods consumed in daily life in Taiwan require the use of cooking oil. Unlike other past events, which only affected certain groups of people or certain kinds of food, this food safety crisis of edible oil, therefore, affected almost all the residents of Taiwan. This crisis led to a general perception among college students in Taiwan that their likelihood of encountering this risk was the same as the risk faced by everyone.

However, unlike college students in Taiwan, college students in mainland China had optimistic bias. This result showed that the students still had cognitive errors [11], and they tended to think that they were unlikely to encounter this risk. To reduce this optimistic bias, correct and sufficient information is highly needed. The intervention of education to make student know more about food safety issue can also improve students' risk perception and confidence toward food [25].

5.2. The Differences of Structural Models between Taiwan and Mainland China

The Taiwan and mainland China models both significantly illustrated the causal relationship between consumption cognition and purchase intention, which echoed the findings of Huang et al. [6]. However, the results of this study are inconsistent with the findings of Elliott and Ellison [58]. They found that food safety awareness among middle school students is that the concept of food safety is equivalent to a food safety scandal. However, the results of this study found that from the perspective of consumer cognition, college students pay more attention to other aspects of food safety in addition to scandals.

Consumption cognition covers how people consider traceability, perceive the risk of food unsafety, and consume foods with organic labels and certification marks. The consumption cognition represented the value consumers place on food and affected their willingness to purchase certified food. Comparing the consumption models of college students in Taiwan and those in mainland China, the consumption cognition of Taiwan respondents had a greater impact on the causality of purchase intention. This result suggests that Taiwan respondents possessed better comprehension toward food consumption so that they relied on themselves and applied their own food knowledge autonomously. On the other hand, a previous study showed that people in mainland China cared more about the production date and shelf life of food product [59], which could result in weak purchase intention of certified food. However, the results of the study also suggested the needs of more active measures taken by the governments and manufacturers to increase consumer's inherent awareness of food selection, preparation, handling, and even cooking and eating. For the college students in both Taiwan and mainland China, imparting more knowledge about food to increase awareness may cause people to value food and make them want to buy certified food. Despite some gaps in knowledge about certified food, people are still willing to buy it and are motivated by health and environmental reasons [42].

Another difference between Taiwanese and mainland Chinese college students was the impact of news attention on purchase intention. In Taiwan, there was no significant causal relationship between news attention and willingness to buy, but news attention had an effect in mainland China. People are eager for more information because they are aware of the risks and want to take actions to reduce their uncertainty about the risks [60]. However, young adults often lack the knowledge or motivation to use their ability to solve food safety issues, and this information behavior is temporary [15,30]. To develop lasting habits, this study considered that education on food media literacy and drawing attention to food safety issues are very important. In addition, mobile media have received a lot of attention in mainland China, especially because information on the food scandals was widely disseminated by the public on social media [61], and most users take complete advantages of the multimedia and interaction functions available on social media [34]. Such new media types might continue to remind people to pay attention to food safety.

In both the Taiwan and mainland China samples, the causal effects of information credibility and social trust on purchase intention were not significant. Lu et al. [18] pointed out that if the news media continue to expose negative scandals about these events and if the government does not explain them in due course, the public will no longer trust the government. Therefore, the results of this study indicated that neither the Taiwanese nor the mainland Chinese government has responded appropriately on food safety issues. Because government-certified foods lack the public's social trust, people do not consider information about food and social trust when deciding whether to buy food with a certification mark. To make communication more effective, regulators need to reduce consumer-perceived information overload, provide targeted consumers with the right information, and reduce the gap between consumers and regulators [10,38].

6. Policy Implications

6.1. The Development of Food Certification Marks

Since 2013, the State Council of the People's Republic of China has formulated 11 regulations and several documents on food safety. In 2015, the "Food Safety Law" was promulgated and implemented. To improve risk management of food safety, the government tried to revolutionize the organizations [62]. Since there are several related agencies responsible for food safety [63], the government has attempted to cooperate with these agencies. Overlapping government agencies include the Ministry of Commerce, the Ministry of Industry and Information Technology, the Ministry of Public Security, the Ministry of Agriculture and Rural Affairs, the Food and Drug Administration, the Quality Supervision Administration, and the National Security Administration [7].

On the other hand, after the cooking oil scandal in Taiwan in 2014, the public's anger at Taiwan's harsh food safety environment reached its highest point. As a result, the Taiwanese government established the "Food Safety Office of the Executive Yuan" to help liaise with interministerial food safety policies including the Agricultural Committee, Environmental Protection Agency, and Ministry of Health and Welfare. At the same time, the Taiwan government formulated the "Act Governing Food Safety and Sanitation" at the end of 2014 as the highest guiding law for the management of food safety and hygiene.

Section 2.4 of this paper summarizes the certification marks currently used by both governments. The legal sources of food certification marks in mainland China include the Regulation of the People's Republic of China on the Administration of Production License for Industrial Products and the Trademark Law of the People's Republic of China. In order to make it easier for consumers to choose safe food, the Taiwan government has issued about 20 different certification marks. These certification marks have different legal sources and are issued by different units. It includes Agricultural Production and Certification Act, Agricultural Food Control Act, Act Governing Food Safety and Sanitation, etc.

In addition, in the context of promoting cross-strait exchanges, in November 2008, the Taiwan government and the Chinese government signed the "Cross-Strait Food Safety Agreement"; the purpose is to enhance cross-strait food safety communication and mutual trust. This agreement was negotiated between the Straits Exchange Foundation and the Association for Relations across the Taiwan Straits.

6.2. Food Safety Incidents and Government Response

Over the past 20 years, there have been many serious food safety incidents in mainland China and Taiwan. In China, in 2008, melamine-contaminated milk powder caused babies with kidney stone disease. The milk powder producer Sanlu also lost public trust [63]. In 2010, another gutter oil scandal occurred in China. Between 2005 and 2011, news reports indicated that three leather milk incidents occurred. Milk manufacturers use waste leather to squeeze leather protein powder and add it to milk to increase the ratio of milk protein. In 2015, "zombie meat" was illegally imported into China, and these meats are all frozen meat that has expired. As food delivery services have become more and more popular in recent years, many related food safety scandals have also appeared in 2016.

Faced with so many food scandals, starting from 2011, the Food Safety Committee of the State Council of China held the "China Food Safety Promotion Week" every June. The project is seen as an educational activity for the public to popularize scientific knowledge and improve public perception of risk. Until 2019, it has been held about 8 times. In addition, at the end of 2019, the new coronavirus disease affected the entire world. Under such circumstances, food safety has also become a worrying issue. In China, traditional eating habits may increase the possibility of infection. In particular, February 2020 is the Chinese New Year, and the family's return to their hometowns also makes the virus more likely to spread more quickly. Therefore, the habit of eating together at the table should be different from the past [64].

As for Taiwan, in 2009, a large number of ducks contaminated with dioxin died. In 2011, unscrupulous manufacturers used cheap industrial plasticizers to replace normal food additives. Food and beverages containing plasticizers such as bis(2-ethylhexyl) phthalate (DEHP) were sold in stores. These foods caused harm to the human body, affecting the people not only of Taiwan but also of Hong Kong and China. In 2014, cooking oil recovered from restaurant waste almost destroyed many food manufacturers in Taiwan. In 2015 and 2016, there were also scandals of unsafe food and restaurants selling food with pesticides. As a result of the food security scandal in 2014, the Taiwanese government has established important organizations and government agencies since then, and also formulated special laws to manage it.

The perpetrators of the most serious food safety scandals in mainland China and Taiwan have been sentenced to heavy penalties. Taiwan Dingxin Enterprise has been fined US $3 million, and the chairman was sentenced to 22 years in prison. Sanlu, Zhang Yujun, and Geng Jinping of mainland China have all passed away [65].

6.3. The Role of Social Trust and Information Credibility

From the foregoing discussion, it can be seen that the government plays an important role in dealing with food safety issues. In spite of the availability of more useful information, still the general public may be subjective and the information would be inefficient [20]. Therefore, if it can increase the public's trust in the government and the credibility of disseminating information, it should be beneficial to public communication and education efficiently [20,25]. Liu et al. [66] also indicated that to raise public concern, one cannot just reply in the media. Although the media continues to report food safety news, it still has no significant relationship with the public's concern. However, the government needs to pay more attention to the public's views and make the public more concerned about food safety. Meanwhile, it is necessary to strengthen risk communication between the government and the public, and in the future, the news media should be used in more efficient way [22].

Food safety labeling and certification systems can be a simple and effective way to inform consumers that food products with a mark are safe and trustworthy [7,20]. In order to make certification more reliable, it is necessary to trace back to the public's trust in the government's supervision of food safety. The model proposed by this current research takes into account the purchase intention of certified food and thus can help to understand the relationship between consumer consumption decisions and government policies.

Liu et al. [63] recommended that public authorities need to provide the public with detailed information such as traceability information, food certification, and other food quality standard information. The results of this study showed that people tend to focus on their consumer perceptions and there is no significant relationship between media attention and information. The result also echoed Zhang et al. [48], which indicated that the more the information access people have, the less chances they would examine food safety. It is recommended that the government need to make more efforts in the future to maintain its credibility and provide more reliable information in order to facilitate more positive and appropriate public perception of food safety.

7. Conclusions

This study proposed factors that influence the purchase intention of certified foods, including optimistic bias, consumption cognition, news attention, information credibility and social trust. The results of this study revealed that students' optimistic bias can be reduced in different social or cultural contexts and that the Taiwanese model and the mainland Chinese model proposed in this study were indeed different. Through empirical research, the results confirmed once again that people do have optimistic bias on food safety issues. College students tend to think that they are less likely to eat unsafe food than others, and this bias becomes a systemic bias. The results show that the optimism bias of the two groups of students is different, which may be due to differences in personal experience and social context, leading to different degrees of cognitive biases on food safety between Taiwanese students and mainland Chinese students.

In this study, consumer cognition significantly affected the willingness of Taiwanese and mainland Chinese college students to purchase food with certification marks. News attention significantly influenced only the willingness of mainland Chinese students to purchase certified food, not that of Taiwan students. Meanwhile, the other variables in the two models had no significant relationships with purchase intention. The results of this empirical study showed that people tended to believe in themselves and isolated themselves from other sources of information. In addition, consumers valued the product itself, indicating that their consumption perception comes from the consumption process. For this reason, it is difficult for the media and government to exert influence on consumers. In addition, the media can have a small impact on consumer behavior only when consumers are actively concerned about the issue. Therefore, it is necessary to improve consumers' impressions of products, their governments and even media resources. The quality of news from the government, the media and manufacturers should also be improved.

Since data were collected only from Taipei and Beijing, which is an important limitation of the study, it must be noted that the results cannot be generalized to other populations. However, this empirical study also specifically discusses the impact of social context. The governments in Taiwan and mainland China implemented different measures on food safety management, certification marks, and policy communication, so college students on the two sides of the Taiwan Strait showed different purchasing behaviors in the two models. The different information behaviors adopted by college students in Taiwan and mainland China may cause them react differently to food risk events. It is recommended that future research should re-examine the impact of government policies, and to what extent consumers may only believe in themselves, rather than rely on the credibility and trust of the government and food manufacturers.

Furthermore, people's attitudes towards food safety issues may change over time, and more social context variables may need to be considered. In this current study, the food safety cognition mainly focused on people's views on food safety and their information processing behavior. While the subjects recruited were only college students, it is suggested that future research can be expanded to include a wider sample or consider more variables to improve the generalizability of the model. Moreover, this study found that optimistic bias was not the main factor affecting purchase intention, and it showed different results in the two models of college students in Taiwan and mainland China. This weak explanatory power reflects the need to explore other food safety risk assessment factors. Future research should consider time effects, subject areas, and other risk assessment variables, and more psychological and social factors can be added to better understand food safety risk assessment issues.

Author Contributions: Conceptualization, H.-P.Y. and G.-Y.W.; methodology, H.-P.Y. and G.-Y.W.; software, G.-Y.W.; validation, H.-P.Y.; formal analysis, G.-Y.W.; investigation, G.-Y.W.; resources, H.-P.Y. and G.-Y.W.; data curation, G.-Y.W.; writing—original draft preparation, G.-Y.W.; writing—review and editing, H.-P.Y.;

visualization, G.-Y.W.; supervision, H.-P.Y.; project administration, G.-Y.W. All authors have read and agreed to the published version of the manuscript.

Funding: This research was funded by the Ministry of Science and Technology, R. O. C., grant number 105-2815-C-002-186-U.

Acknowledgments: We would like to appreciate the participants who took our questionnaire survey.

Conflicts of Interest: The authors declare no conflict of interest.

References

1. FAO/WHO. *Risk Managemtent and Food Safety*; FAO/WHO: Rome, Italy, 1997.
2. FAO. *Risk Communication Applied to Food Safety Handbook*; FAO: Rome, Italy, 2016.
3. Griffith, C.J. Food safety: Where from and where to? *Br. Food J.* **2006**, *108*, 6–15. [CrossRef]
4. Mortlock, M.; Peters, A.; Griffith, C.; Lloyd, D. Evaluating impacts of food safety control on retail butchers. In *Interdisciplinary Food Safety Research*; Hooker, N.H., Murano, E.A., Eds.; CRC Press: Boca Raton, FL, USA, 2000; pp. 159–182.
5. Weng, S.-T.; Lu, H.-Y.; Hou, H.-Y. Factors influencing intentions to consume food made in China: A study on women in Kinmen. *J. Health Promot. Health Educ. Contents* **2012**, *37*, 79–99.
6. Huang, C.-H.; Lin, C.-S.; Kuo, T.-I.; Lee, N.-H. Perceiving and segmenting for consuming the certified label of agricultural products in Kinmen area. *Xing Xiao Ping Lun* **2011**, *8*, 315–330.
7. Liu, R.; Gao, Z.; Nayga, R.M., Jr.; Snell, H.A.; Ma, H. Consumers' valuation for food traceability in China: Does trust matter? *Food Policy* **2019**, *88*, 101768.
8. Wang, Z.-G.; Yang, Y.-X.; Liu, H.; Lin, M.-Y. An analysis of consumers' cognition, referential behavior and benefits from agricultural products certification mark based on survey in 20 provinces, cities and autonomous regions. *Agric. Econ. Manag.* **2013**, *6*, 38–59.
9. Zhao, Y.; Tang, J.-S.; Li, F.-F. A research on risk perception and information demand of the public in food security crisis. *Mod. Financ. Econ. J. Tianjin Univ. Financ. Econ.* **2012**, *6*, 61–70.
10. Van Kleef, E.; Frewer, L.J.; Chryssochoidis, G.M.; Houghton, J.R.; Korzen-Bohr, S.; Krystallis, T.; Lassen, J.; Pfenning, U.; Rowe, G. Perceptions of food risk management among key stakeholders: Results from a cross-European study. *Appetite* **2006**, *47*, 46–63. [CrossRef] [PubMed]
11. Weinstein, N.D. Unrealistic optimism about future life events. *J. Personal. Soc. Psychol.* **1980**, *39*, 806–820. [CrossRef]
12. Miles, S.; Scaife, V. Optimistic bias and food. *Nutr. Res. Rev.* **2003**, *16*, 3–20. [CrossRef]
13. Clarke, V.A.; Lovegrove, H.; Williams, A.; Machperson, M. Unrealistic optimism and the health belief model. *J. Behav. Med.* **2000**, *23*, 367–376. [CrossRef]
14. Salmon, C.T.; Park, H.S.; Wrigley, B.J. Optimistic bias and perceptions of bioterrorism in Michigan corporate spokespersons, fall 2001. *J. Health Commun.* **2003**, *8* (Suppl. S1), 130–143. [CrossRef] [PubMed]
15. Abbot, J.M.; Byrd-Bredbenner, C.; Schaffner, D.; Bruhn, C.M.; Blalock, L. Comparison of food safety cognitions and self-reported food-handling behaviors with observed food safety behaviors of young adults. *Eur. J. Clin. Nutr.* **2009**, *63*, 572–579. [CrossRef] [PubMed]
16. Wei, R.; Lo, V.-H.; Lu, H.-Y. Reconsidering the relationship between the third-person perception and optimistic bias. *Commun. Res.* **2007**, *34*, 665–684.
17. Lu, H.-Y.; Hou, H.-Y.; Chen, T.-P.; Lin, M.-C.; Lee, C.-C. Internet use motivations, perceived credibility of online information, and decision-making. *Chin. J. Commun. Res.* **2009**, *16*, 255–285.
18. Lu, H.-Y.; Hsu, F.-S.; Hou, H.-Y. Optimistic bias, self-efficacy, social trust, and intentions to receive H1N1 influenza vaccine. *Commun. Soc.* **2012**, *22*, 135–158.
19. Zhang, Y.; Jing, N.; Song, M. Food quality information cognition and public purchase decisions: Research from China. *Qual. Assur. Saf. Crop. Food* **2019**, *11*, 647–657. [CrossRef]
20. Zhang, B.; Fu, Z.; Huang, J.; Wang, J.; Xu, S.; Zhang, L. Consumers' perceptions, purchase intention, and willingness to pay a premium price for safe vegetables: A case study of Beijing, China. *J. Clean. Prod.* **2018**, *197*, 1498–1507. [CrossRef]

21. Redmond, E.C.; Griffith, C.J. Factors influencing the efficacy of consumer food safety communication. *Br. Food J.* **2005**, *107*, 484–499. [CrossRef]
22. Sun, X. Investigation and analysis of consumers' food safety cognition in Shandong Province. *IOP Conf. Ser. Earth Environ. Sci.* **2019**, *252*, 042049. [CrossRef]
23. Tsai, H.-T.; Hsiao, L.; Hong, J.-T.; Chen, Y.-C. An empirical study on Taiwan's agriculture and food traceability policy. *Public Admin. Policy* **2012**, *55*, 67–108.
24. Yang, S.; Leff, M.G.; McTague, D.; Horvath, K.A.; Jackson-Thompson, J.; Murayi, T.; Boeselager, G.K.; Melnik, T.A.; Gildemaster, M.C.; Ridings, D.L. Multistate surveillance for food-handling, preparation, and consumption behaviors associated with foodborne diseases: 1995 and 1996 BRFSS food-safety questions. *Morb. Mortal. Wkly. Rep. CDC Surveill. Summ.* **1998**, *47*, 33–57.
25. Barrett, T.; Feng, Y. Evaluation of food safety curriculum effectiveness: A longitudinal study of high-school-aged youths' knowledge retention, risk-perception, and perceived behavioral control. *Food Control* **2021**, *121*, 107587. [CrossRef]
26. Fan, T.S.; Li, Y.C. Consumers perception and consumption behavior on organic products for Pingtung city. *J. Agric. Assoc. Taiwan* **2009**, *10*, 504–519.
27. Cope, S.; Frewer, L.; Houghton, J.; Rowe, G.; Fischer, A.; de Jonge, J. Consumer perceptions of best practice in food risk communication and management: Implications for risk analysis policy. *Food Policy* **2010**, *35*, 349–357. [CrossRef]
28. Houghton, J.R.; Rowe, G.; Frewer, L.J.; Van Kleef, E.; Chryssochoidis, G.; Kehagia, O.; Korzen-Bohr, S.; Lassen, J.; Pfenning, U.; Strada, A. The quality of food risk management in Europe: Perspectives and priorities. *Food Policy* **2008**, *33*, 13–26. [CrossRef]
29. Fleming, K.; Thorson, E.; Zhang, Y. Going beyond exposure to local news media: An information-processing examination of public perceptions of food safety. *J. Health Commun.* **2006**, *11*, 789–806. [CrossRef] [PubMed]
30. Rieger, J.; Weible, D.; Anders, S. "Why some consumers don't care": Heterogeneity in household responses to a food scandal. *Appetite* **2017**, *113*, 200–214. [CrossRef]
31. Lu, H.-Y.; Dzwo, T.-H.; Hou, H.-Y.; Andrews, J.E. Factors influencing information-seeking intentions and support for restrictions: A study on an arsenic-contaminated frying oil event. *Br. Food J.* **2011**, *113*, 1439–1452. [CrossRef]
32. Wei, R.; Lo, V.-H.; Lu, H.-Y. The third-person effect of tainted food product recall news: Examining the role of credibility, attention, and elaboration for college students in Taiwan. *J. Mass Commun. Q.* **2010**, *87*, 598–614. [CrossRef]
33. Liao, L.X.; Wang, J.X.; Huang, X.R.; Li, L.M. Study in food safety management and standard between the two sides of the Taiwan Straits. *J. Insp. Quar.* **2012**, *4*, 68–72.
34. Hu, Y.-H.; Yueh, H.-P. Food safety risk communication behavior on social media: The case of Sina Weibo. *J. Libr. Inf. Stud.* **2019**, *17*, 151–183.
35. Yin, S.; Li, Y.; Chen, Y.; Wu, L.; Yan, J. Public reporting on food safety incidents in China: Intention and its determinants. *Br. Food J.* **2018**, *120*, 2615–2630. [CrossRef]
36. McCroskey, J.C.; Teven, J.J. Goodwill: A reexamination of the construct and its measurement. *Commun. Monogr.* **1999**, *66*, 90–103. [CrossRef]
37. Westerman, D.; Spence, P.R.; Van Der Heide, B. Social media as information source: Recency of updates and credibility of information. *J. Comput. Mediat. Commun.* **2014**, *19*, 171–183. [CrossRef]
38. Wilson, A.M.; Meyer, S.B.; Webb, T.; Henderson, J.; Coveney, J.; McCullum, D.; Ward, P.R. How food regulators communicate with consumers about food safety. *Br. Food J.* **2015**, *117*, 2129–2142. [CrossRef]
39. Bu, F.; Zhu, D.; Wu, L. Research on the consumers' willingness to buy traceable pork with different quality information: A case study of consumers in Weifang, Shandong Province. *Asian Agric. Res.* **2013**, *5*, 121–124.
40. Yin, S.; Wu, L.; Du, L.; Chen, M. Consumers' purchase intention of organic food in China. *J. Sci. Food Agric.* **2010**, *90*, 1361–1367. [CrossRef]
41. Coveney, J. Food and trust in Australia: Building a picture. *Public Health Nutr.* **2008**, *11*, 237–245. [CrossRef]
42. McCarthy, B.L. Trends in organic and green food consumption in China: Opportunities and challenges for regional Australian exporters. *J. Econ. Soc. Policy* **2015**, *17*, 6–31.
43. Lobb, A.E.; Mazzocchi, M.; Traill, W.B. Modelling risk perception and trust in food safety information within the theory of planned behaviour. *Food Qual. Prefer.* **2007**, *18*, 384–395. [CrossRef]

44. Van Dijk, H.; Houghton, J.; Van Kleef, E.; Van Der Lans, I.; Rowe, G.; Frewer, L. Consumer responses to communication about food risk management. *Appetite* **2008**, *50*, 340–352. [CrossRef]
45. Rothstein, B.; Uslaner, E.M. All for all: Equality, corruption, and social trust. *World Polit.* **2005**, *58*, 41–72. [CrossRef]
46. Lu, H.-Y.; Hou, H.-Y.; Dzwo, T.-H.; Wu, Y.-C.; Andrews, J.E.; Weng, S.-T.; Lin, M.-C.; Lu, J.-Y. Factors influencing intentions to take precautions to avoid consuming food containing dairy products: Expanding the theory of planned behaviour. *Br. Food J.* **2010**, *112*, 919–933. [CrossRef]
47. Moy Foong, M.; Alias Abd, A.; Jani, R.; Abdul Halim, H.; Low Wah, Y. Determinants of self-reported food safety practices among youths: A cross-sectional online study in Kuala Lumpur, Malaysia. *Br. Food J.* **2018**, *120*, 891–900.
48. Zhang, J.; Cai, Z.; Cheng, M.; Zhang, H.; Zhang, H.; Zhu, Z. Association of Internet use with attitudes toward food safety in China: A cross-sectional study. *Int. J. Environ. Res. Public Health* **2019**, *16*, 4162. [CrossRef] [PubMed]
49. Ma, Q.J. On cross-straits mechanisms and oversight systems for food safety. *J. Beijing Adm. Coll.* **2014**, *6*, 47–52.
50. Anderson, J.C.; Gerbing, D.W. Structural equation modeling in practice: A review and recommended two-step approach. *Psychol. Bull.* **1988**, *103*, 411–423. [CrossRef]
51. Bagozzi, R.P.; Yi, Y. On the evaluation of structural equation models. *J. Acad. Mark. Sci.* **1988**, *16*, 74–94. [CrossRef]
52. Browne, M.W.; Cudeck, R. Alternative ways of assessing model fit. *Sociol. Methods Res.* **1992**, *21*, 230–258. [CrossRef]
53. Koufteros, X.; Marcoulides, G.A. Product development practices and performance: A structural equation modeling-based multi-group analysis. *Int. J. Prod. Econ.* **2006**, *103*, 286–307. [CrossRef]
54. Vandenberg, R.J.; Lance, C.E. A review and synthesis of the measurement invariance literature: Suggestions, practices, and recommendations for organizational research. *Organ. Res. Methods* **2000**, *3*, 4–70. [CrossRef]
55. Cheung, G.W.; Rensvold, R.B. Evaluating goodness-of-fit indexes for testing measurement invariance. *Struct. Equ. Model.* **2002**, *9*, 233–255. [CrossRef]
56. Teng, C.-C.; Lu, C.-H. Organic food consumption in Taiwan: Motives, involvement, and purchase intention under the moderating role of uncertainty. *Appetite* **2016**, *105*, 95–105. [CrossRef] [PubMed]
57. Bentler, P.; Yuan, K.-H. Structural equation modeling with small samples: Test statistics. *Multivar. Behav. Res.* **1999**, *34*, 181–197. [CrossRef]
58. Elliott, C.; Ellison, K. Negotiating choice, deception and risk: Teenagers' perceptions of food safety. *Br. Food J.* **2018**, *120*, 2748–2761. [CrossRef]
59. Dai, Y.; Li, Q.; Liu, W.; Zhang, H.; Hao, J.; Dong, Y.; Liu, P.; Duan, M.; Fan, X. Research on consumers' cognition and demand for food label information. *IOP Conf. Ser. Earth Environ. Sci.* **2020**, *512*, 012071.
60. Lachlan, K.A.; Westerman, D.K.; Spence, P.R. Disaster news and subsequent information seeking: Exploring the role of spatial presence and perceptual realism. *Electron. News* **2010**, *4*, 203–217. [CrossRef]
61. Peng, Y.; Li, J.; Xia, H.; Qi, S.; Li, J. The effects of food safety issues released by we media on consumers' awareness and purchasing behavior: A case study in China. *Food Policy* **2015**, *51*, 44–52. [CrossRef]
62. Wu, L.-H.; Yin, S.-J.; Chen, X.-J.; Pu, X.-J.; Wang, J.-H. How to ensure "Safety of Every Bite of Food" from the farmland to the table: China's food safety risk management and situation analysis. *China Food Saf. Manag. Rev.* **2018**, *1*, 1–10.
63. Tainted-Baby-Milk Scandal in China. Available online: http://content.time.com/time/world/article/0,8599,1841535,00.html (accessed on 2 September 2020).
64. Chang-Xue Zhu, President of China Food Safety News, Said "Three Guarantees' Action Has Rooted in People's Hearts Food Consumption Habits Change". Available online: http://www.cfsn.cn/front/web/site.searchshow?pdid=0&id=1106 (accessed on 2 September 2020).

65. Re-judgment of Taiwan's Qiang-Guan Incident Has Implications For Food Safety Across the Taiwan Straits. Available online: https://www.bbc.com/zhongwen/trad/chinese-news-41292774 (accessed on 2 September 2020).
66. Liu, P.; Ma, L. Food scandals, media exposure, and citizens' safety concerns: A multilevel analysis across Chinese cities. *Food Policy* **2016**, *63*, 102–111. [CrossRef]

Article

A Closer Look at Changes in High-Risk Food-Handling Behaviors and Perceptions of Primary Food Handlers at Home in South Korea across Time

Tae Jin Cho [1,†], Sun Ae Kim [2,†], Hye Won Kim [3] and Min Suk Rhee [3,*]

1 Department of Food and Biotechnology, College of Science and Technology, Korea University, 2511, Sejong-ro, Sejong 30019, Korea; microcho@korea.ac.kr
2 Department of Food Science and Engineering, Ewha Womans University, Seoul 03760, Korea; sunaekim@ewha.ac.kr
3 Department of Biotechnology, College of Life Sciences and Biotechnology, Korea University, 145, Anam-ro, Seongbuk-gu, Seoul 02841, Korea; kgpdnjs@korea.ac.kr
* Correspondence: rheems@korea.ac.kr; Tel.: +82-2-3290-3058
† These authors contributed equally to this work.

Received: 14 September 2020; Accepted: 6 October 2020; Published: 13 October 2020

Abstract: Food-handling behaviors and risk perceptions among primary food handlers were investigated by consumer surveys from different subjects in 2010 (N = 609; 1st survey will be called here "Year 2010") and 2019 (N = 605; 2nd survey will be called here "Year 2019"). Year 2010 was characterized by consumers' risk perception-behavior gap (i.e., consumers knew safe methods for food-handling, but responses regarding the behaviors did not support their confidence in food safety): they (1) did not wash/trim foods before storage, (2) thawed frozen foods at room temperature, and (3) exposed leftovers to danger zone temperatures. These trends were not improved and the gaps in Year 2010 remained in Year 2019. Year 2010 was also characterized by other common high-risk behaviors improved during 8 years for the following aspects: (1) 70.0% of consumers divided a large portion of food into smaller pieces for storage, but few consumers (12.5%) labeled divided foods with relevant information, and (2) they excessively reused kitchen utensils. Whereas in Year 2019, more consumers (25.7%) labeled food and usage periods for kitchen utensils were shortened. Consumers usually conformed to food safety rules in both Year 2010 and 2019: (1) separate storage of foods, (2) storage of foods in the proper places/periods, (3) washing fruits/vegetables before eating, (4) washing hands after handling potentially hazardous foods, and (5) cooking foods and reheating leftovers to eat. Our findings provided resources for understanding consumers' high-risk behaviors/perceptions at home, highlighting the importance of behavioral control.

Keywords: consumer survey; food safety; food hygiene; food handling; consumer behavior; risk perception; healthy food consumption; cultural consumer context; microbiological risk; health

1. Introduction

Food safety is one of the most important global public health issues which has repeatedly created social anxiety and resulted in the economic loss of many countries [1–3]. Most people commonly know that foodborne diseases are mainly associated with foods consumed outside the home, however, the private home is also a crucial site where foodborne illnesses are engendered [4–6]. The World Health Organization (WHO) [7] estimated that approximately 40% of foodborne illnesses have been associated with food prepared at home, and Redmond and Griffith [8] also reported that 50–87% of foodborne diseases were attributed to the consumption of food at home. Researchers assumed that the number of actual foodborne diseases caused at home might be much higher than reported since most foodborne illnesses were unreported and/or unconfirmed [4]; Redmond and Griffith [9] estimated that this number could reach 95%. In South Korea, it is also assumed that the number of foodborne diseases has been underestimated (i.e., most cases are expected to be unreported); Ministry of Food and Drug Safety reported that only 1.32% outbreaks and 0.36% cases were attributed to foods prepared at private home [10].

Mishandling of foods could frequently occur from preparation, handling, cooking, and storing in the home [11]. Byrd-Bredbenner et al. [4] summarized the reasons why the home was a risky place for foodborne disease as follows: (1) the greatest portion of foods is prepared at home; (2) there are many consumers at home in high risk groups for health problems (YOPI: young/old/pregnant/immunocompromised people); (3) people even in the YOPI group do not perceive themselves as susceptible to illness or do not follow recommended practices for food safety, and (4) home-prepared food can be served to a wider community (e.g., school picnics, lunch boxes, and bake sales).

Risk perception on food hygiene has been regarded as one of the most important topics for estimating the actual risk levels of food safety for lay public [12]. Consumer surveys can be used to explore the underreported and/or underrecognized risk perceptions linked to the improper behaviors with the perspective to domestic food safety, which could also support to overcome the limitation of current conventional consumer guidelines [13,14]. Moreover, understanding national differences in risk perceptions should be the pre-requisites for the establishment of internationally-recognized interventions (e.g., the education, public campaign, etc.) against the major risk factors [15]. Especially previous research has focused on the gap between risk perception and actual practices of food handlers as the key clues for the improvement of both the proper perception and behaviors for food safety [16–18]. In the case of domestic food safety, the importance of the risk perception as the background knowledge for the cause of improper behaviors of food handlers at home has been highlighted [3,4].

Hygienic handling practices based on proper knowledge and risk perception are essential to prevent diseases at the home [3,8,19]. However, many consumers have been improperly informed about the methods required to prevent foodborne disease at home [20]. International concern about food safety has prompted considerable studies to gain insights into domestic food-handling practices and risk perceptions mainly in the Western countries (including the United Kingdom, Northern Ireland, Canada, Germany, and Belgium) and the United States [2,8,21–28]. However, there has been only a limited number of studies in South Korea. In this study, we conducted a nationwide survey of primary food handlers to investigate consumers' behaviors as well as risk perceptions at home in South Korea. Since food-handling behaviors and perceptions of consumers could be altered by trends over time, the surveys were conducted in both 2010 (N = 609; 1st survey will be called here "Year 2010") and 2019 (N = 605; 2nd survey will be called here "Year 2019") with the same questionnaires to track changes in behaviors and perceptions.

Although consumers' perception has been revealed as the motivation behind high-risk behaviors associated with domestic food safety, a lack of empirical data demonstrating the unchanged perception-behavior gap acts as a major hurdle for conventional intervention strategies (e.g., advertising, providing guidelines, publishing pamphlets, providing education, etc.) to drive the alteration of the behaviors of the lay public. Since the previous research on consumers' risk perception and/or high-risk behaviors regarding the domestic food safety have been conducted by the cross-sectional analyses [2,21,25–28], we expected that unchanged and/or emerging high-risk behaviors can be identified by the comparative analysis of two individual consumer surveys using sociodemographic characteristics of respondents on a decade basis. The major aim of this research is to identify the unchanged gap between the major risk perception and behaviors of consumers over time. The survey on the consumers' practices linked to the risk perception was expected to identify the self-reported behaviors which can be regarded as the potential for the deviate behaviors [29–31]. In this study, we conducted a nationwide trend survey of primary food handlers (i.e., food consumers who are the main people involved in food preparation at home) to investigate consumers' behaviors as well as risk perceptions at home in South Korea.

2. Materials and Methods

2.1. Questionnaires for Food-Handling Behaviors and Perception

To develop questionnaires, experts in different fields, representing the South Korean government (Ministry of Food and Drug Safety, Cheongju-si, Korea), food safety laboratories (Korea University, Seoul, Korea), consumer organizations, and a professional market research company (Gallup Korea, Seoul, Korea) formed a consulting committee. These experts compiled a draft questionnaire. The questions were determined based on the general food safety guidelines for the home provided by health authorities, including the US Food and Drug Administration (US FDA) and Food Safety and Inspection Service in the US Department of Agriculture [32,33]. Only general guidelines were included in a draft questionnaire while guidelines, not culturally applicable to Korean consumers, were excluded. A draft questionnaire was inspected with a consulting committee and revised to develop the final questionnaire.

A food safety perception questionnaire was designed to obtain information about food handlers' perceptions about hazardous behaviors that frequently occurred at home. Detailed questions are shown in Table 1.

Table 1. Composition of the questionnaire.

Topic	Questions [1]	Answer Options [2]
Consumer's perception	Q1. Do you think storing raw materials including fruits, vegetables, meat, and fish/shellfish without preparation (e.g., washing or trimming) is hazardous?	Completely hazardous Mostly hazardous Moderate Mostly safe Completely safe
	Q2. Do you think thawing frozen foods at room temperature is hazardous?	
	Q3. Do you think to touch cooked foods after handling raw materials without washing your hands is hazardous?	
	Q4. Do you think to incompletely cook meat, fish/shellfish, and eggs is hazardous?	
	Q5. Do you think storing leftovers at room temperature is hazardous?	
	Q6. Do you think to eat leftover without re-heat is hazardous?	
Food storage behavior	Q7. Do you wash or trim meat, fish, fruits, and vegetables before storage?	Always Frequently Sometimes Seldom Never
	Q8. When you buy a large portion of food, do you divide it into small portions for storage?	
	Q9. When you store a large portion of food in smaller portions, do you label the small portions with information about those foods?	
	Q10. Where do you store foods (meat, chicken, fish/shellfish, fruit/vegetables, eggs, milk, and frozen processed foods) among refrigerator, freezer, or other environments at room temperature? Then, how long do you store those foods in the selected storage place? [3]	<Answer options for the storage place> Refrigerator Freezer Other environments at room temperature Do not purchase <The storage period was asked as an open question>
Preparing and cooking behavior	Q11. Do you wash your hands to handle other foods after handling meat, fish/shellfish, or eggs?	Always Frequently Sometimes Seldom Never
	Q12. How do you defrost frozen foods?	Do not thaw Place in the refrigerator for 1–2 days before it is cooked Use a microwave oven Dip in the water at room temperature Place on the counter at room temperature
	Q13. Do you thaw only a portion of food as much as you intend to cook rather than thawing all of the foods?	Always Frequently Sometimes Seldom Never
	Q14. Do you use plastic gloves when you handle meat, fish/shellfish, or eggs?	

Table 1. *Cont.*

Topic	Questions [1]	Answer Options [2]
	Q15. What do you do when you have a slight wound on your hand during cooking?	Keep cooking after the wound has been treated Keep cooking without any treatment Do not cook
Eating behavior	Q16. Do you fully cook foods including meat, chicken, and fish/shellfish for eating?	Always Frequently Sometimes Seldom Never
	Q17. Do you wash fruits and vegetables before eating?	
Management of leftovers	Q18. How do you store hot and leftover foods?	Store immediately in the refrigerator Store in refrigerator after chilling in cold water Store at room temperature Store in refrigerator after chilling at room temperature
	Q19. Do you re-heat leftovers for eating?	Always Frequently Sometimes Seldom Never
	Q20. Do you store remaining raw foodstuffs with sealing after cooking?	
Management of domestic kitchen utensils	Q21. How long do you use a cutting board?	Less than 6 months 6 months–1 year 1–2 years 2–3 years More than 3 years Other
	Q22. Do you sanitize your cutting board?	Yes No
	Q23. How often do you replace your kitchen cloth?	1 or more than in a week 1–2 times in a month Once in 3 months Once in 6 months Once in a year or more than a year Do not replace
	Q24. Do you sanitize your kitchen cloth?	Yes No

[1] All questions except for the "Q10. How long do you store those foods in the selected storage place?" (an open question) were in multiple-choice questions with a single-select answer option. All answer options (i.e., choices) for each question are provided in Tables and Figures of this study. [2] All questions include the "Do not know/no response" as an answer option (this option is omitted in this Table to avoid the repetition). [3] Respondents were asked to select the storage place (refrigerator, freezer, or other environments at room temperature) for each food (meat, chicken, fish/shellfish, fruit/vegetables, eggs, milk, and frozen processed foods) and to answer the storage period as a unit of "days" for each food in the selected storage place.

2.2. Pilot Test of Questionnaires

To confirm the clarity of the draft questionnaire a pilot test, using a draft questionnaire, was conducted with 15 randomly selected consumers and 15 expert researchers before performing the main survey to confirm the clarity of the draft questionnaire. The pretest consumers were asked to respond to the following questions: (1) Did you clearly understand the terminology used? (2) How much time did it take you to respond to all the questions? (3) Did the questionnaire contain unclear expressions? (4) Is there any question that is difficult to interpret? (5) Do any of the terms need to be clarified? (6) Did you feel displeasure or resistance when you responded to the questionnaire? (7) Did you have any opinions that differed from the existing response options? The questionnaires were revised based on the results of the pretest, specifically focusing on the understandability of questions. The answers of the pre-testers were not included in the survey results.

2.3. Surveys

This survey targeted adult consumers (>18 years old), mainly the primary food handlers in their homes, from South Korea. Before the survey, consumers (N = 609) were pre-allocated (i.e., quota sampling) according to population data from the statistical yearbook of South Korea using the multistage stratified systematic sampling method [34,35]. Participants of the survey were asked for their sociodemographic characteristics to obtain responses according to the pre-allocated population composition. A survey was conducted in households or shopping centers from various locations throughout Korea, including large cities (Busan, Daegu, Daejeon, Gwangju, Incheon, Seoul, and Ulsan), small and medium cities, and country towns. The sampling fraction used for the geographic location was proportional to the total population. All the respondents were interviewed face-to-face by trained panels (Gallup Korea). The instructions explaining the purpose of the study were displayed at the front of the questionnaire. The investigator briefly explained the purpose and nature of the present study. The questionnaires (a total of 24 questions) as described in Table 1 were used for both surveys. A survey in 2019 was conducted for targeted adult consumers (N = 605) using the same questionnaire and method for Year 2010. They were recruited by a multistage stratified systematic sampling method for the homogeneity of sociodemographic characteristics both within and between surveys to collect comparable responses. The average duration for the data collection process of both surveys was ca. 3 months for each time-point (2010 or 2019). As consumers' responses for the risk perception and food handling practices in both surveys were obtained by same questionnaires organized as multiple-choice questions with a single-select answer option (except for the "Q10. How long do you store those foods in the selected storage place?" (an open question)), comparative analysis on the results from two time period (2010 and 2019) could be conducted. The sociodemographic characteristics of the participants (e.g., gender, age, location, level of education, number of family members, and average monthly income) are shown in Table 2. To evaluate the effect of the year on the survey results, Pearson chi-square tested the associations of each sample characteristic between the time-points of Year 2010 and Year 2019. There was no effect of the year for all variables (sociodemographic characteristics) considered in this study (i.e., gender, age (years), location, level of education, number of family members).

Table 2. Sociodemographic characteristics of the respondents and chi-square test results for each variable.

Variables	Number of Respondents to Year 2010 (n = 609) [1]	Number of Respondents to Year 2019 (n = 605) [2]	χ^2	p-Value
Gender			0.005	0.943
Male	49	48		
Female	560	557		
Age (years)			0.604	0.963
19–29	67	74		
30–39	144	144		
40–49	148	145		
50–59	112	112		
>60	138	130		
Location			0.003	0.998
Large city	270	269		
Small or medium city	271	269		
Country town	68	67		
Level of education			2.003	0.572
Less than high school	122	120		
High school	231	231		
University	254	254		
No response	2	-		
Number of family members			3.620	0.306
One person	64	47		
2–3 persons	246	267		
4–5 persons	266	257		
More than 6 persons	33	34		

[1] 1st survey conducted in 2010 will be called here "Year 2010". [2] 2nd survey conducted in 2019 will be called here "Year 2019".

2.4. Data Analysis

All the questions and responses were manually coded by assigning a unique number with the sui generis data coding system used by Gallup Korea for each response, and the codes were entered into a multivariate Excel spreadsheet. As shown in Section 2.3, the Pearson chi-square test was conducted to evaluate the effect of the year to the survey results by the analysis of the association of each sociodemographic characteristic between Year 2010 and Year 2019. Kruskal–Wallis test method was used to evaluate the changes in common high-risk behaviors between surveys by the analysis of the significant differences ($p < 0.05$) of responses from Year 2010 and Year 2019. All statistical analyses (i.e., Pearson chi-square test, Kruskal–Wallis test) for the data were conducted by using the SPSS statistical package (Statistical Package for the Social Sciences, version 12.0, SPSS Inc. Chicago, IL, USA).

3. Results

3.1. Discordance Between Consumers' Food Safety Perceptions and Behaviors (Risk Perception-Behavior Gap)

Figures 1–3 showed that there were large gaps between consumers' food safety perceptions and behaviors in surveys performed in 2010 and 2019, respectively. In general, a similar tendency was observed in both surveys, indicating that the gaps between safety knowledge and behaviors were not narrowed, even after 8 years. Detailed data for Figures 1–3 with the results of statistical analysis were also indicated in Table S1.

3.1.1. Storage of Perishable Foods Without Washing and Trimming

As shown in Figure 1, many consumers (71.8% and 67.8% in the Year 2010 and Year 2019, respectively) perceived the storage of raw food materials (e.g., fruits, vegetables, meat, fish/shellfish) without any preparation, including washing/trimming, as hazardous. A considerable number of consumers (32.5%) did not wash or trim food before storage in Year 2010, and 33.7% of consumers did not wash perishables in Year 2019. Moreover, significant differences ($p < 0.05$) represented by the decreases in the response to proper behavior were observed: from 22.5% in Year 2010 to 7.9% in Year 2019 for "completely yes" (Figure 1); 51.7% in Year 2010 to 35.7% in Year 2019 for "always + frequently" (Table S1). The response "moderate" (from 15.8% in Year 2010 to 30.6% in Year 2019) was also increased ($p < 0.05$).

3.1.2. Thawing Foods at Room Temperature

Figure 2 showed the risk perception on thawing frozen foods at room temperature and practices with perspective to various thawing methods (e.g., using a refrigerator or microwave, placed at room temperature). Results of Year 2010 indicated that many consumers (63.2%) were knowledgeable about the hazard of thawing frozen foods at room temperature (mostly hazardous = 36.8%; completely hazardous = 26.4%). However, the interviewed consumers (53.5%) thawed frozen foods at room temperature, 36.9% on the countertop, and 16.6% at room temperature water. Only 42.1% of consumers properly thawed foods; 25.0 and 17.1% thawed them in the microwave oven and in the refrigerator for 1–2 days before use, respectively. In the case of Year 2019, 58.5% of the consumers thawed them at room temperature although approximately half of the consumers (48.1%) were knowledgeable about the relevant hazard. Especially responses to "Place on counter at room temperature" increased ($p < 0.05$) from 36.9% to 43.6%. Whereas significant decreases in the proper risk perception (i.e., the correct response in line with conventional food safety guidelines) in Year 2019 were noticeable, implying that the risk perception worsened between studies: from 26.4% to 17.4% ($p < 0.05$), from 36.8% to 30.7% ($p < 0.05$), and from 63.2% to 48.1% for the responses "completely hazardous", "mostly hazardous", and "completely hazardous + mostly hazardous", respectively (Table S1). The percentage of respondents who properly thawed food (using a microwave or placing foods in a refrigerator 1–2 days before use) was 38.5%.

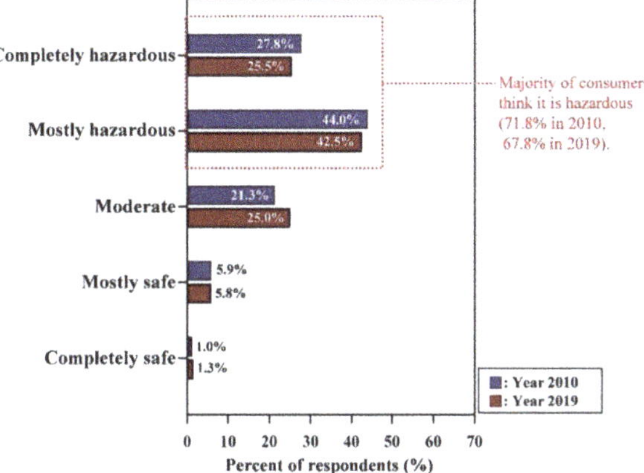

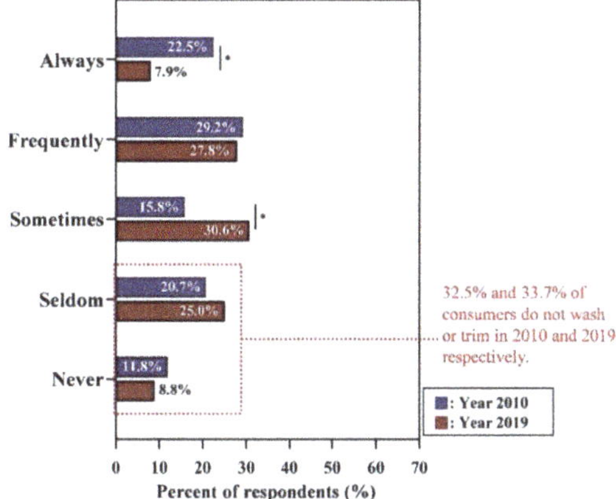

Figure 1. Gap between the consumers' perceptions and behaviors regarding the storage of perishable foods without washing and trimming. Asterisk (*) indicated between graphs means the significant differences of responses (the Kruskal-Wallis test; $p < 0.05$) in each answer option from Year 2010 and Year.

3.1.3. Improper Handling and the Storage of Leftovers

The gap between the risk perception and potential for deviant behavior regarding the leftovers was shown in Figure 3. In terms of Year 2010, although the respondents (64.0%) perceived the hazard of storing leftover food at room temperature, 95.6% of consumers (56.0% stored leftover foods in the refrigerator after chilling at room temperature, and 39.6% stored foods at room temperature) exposed leftovers to room temperature conditions. Only a few respondents properly handled leftovers: refrigerating them immediately (1.6%) or after chilling them in cold water (2.5%). Survey results in 2019 showed that 60.8% of the respondents perceived the hazard of storing leftovers at room temperature. While the increase in the response to "moderate" (from 23.3% to 32.4%) ($p < 0.05$) and the decrease in the response to "mostly safe" (from 11.2% to 6.3%) ($p < 0.05$) for risk perception resulted in the overall decrease in the proper risk perception (from 12.6% to 6.8% for the response to "completely safe + mostly safe") ($p < 0.05$) (Table S1). Most consumers (93.0%) exposed leftovers to room temperature conditions, and only a small percentage of respondents properly handled leftovers, including placing them in a refrigerator after chilling in cold water (3.5%) and immediately placing them in a refrigerator (3.5%). Distinct increases in the responses to "store in the refrigerator after chilling at room temperature" were analyzed between survey rounds (from 56.0% to 68.9%; $p < 0.05$) despite the decreases in the responses to "store at room temperature" between survey rounds (from 39.6% to 24.1%; $p < 0.05$).

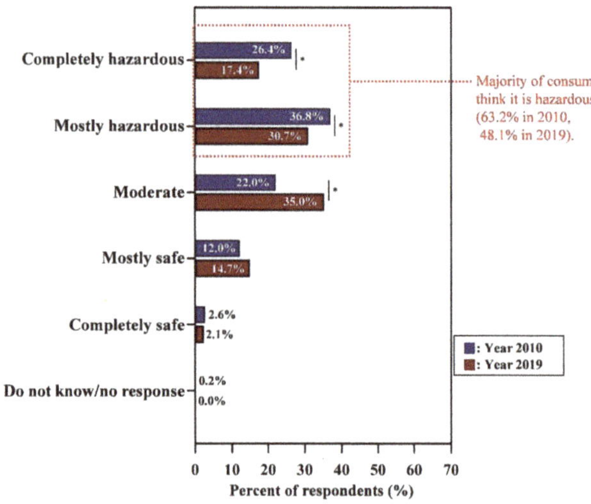

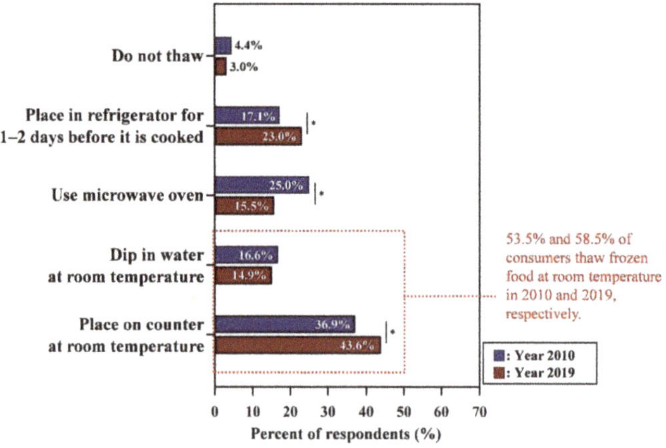

Figure 2. The gap between the consumers' perceptions and behaviors regarding frozen foods. Asterisk (*) indicated between graphs means the significant differences of responses (the Kruskal–Wallis test; $p < 0.05$) in each answer option from Year 2010 and Year 2019. Red letters indicated the improper risk perceptions (i.e., the response in opposition to conventional food safety guidelines) and high-risk behaviors of consumers. All data regarding the percentage of the combined answer options (e.g., Always + Frequently) are provided in Table S1.

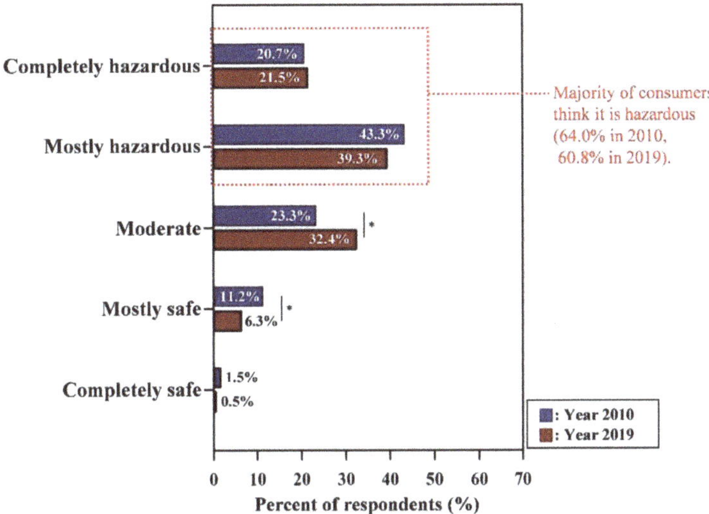

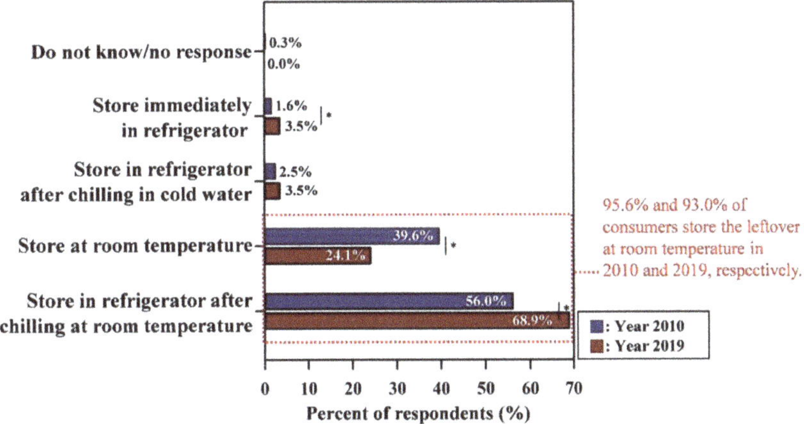

Figure 3. The gap between the consumers' perceptions and behaviors regarding the leftovers. Asterisk (*) indicated between graphs means the significant differences of responses (the Kruskal–Wallis test; $p < 0.05$) in each answer option from Year 2010 and Year 2019. Red letters indicated the improper risk perceptions (i.e., the response in opposition to conventional food safety guidelines) and high-risk behaviors of consumers. All data regarding the percentage of the combined answer options (e.g., Always + Frequently) are provided in Table S1.

3.2. Changes in Common High-Risk Behaviors Between Surveys

Other common high-risk behaviors of consumers in 2010 were summarized in Table 3: (1) Packaging and storage of foods, (2) Management of cutting board, (3) Management of kitchen cloth.

Table 3. The common high-risk behaviors of consumers were observed from surveys performed in 2010 and 2019.

Topic	Questions and Choices	Percent of Respondents or Answers		p-Value [1]
		Year 2010 (N = 609)	Year 2019 (N = 605)	
Packaging and storage of foods	Q8. When you buy a large portion of food, do you divide it into small portions for storage?			
	Never	5.1%	5.8%	0.594
	Seldom	11.8%	16.0%	0.034
	Sometimes	13.1%	20.3%	0.001
	Frequently	35.5%	40.0%	0.103
	Always	34.5%	17.9%	0.000
	<Analysis of the combined responses> [2]			
	Never + Seldom	16.9%	21.8%	0.031
	Always + Frequently	70.0%	57.9%	0.000
	Q9. When you store a large portion of food in smaller portions, do you label the small portions with information about those foods?			
	Never	36.1%	17.0%	0.000
	Seldom	39.1%	31.2%	0.000
	Sometimes	12.3%	26.1%	0.000
	Frequently	8.9%	20.7%	0.000
	Always	3.6%	5.0%	0.247
	<Analysis of the combined responses> [2]			
	Never + Seldom	75.2%	48.3%	0.000
	Always + Frequently	12.5%	25.6%	0.000
Management of cutting board	Q21. How long do you use a cutting board?			
	Less than 6 months	0.5%	3.8%	0.000
	6 months–1 year	2.3%	6.1%	0.001
	1–2 years	12.0%	30.6%	0.000
	2–3 years	13.6%	18.2%	0.03
	More than 3 years	69.3%	41.3%	0.000
	Other	2.3%	0.0%	0.000
	Q22. Do you sanitize your cutting board?			
	Yes	68.0%	44.8%	0.000
	No	32.0%	55.2%	0.000
Management of kitchen cloth	Q23. How often do you replace your kitchen cloth?			
	1 or more than in a week	19.4%	26.6%	0.003
	1–2 times in a month	31.9%	35.9%	0.140
	Once in 3 months	33.7%	25.3%	0.001
	Once in 6 months	10.7%	8.1%	0.124
	Once in a year or more than a year	4.3%	4.1%	0.905
	Do not replace	0.2%	0.0%	0.319
	Q24. Do you sanitize your kitchen cloth?			
	Yes	87.7%	59.3%	0.000
	No	12.3%	40.7%	0.000

[1] *p*-value was provided according to the results of the Kruskal–Wallis test conducted for the analysis on the differences of responses in each answer option between Year 2010 and Year 2019. [2] To identify the difference between positive responses (i.e., Always + Frequently) and negative responses (i.e., Never + Seldom), answer options were combined.

Common risk food handling practices of consumers could be identified by the analysis of results from Year 2010. When purchased foods are too large to cook or eat, cutting or dividing them into smaller pieces is effective for handling and storage. Most consumers (mostly yes, 35.5%; completely yes, 34.5%) responded that they divided large portion of foods into smaller pieces for storage; however, only 12.5% (mostly yes, 8.9%; completely yes, 3.6%) of the consumers wrote information (e.g., product name, shelf life, proper storage method) on the divided foods). Therefore, a considerable number of consumers could not obtain information about divided foods after storage, which could lead to the mishandling of foods. When consumers were asked how long they used cutting boards, only 2.8% of consumers used cutting boards for < 1 year, while 69.3% reported using them for > 3 years. In terms of the question for the sanitization of kitchen utensils, 68.0% of respondents sanitized the cutting board. In the case of kitchen cloths, respondents usually replaced them once for every 3 months (33.7%), 1–2 times in a month (31.9%), and once or more in a week (19.4%). The percentage of respondents who sanitized kitchen cloths was 87.7%.

According to the results of Year 2019, common high-risk behaviors in Year 2010 were improved after 9 years (Table 3). Although both the increases ($p < 0.05$) and decreases ($p < 0.05$) in the negative responses (i.e., Never + Seldom) and the positive responses (i.e., Always + Frequently) were observed, respectively, more than half of consumers (57.9%) responded that they divided the large portion of foods into smaller portions. Although a substantial percentage (48.3%) of respondents did not write information, more consumers (25.6%) compared to Year 2010 ($p < 0.05$) indicated the information on divided foods. Over time, the period of using cutting board was shortened; distinct increases ($p < 0.05$) in the responses to 1–2 years (30.6%) and decreases ($p < 0.05$) in > 3 years (41.3%). The usage period of kitchen cloth was also shortened in Year 2019 compared to Year 2010; more consumers ($p < 0.05$) replaced it once or more than in a week (26.6%), and fewer consumers ($p < 0.05$) replaced it once in 3 months (25.3%). The percentages of respondents who sanitized cutting board and kitchen cloth decreased (from 68.0% to 44.8% and from 87.7% to 59.3%, respectively; $p < 0.05$), which was likely due to the frequent replacement of kitchen utensils rather than sanitizing them.

3.3. *Proper Behaviors of Consumers (Reported to Practice Proper Food Handling in Line with Guidance)*

Figures 4–6 showed that consumers practiced proper behaviors in accordance with their perceptions in both surveys (Detailed data with the results of statistical analysis were also indicated in Table S2).

3.3.1. Risk Perception and Behaviors of Consumers

The majority of consumers (83.4% for Year 2010, 77.8% for Year 2019) perceived the hazard of touching cooked foods after handling raw material (meat, fish/shellfish, eggs) without washing hands, and substantial consumers (74.3% for Year 2010, 67.1% for Year 2019) washed hands after handling potentially hazardous foods, reducing the risk of cross-contamination (Figure 4). Although there were decreases in the responses to proper risk perceptions (from 83.4% in Year 2010 to 77.9% in Year 2019 for the response "completely hazardous + mostly hazardous") and hygienic behaviors (from 74.2% in Year 2010 to 67.1% in Year 2019 for the response "always + frequently"), distinct trends that consumers confirmed the food safety rule with proper risk perception were maintained in both Year 2010 and Year 2019. Increases in the responses as "moderate" were also observed in both risk perception (from 13.3% to 17.4%; $p < 0.05$) and behavior (from 8.7% to 17.5%; $p < 0.05$) (Figure 4).

Of the respondents, most participants perceived the hazard of not fully cooking foods (90.0% and 80.0% in Year 2010 and Year 2019, respectively) with their self-reported behaviors supporting those risk perceptions (94.1% and 85.6% of consumers fully cook foods in Year 2010 and Year 2019, respectively) (Figure 5).

Generally, consumers reheated leftovers before eating them (91.9% for Year 2010, 85.0% for Year 2019), and most consumers (76.8% for Year 2010, 66.8% for Year 2019) were knowledgeable that eating leftovers without reheating was hazardous (Figure 6). Responses from Year 2010 to Year 2019 regarding the proper risk perception ((from 50.9% to 43.5% for "mostly hazardous" ($p < 0.05$), from 76.8% to 66.8% for "completely hazardous + mostly hazardous") ($p < 0.05$)) and proper behavior (from 45.6% to 34.4% for "frequently" ($p < 0.05$), from 92.0% to 85.0% for "always + frequently" ($p < 0.05$)) decreased by the increases in the responses "moderate" for both the risk perception (from 17.4% to 26.6%; $p < 0.05$) and behavior (from 6.7% to 12.2%; $p < 0.05$) of consumers (Table S2).

3.3.2. Storage Places/Periods for Food

Table 4 showed the storage places/periods for food in 2010 and 2019, respectively. Food safety authorities recommended that fresh eggs should be stored in refrigerators for 3–5 weeks [36], and consumers usually kept them in refrigerators (96.9 and 90.6% in Year 2010 and Year 2019, respectively) for fewer days than recommend (11.7 and 14.5 days on average, respectively). The US FDA (2018) also proposed the proper storage length and location for meat, chicken, fish, and shellfish as follows: 3–5 days in the refrigerator and 6–12 months in the freezer for meat (steak), 1–2 days in the refrigerator and 1 year in the freezer for chicken (whole chicken or turkey), 1–2 days in the refrigerator and 6-8 months in the freezer for fish (lean fish), and 1–2 days in the refrigerator and 3–6 months in the freezer for shrimp. Most consumers generally followed the rules for food storage. In terms of the storage places, significant differences ($p < 0.05$) in the responses on the preferred place for each food item between Year 2010 and Year 2019 were mainly observed (except for refrigerator and freezer for chicken and fish; other environments at room temperature for meat, shellfish, and milk; do not know/no response for all food items), but distinct trends for the preference were maintained as follows: meat for freezer, chicken for refrigerator or freezer, fish for freezer, shellfish for freezer, fruit and vegetables for refrigerator, eggs for refrigerator, milk for refrigerator, and frozen processed foods for freezer.

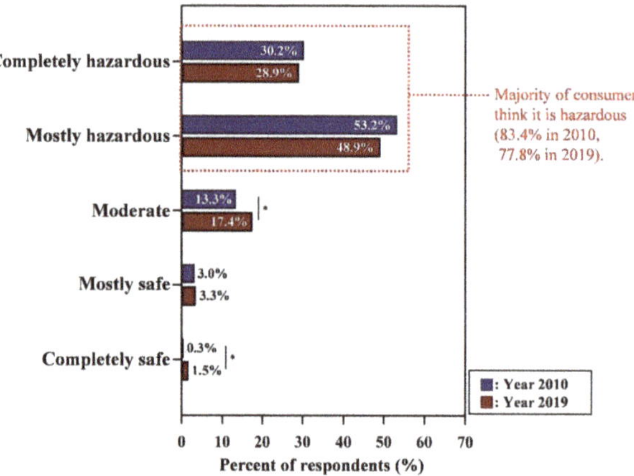

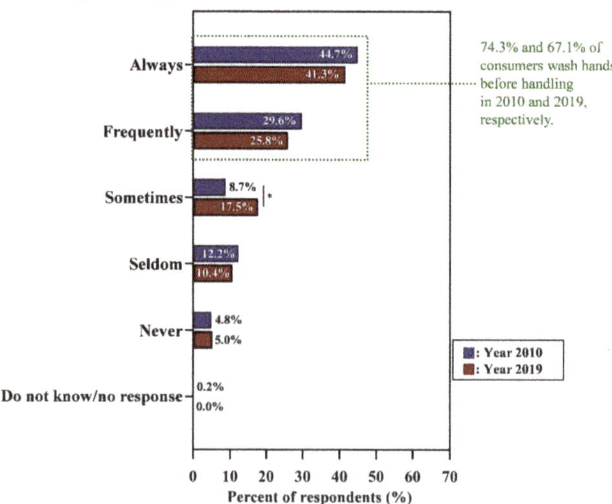

Figure 4. Proper behaviors of consumers in accordance with their perceptions regarding the washing hands. Asterisk (*) indicated between graphs means the significant differences of responses (the Kruskal–Wallis test; $p < 0.05$) in each answer option from Year 2010 and Year 2019. Green letters indicated the proper risk perceptions (i.e., the correct response in line with conventional food safety guidelines) and behaviors of consumers. All data regarding the percentage of the combined answer options (e.g., Always + Frequently) are provided in Table S2.

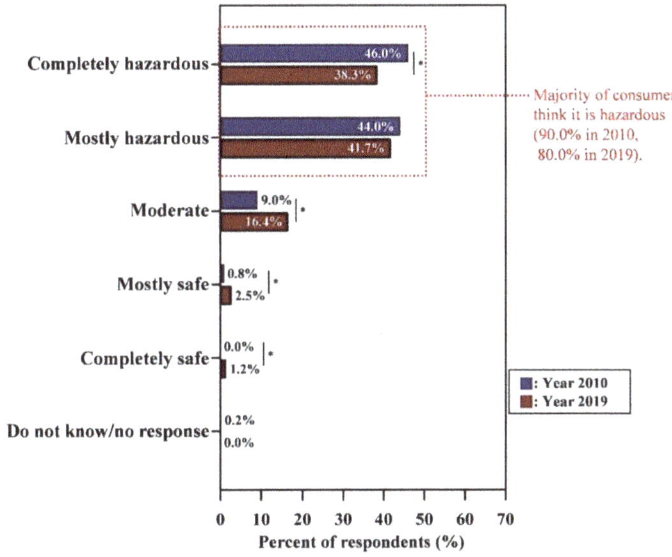

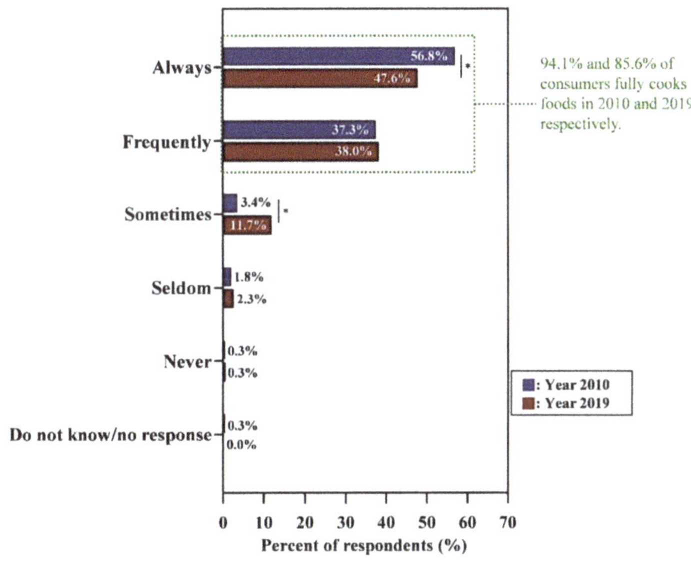

Figure 5. Proper behaviors of consumers in accordance with their perceptions regarding incomplete cooking. Asterisk (*) indicated between graphs means the significant differences of responses (the Kruskal–Wallis test; $p < 0.05$) in each answer option from Year 2010 and Year 2019. Green letters indicated the proper risk perceptions (i.e., the correct response in line with conventional food safety guidelines) and behaviors of consumers. All data regarding the percentage of the combined answer options (e.g., Always + Frequently) are provided in Table S2.

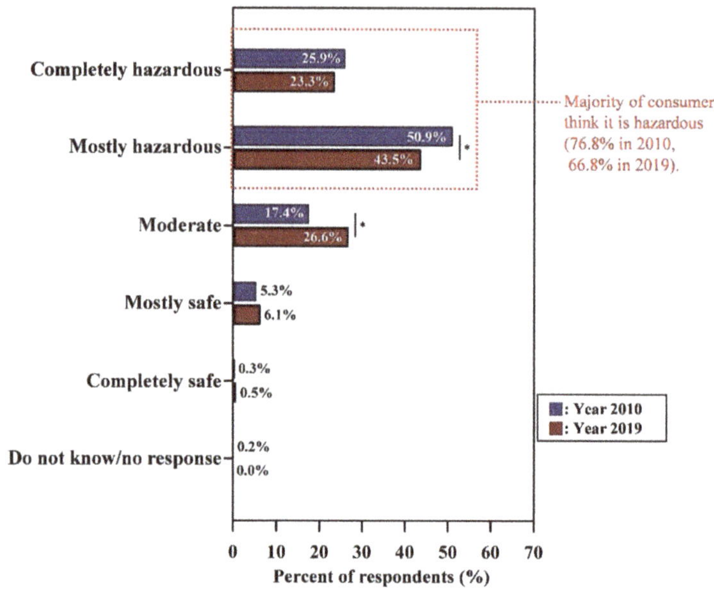

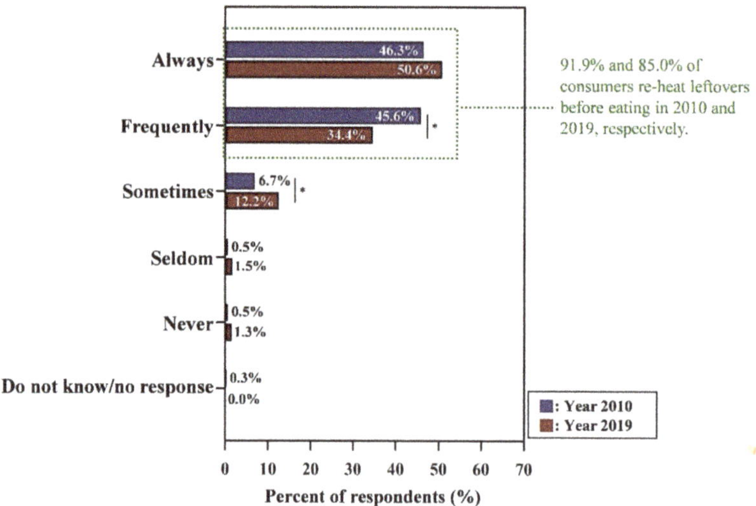

Figure 6. Proper behaviors of consumers in accordance with their perceptions regarding reheating leftovers. Asterisk (*) indicated between graphs means the significant differences of responses (the Kruskal–Wallis test; $p < 0.05$) in each answer option from Year 2010 and Year 2019. Green letters indicated the proper risk perceptions (i.e., the correct response in line with conventional food safety guidelines) and behaviors of consumers. All data regarding the percentage of the combined answer options (e.g., Always + Frequently) are provided in Table S2.

Table 4. The storage place and periods for specific foods in Year 2010 and Year 2019.

Q10. Where do you store foods (meat, chicken, fish/shellfish, fruit/vegetables, eggs, milk, and frozen processed foods) among refrigerator, freezer, or other environments at room temperature? Then, how long do you store those foods in the selected storage place?

	Refrigerator			Freezer			Other Environments at Room Temperature			Do Not Know /No Response			Do Not Purchase		
	Year 2010	Year 2019	p-Value [1]	Year 2010	Year 2019	p-Value	Year 2010	Year 2019	p-Value	Year 2010	Year 2019	p-Value	Year 2010	Year 2019	p-Value
Meat (pork and beef)															
Storage place (storage period; day)	20.9% (2.5)	30.3% (3.8)	0.000	77.0% (11.6)	67.9% (14.2)	0.000	1.8% (1.3)	1.8% (4.7)	0.988	0.3% (4.5)	-	0.158	-	-	-
Chicken															
Storage place (storage period; day)	44.8% (1.7)	48.1% (3.0)	0.253	48.3% (10.0)	50.2% (12.3)	0.492	6.6% (1.4)	1.7% (2.9)	0.000	0.3% (4.5)	-	0.158	-	-	-
Fish															
Storage place (storage period; day)	26.9% (2.2)	26.5% (3.6)	0.901	72.6% (13.0)	72.3% (18.5)	0.893	0.2% (2.0)	1.2% (2.7)	0.033	0.3% (8.5)	-	0.158	-	-	-
Shellfish															
Storage place (storage period; day)	37.4% (2.5)	30.5% (3.2)	0.012	59.3% (11.4)	67.5% (15.4)	0.003	2.0% (2.2)	2.0% (2.9)	0.987	0.3% (8.5)	-	0.158	1.0% (-)	-	0.014
Fruit and vegetables															
Storage place (storage period; day)	96.9% (5.7)	90.7% (7.8)	0.000	0.3% (7.0)	3.1% (4.1)	0.000	2.5% (3.6)	6.1% (5.1)	0.002	0.3% (5.0)	-	0.158	-	-	-
Eggs															
Storage place (storage period; day)	96.9% (11.7)	90.6% (14.5)	0.000	-	3.1% (8.8)	0.000	3.1% (13.3)	6.3% (12.8)	0.009	-	-	-	-	-	-
Milk															
Storage place (storage period; day)	99.5% (3.5)	95.2% (5.8)	0.000	-	4.0% (5.0)	0.000	0.5% (2.3)	0.8% (6.4)	0.472	-	-	-	-	-	-
Frozen processed foods															
Storage place (storage period; day)	5.1% (7.8)	9.6% (16.0)	0.003	92.6% (14.7)	88.9% (23.2)	0.027	-	1.5% (16.2)	0.003	-	-	-	2.3% (-)	-	0.000

[1] p-value was provided according to the results of the Kruskal–Wallis test conducted for the analysis on the differences of responses in each answer option (only for storage place, not for storage period) between Year 2010 and Year 2019.

3.3.3. Other Behaviors of Consumers

Other consumers' behaviors were also shown in Table 5. In short, the most consumers properly followed the food safety rules (at least 70% of the consumers): (1) store foods separately (thaw the only portion of frozen foodstuffs for cooking, store remaining raw foodstuffs with sealing after cooking), (2) store foods in the proper place and for the proper period, (3) wash fruits and vegetables properly before eating, (4) wash hands after handling potentially hazardous foods, (5) cook food to eat, and (6) reheat leftovers to eat. Although some answer options showed changes over time (i.e., a significant decrease of increase of the responses ($p < 0.05$)), distinct differences between the answer options within the results from each survey were obviously observed.

Table 5. Consumer behaviors for food purchasing, preparing, cooking, eating, leftovers management, and kitchen utensils.

Questions and Choices	Percent of Respondents or Answers		p-Value [1]
	Year 2010 (N = 609)	Year 2019 (N = 605)	
Q13. Do you thaw only a portion of food as much as you intend to cook rather than thawing all of the foods?			
Never	1.3%	1.0%	0.6
Seldom	3.9%	3.8%	0.9
Sometimes	10.8%	19.0%	0
Frequently	50.2%	45.1%	0.074
Always	33.2%	31.1%	0.435
Do not know/no response	0.5%	–	0.084
<Analysis of the combined responses> [2]			
Never + Seldom	5.3%	4.8%	0.713
Always + Frequently	83.4%	76.2%	0.002
Q14. Do you use plastic gloves when you handle meat, fish/shellfish, or eggs?			
Never	8.9%	6.1%	0.069
Seldom	16.4%	21.2%	0.035
Sometimes	17.6%	22.5%	0.033
Frequently	30.0%	30.6%	0.841
Always	27.1%	19.7%	0.002
<Analysis of the combined responses> [2]			
Never + Seldom	25.3%	27.3%	0.432
Always + Frequently	57.1%	50.2%	0.016
Q15. What do you do when you have a slight wound on your hand during cooking?			
Keep cooking after the wound has been treated	74.7%	71.7%	0.242
Keep cooking without any treatment	19.7%	24.6%	0.039
Do not cook	5.6%	3.6%	0.018
Q17. Do you wash fruits and vegetables before eating?			
Never	0.3%	0.5%	0.649
Seldom	0.3%	1.8%	0.012
Sometimes	1.8%	12.6%	0
Frequently	30.2%	30.4%	0.94
Always	67.2%	54.7%	0
<Analysis of the combined responses> [2]			
Never + Seldom	0.7%	2.3%	0.017
Always + Frequently	97.4%	85.1%	0
Q20. Do you store remaining raw foodstuffs with sealing after cooking?			
Never	0.3%	0.3%	0.995
Seldom	2.3%	2.6%	1
Sometimes	3.9%	16.4%	0
Frequently	46.3%	39.3%	0.014
Always	47.0%	41.3%	0.048
<Analysis of the combined responses> [2]			
Never + Seldom	2.6%	3.0%	0.713
Always + Frequently	93.3%	80.7%	0

[1] p-value was provided according to the results of the Kruskal–Wallis test conducted for the analysis on the differences of responses in each answer option between Year 2010 and Year 2019. [2] To identify the difference between positive responses (i.e., Always + Frequently) and negative responses (i.e., Never + Seldom), answer options were combined.

4. Discussion

In the present study, we conducted multiple individual surveys for the demonstration of the distinct/obvious gap between risk perception and practices regardless of the time-point for the surveys. This novel approach can contribute to the body of the knowledge on consumers' risk perception-behaviors by overcoming a limitation of relevant research which have been conducted as the singular cross-sectional study for specific time-point [2,21,25–28]. Comparative analysis of multiple individual surveys conducted in different time-points was rarely reported from the research not only for consumers [1,21,37–43] but also for food handlers in food services [44–51]. This is the first study regarding the comparative analysis of the surveys from multiple time-points to identify the distinct/obvious gap between consumers' risk perception and practices which were not improved over time. Previous consumer surveys regarding the domestic food safety identify the risk perception and behaviors in the specific time-point [1,21,37–43], however, the results from research cannot demonstrate whether the risk perception and/or behaviors will be changed over time. This limitation implies that the risk perception-behavior gap described by the survey conducted in a singular specific time-point cannot represent the general responses due to the time-dependent variability of consumers' perception and behaviors. Whereas most previous research have mainly conducted not only in the specific time-point but also from the specific region as follows: Belgium [41], China [40], Island of Ireland [21], Poland and Thailand [39], Republic of Ireland [37], Slovenia [1], Trinidad [43], Turkey [38], and USA [42]. Although previous research for the survey of respondents from multiple nations have been reported [39,48], comparative analysis between the results of research from other countries was rarely conducted. Since the risk perception and/or behaviors are different according to the regions and changeable over time, both regions and time-points can act as determinant factors for the survey results. Thus, the comparative analysis among the survey results with the perspectives to the region (i.e., countries) was rarely conducted even for the integrative reviews regarding the consumer surveys on food safety [4,52]. While the impact of the regional factor can be estimated by the comparative analysis of research which adopted the multiple individual surveys for general consumers (i.e., respondents selected by considering the homogeneity of sociodemographic characteristics) from various time-points because the longitudinal study can support the generalizability of the results with the perspectives to the time as a determinant factor. Consequently, further survey research based on our study design with the consideration of the time-points in other countries are expected to identify the most important food handling deficiencies by the analysis of the unchanged risk perception-behaviors over time regardless of the nations. Moreover, the accumulation of the survey data feasible to the comparative analysis between relevant research can also be useful for the modeling among the knowledge-attitude-practice regarding the food safety to predict the level of the risks and to establish the risk management strategies [53,54].

This study attempted to provide empirical data about food-handling behaviors as well as perceptions of food safety at home through consumer surveys focused on primary food handlers in 2010 and 2019. From the present study, we could follow the major trends for food safety knowledge and provide practical information about the high risk or proper behaviors implemented at home. Major practices properly implemented in accordance with consumers' perceptions and high-risk behaviors that did not support consumers' confidence in food safety were identified. Although the improvement in high-risk behaviors over a decade (from Year 2010 to Year 2019) and proper behaviors observed in both surveys were also noticed, several common high-risk behaviors were not corrected. The unchanged gap between risk perception and practices of consumers should be considered as the endemic problems for domestic food safety and the blind spots of the current intervention strategies against high-risk behaviors which required novel countermeasures. There were large gaps of perception-behaviors in 2010, including (1) storing perishable foods without any preparation (washing or trimming), (2) thawing foods at room temperature, and (3) exposing leftovers to danger zone temperatures. These gaps between consumer perceptions and behaviors in 2010 remained in 2019.

Since the internationally-recognized consumer guidelines suggested cleaning as a basic step to food safety at home [32], previous research regarding the washing perishables have consistently reported the proper perceptions with actual practices [39,40] whereas there was a lack of survey research washing raw materials prior to storage. However, in this study, the storage of perishable foods without washing and trimming was analyzed as a major potential for deviant behaviors in both surveys despite consumers' proper safety perception/knowledge. Preparation of perishables before storage can prevent not only the risk of foodborne diseases by the removal of contaminants attached to foods but also the cross-contamination during the storage of various foodstuffs [55,56]. These contributions to food safety should be emphasized by the educational tools (e.g., guidelines, programs, leaflets) regarding the preparation followed by the storage of perishables. The importance for washing and trimming of foods mainly for eaten as raw (e.g., fruits, vegetables, etc.) has been highlighted by the previous research on the effectiveness of the decontamination of microbiological risk factors (i.e., pathogens) [57,58] and/or the validation of the cross-contamination [59]. Moreover, the establishment and the improvement of the knowledge basis regarding proper packaging methods (e.g., suggestions on how not to group a large amount of food together, how to separate large pieces in individual packages as an alternative to cutting them into smaller pieces, and how to place them properly in the freezer to ensure optimum quick freezing, etc.) should also be followed to alter the consumers' high-risk behaviors.

Frozen foods should not be thawed at room temperature to avoid the exposure of foods to "danger zone" (4.4–60.0 °C; 40–140 °F), the temperature range that can cause the growth of foodborne bacteria in foods [60,61]. Thawing frozen foodstuffs at room temperature have been reported as general behaviors or underrecognized risk perceptions of consumers in other relevant research [21,41,43]. Thawing in a refrigerator or in cold water with the time-temperature control has not been generally preferred than the exposure to room temperate [43,62], and other inadequate methods using tap water [63] or immersing in warm water [64] were also frequently used. Whereas a direct gap of perceptions-behaviors observed in this study has been rarely highlighted by previous studies. Low-risk perception on the temperature control has been suggested as the major cause for the high-risk behaviors of thawing frozen foods from relevant studies [39,40], however, this research implied that high-risk behavior could be occurred by consumers with proper risk perception.

Leaving food out at an improper temperature is one of the main factors commonly associated with foodborne disease at home [3]. The government recommended the following practice for leftovers: to keep food out of the "danger zone", wrap leftovers well, store leftovers safely, thaw frozen leftovers safely, reheat leftovers without thawing, and reheat leftovers [65]. Leftovers should be refrigerated or frozen quickly as soon as possible to prevent the exposure of foods under favorable conditions for microbial growth [1]. However, in this research, exposure leftovers at room temperature have been reported as one of the common high-risk behaviors of primary food handlers. This risk factor has been also highlighted in recent studies on consumer surveys which reported the preference for cooling leftovers at room temperature before the storage in the refrigerator [54,62,66].

Whereas the improvement on consumers' high-risk behaviors (Section 3.2) or proper behaviors observed in both surveys (Section 3.3) can lower the microbiological risks of foodborne diseases, however, the importance of overall procedures for consumers' food handling should not be underestimated. Investigations for the identification of underreported potential risks in domestic food safety linked to the consumers' perceptions and behaviors should be persistently conducted. In the case of handling kitchen utensils, although increases in proper practices for the management of the cutting board and kitchen cloths were observed in Year 2019 (Table 3), improper behaviors of consumers for the use of those kitchen utensils to prepare foods could result in the foodborne illnesses despite the hygienic management because cross-contamination can occur when harmful bacteria were transferred to food from kitchen utensils [67]. Especially reusing the same utensils for various kinds of foods including raw materials and the ready-to-eat products has been reported as representative practices of poor sanitation procedures which could result in cross-contamination [43].

Moreover, previous studies on consumer surveys for domestic food safety also highlighted various other cross-contamination routes from raw materials to cooked foods (e.g., via knives, cutting boards, and/or plates, etc.), highlighting the necessities on the systematic intervention structures for kitchen utensils [41]. As both risk perception and behaviors regarding the separated use of kitchen utensils is regarded as a major risk factor [1,40], internationally-recognized consumer guidelines have suggested "separate" as keywords for food safety in the kitchen [32,68]. A recent report reported that most of the respondents from the survey on food handlers' behaviors with the perspective of the meal preparation at home declared that they mainly separated kitchen utensils including the cutting boards for raw and cooked foods [54]. As shown in this study, a significant number of respondents from the relevant previous research also reported consumer preference for washing utensils in hot water with detergent to cut various kinds of raw materials by using the same cutting board [21,66]. Thus, the channels and methods of the delivery of proper information for the effective conditions to eliminate the major microbiological factors present on utensils should be established [69,70] because of the incomplete sterilization of cutting board by using hot water and detergents has been reported [71,72].

The results obtained from 2010 and 2019 suggested that behaviors did not support consumers' confidence in food safety. These gaps between perceptions and behaviors might be significant risk factors for foodborne disease and could increase the likelihood of food deterioration/poisoning. The result of this study highlights that education tools for improving risk perceptions and knowledge should have been implemented [73–75] because these high-risk behaviors are generally based on the lack of awareness of domestic food safety. The intervention for the improvement of consumers' poor food safety knowledge and/or perception-behaviors reported from most previous research has been also limited to education [1,21,38–40,42,43]. However, since there is no enforceable regulation for food handling at home, consumers can easily ignore the importance of handling practices for food safety [11]; this can be a key reason why the gaps were not narrowed over a decade in this research. Thus, strategies for the improvement of the effectiveness and efficiencies of the education should also be adopted (e.g., highlighting responsibility for food safety as a primary food handler, informing the susceptibility and the severity of outcomes, building confidence, etc.) [4]. Moreover, the information sources and/or channels have been reported as the significant determinant factors for the educational effects to consumers [38,76,77], highlighting various education tools which have been regarded as effective strategies for behavior intervention and/or information delivery should also be considered (e.g., the social media and web-based communication with consumers) [52,78–80]. To narrow a perception-behavior gap, education should focus on consumers who think of themselves as knowledgeable regarding food safety issues, and communication with those consumers to recognize the gap is expected to contribute to changes in their hygienic practices [81–84]. As a moderate response to the questions of risk perception-behaviors can also be regarded as the potential for behavior that deviates from the best practice guidelines, further study regarding the in-depth examination on the intention of consumers who responded with the "moderate responses" is expected to change moderate to proper responses [29–31]. Moreover, further studies based on the segmentation analysis of consumers are needed to identify the major education targets that demonstrate high levels of knowledge regarding best practice but tendencies to report deviant behaviors and those with low knowledge and therefore likelihood to also demonstrate deviant behaviors [85]. A continuous education program from health authorities on the potential hazards of improper food handling is essential to motivate changes to improper practices at home using a varied approach including practical guidelines, face-to-face education programs, and the distribution of written materials such as leaflets. We also suggest that future food safety management should be further specialized for the perception-practices gap and proper behavior by the behavioral interventions. Even though most consumer studies have been mainly limited to the surveys investigating current status without the application of the interventions [45], findings from the research on the behavior intervention methods for food handlers in food services (e.g., training program, the legislation of policy and/or regulations for food safety, the inspection, the supervision, etc.) [45–47,49,50] can be applied to support the improvement on the educational tools and/or programs for consumers. Advanced strategies for consumer education are expected to develop consumers' recognition of food safety issues and reduce foodborne illness at home.

5. Conclusions

The results of the present study provided comprehensive data about behaviors and perceptions of consumers (particularly primary food handlers at home) and could be used as an information basis for the development of educational tools specified for food consumers. Whereas the limitations for the study design implied the necessity of further research as follows: (1) the consideration of the recent changes in the increases of male food handlers, (2) the analysis of the responses from the same respondents in multiple individual surveys, and (3) in-depth examination of the determinant factors which can induce the changes in risk perception and/or behaviors and the evaluation on the specific intervention against the risk perception-behavior gap (e.g., advertising, providing guidelines, publishing pamphlets, providing education, etc.). Firstly, we unavoidably used a heavily female-biased sample because of the distinct distribution of male and female primary food handlers at home in South Korea, so the further survey with the increases in the distribution of male food handlers in South Korea is likely to identify emerging high-risk behaviors from males. Secondly, direct comparison for the same respondents and two time periods can effectively show the impact of time in the high-risk behaviors and perception of consumers. Whereas as this research aimed to obtain general responses from consumers (i.e., representative sample) in each survey, we recruited the respondents not only from Year 2010 but also Year 2019 to use a representative sample in each time-point of the survey. Thirdly, as this research did not focus on the specific intervention strategy for domestic food safety and determinant factors which might affect consumers' risk perception and/or behaviors, the further examination on the effectiveness of various interventions is expected to find out optimal methods against risk perception-behavior gap observed in this study. In conclusion, our findings supported the understanding of the risks in domestic food safety necessary for the development of effective perception-behavior interventions to narrow the risk perception-behavior gap. This study is expected to act as a leading role of the representative work for the novel research design (i.e., comparative analysis on the individual consumer surveys conducted with same questionnaires at different time-points for identifying unchanged distinct gaps in risk perception-behaviors over time), highlighting the necessity for the following surveys from various regions and time-points to expand the body of knowledge on food safety for consumers.

Supplementary Materials: The following are available online at http://www.mdpi.com/2304-8158/9/10/1457/s1, Table S1: Detailed data regarding the discordance between consumers' food safety perceptions and behaviors (risk perception-behavior gap), Table S2: Detailed data regarding the proper risk perception (i.e., the correct response in line with conventional food safety guidelines) and behaviors of consumers.

Author Contributions: Contributions for each author are as follows: conceptualization, T.J.C., S.A.K., and H.W.K.; investigation, T.J.C., S.A.K., and H.W.K.; writing—original draft preparation, T.J.C. and S.A.K.; writing—review and editing, H.W.K.; supervision, M.S.R. All authors have read and agreed to the published version of the manuscript.

Funding: This research was supported by a Korea University grant.

Acknowledgments: The authors also thank the School of Life Science and Biotechnology of Korea University for BK 21 PLUS and the Institute of Biomedical Science and Food Safety, Korea University Food Safety Hall, for access to equipment and facilities.

Conflicts of Interest: The authors declare no conflict of interest. The funders had no role in the design of the study; in the collection, analyses, or interpretation of data; in the writing of the manuscript, or in the decision to publish the results.

References

1. Jevšnik, M.; Hlebec, V.; Raspor, P. Consumers' awareness of food safety from shopping to eating. *Food Control* **2008**, *19*, 737–745. [CrossRef]
2. Röhr, A.; Lüddecke, K.; Drusch, S.; Müller, M.; Alvensleben, R. Food quality and safety-consumer perception and public health concern. *Food Control* **2005**, *16*, 649–655. [CrossRef]

3. Unusan, N. Consumer food safety knowledge and practices in the home in Turkey. *Food Control* **2007**, *18*, 45–51. [CrossRef]
4. Byrd-Bredbenner, C.; Berning, J.; Martin-Biggers, J.; Quick, V. Food safety in home kitchens: A synthesis of the literature. *Int. J. Environ. Res. Public Health* **2013**, *10*, 4060–4085. [CrossRef]
5. Nesbitt, A.; Majowicz, S.; Finley, R.; Marshall, B.; Pollari, F.; Sargeant, J.; Ribble, C.; Wilson, J.; Sittler, N. High-risk food consumption and food safety practices in a Canadian community. *J. Food Prot.* **2009**, *72*, 2575–2586. [CrossRef]
6. Scott, E. Food safety and foodborne disease in 21st century homes. *Can. J. Infect. Dis.* **2003**, *14*, 277.
7. WHO. Statistical information on foodborne disease in Europe: Microbiological and chemical hazards. In Proceedings of the FAO-WHO Pan-European Conference on Food Safety and Quality, Budapest, Hungary, 25–28 February 2002.
8. Redmond, E.C.; Griffith, C.J. Consumer food handling in the home: A review of food safety studies. *J. Food Prot.* **2003**, *66*, 130–161. [CrossRef]
9. Redmond, E.C.; Griffith, C.J. The importance of hygiene in the domestic kitchen: Implications for preparation and storage of food and infant formula. *Perspect. Public Health* **2009**, *129*, 69–76. [CrossRef]
10. Ministry of Food and Drug Safety. Surveillance for Foodborne Disease Outbreaks. 2020. Available online: https://www.foodsafetykorea.go.kr/portal/healthyfoodlife/foodPoisoningStat.do?menu_no=3724&menu_grp=MENU_NEW02&menu_no=3724&menu_grp=MENU_NEW02 (accessed on 9 August 2020).
11. Karabudak, E.; Bas, M.; Kiziltan, G. Food safety in the home consumption of meat in Turkey. *Food Control* **2008**, *19*, 320–327. [CrossRef]
12. Knox, B. Consumer perception and understanding of risk from food. *Br. Med Bull.* **2000**, *56*, 97–109. [CrossRef]
13. Phang, H.S.; Bruhn, C.M. Burger preparation: What consumers say and do in the home. *J. Food Prot.* **2011**, *74*, 1708–1716. [CrossRef] [PubMed]
14. Woteki, C.E.; Facinoli, S.L.; Schor, D. Keep food safe to eat: Healthful food must be safe as well as nutritious. *J. Nutr.* **2001**, *131*, 502S–509S. [CrossRef] [PubMed]
15. Frewer, L. Risk perception and risk communication about food safety issues. *Nutr. Bull.* **2000**, *25*, 31–33. [CrossRef]
16. Verbeke, W.; Frewer, L.J.; Scholderer, J.; De Brabander, H.F. Why consumers behave as they do with respect to food safety and risk information. *Anal. Chim. Acta* **2007**, *586*, 2–7. [CrossRef]
17. Mullan, B.A.; Wong, C.; Kothe, E.J. Predicting adolescents' safe food handling using an extended theory of planned behavior. *Food Control* **2013**, *31*, 454–460. [CrossRef]
18. Shapiro, M.A.; Porticella, N.; Jiang, L.C.; Gravani, R.B. Predicting intentions to adopt safe home food handling practices. Applying the theory of planned behavior. *Appetite* **2011**, *56*, 96–103. [CrossRef]
19. Kim, N.H.; Kim, M.J.; Park, B.I.; Kang, Y.S.; Hwang, I.G.; Rhee, M.S. Discordance in risk perception between children, parents, and teachers in terms of consumption of cheap and poorly nutritious food sold around schools. *Food Qual. Prefer.* **2015**, *42*, 139–145. [CrossRef]
20. Medeiros, L.; Hillers, V.; Kendall, P.; Mason, A. Evaluation of food safety education for consumers. *J. Nutr. Educ.* **2001**, *33*, S27–S34. [CrossRef]
21. Kennedy, J.; Jackson, V.; Blair, I.; McDowell, D.; Cowan, C.; Bolton, D. Food safety knowledge of consumers and the microbiological and temperature status of their refrigerators. *J. Food Prot.* **2005**, *68*, 1421–1430. [CrossRef]
22. Meysenburg, R.; Albrecht, J.A.; Litchfield, R.; Ritter-Gooder, P.K. Food safety knowledge, practices and beliefs of primary food preparers in families with young children. A mixed methods study. *Appetite* **2014**, *73*, 121–131. [CrossRef]
23. Rimal, A.; Fletcher, S.M.; McWatters, K.; Misra, S.K.; Deodhar, S. Perception of food safety and changes in food consumption habits: A consumer analysis. *Int. J. Consum. Stud.* **2001**, *25*, 43–52. [CrossRef]
24. Roseman, M.; Kurzynske, J. Food safety perceptions and behaviors of Kentucky consumers. *J. Food Prot.* **2006**, *69*, 1412–1421. [CrossRef] [PubMed]
25. Murray, R.; Glass-Kaastra, S.; Gardhouse, C.; Marshall, B.; Ciampa, N.; Franklin, K.; Hurst, M.; Thomas, M.K.; Nesbitt, A. Canadian consumer food safety practices and knowledge: Foodbook study. *J. Food Prot.* **2017**, *80*, 1711–1718. [CrossRef] [PubMed]

26. Patil, S.R.; Cates, S.; Morales, R. Consumer food safety knowledge, practices, and demographic differences: Findings from a meta-analysis. *J. Food Prot.* **2005**, *68*, 1884–1894. [CrossRef]
27. Terpstra, M.; Steenbekkers, L.; De Maertelaere, N.; Nijhuis, S. Food storage and disposal: Consumer practices and knowledge. *Br. Food J.* **2005**, *107*, 526–533. [CrossRef]
28. Verbeke, W.; Sioen, I.; Pieniak, Z.; Van Camp, J.; De Henauw, S. Consumer perception versus scientific evidence about health benefits and safety risks from fish consumption. *Public Health Nutr.* **2005**, *8*, 422–429. [CrossRef]
29. De Boer, M.; McCarthy, M.; Brennan, M.; Kelly, A.L.; Ritson, C. Public understanding of food risk issues and food risk messages on the island of Ireland: The views of food safety experts. *J. Food Saf.* **2005**, *25*, 241–265. [CrossRef]
30. McCarthy, M.; Brennan, M.; Kelly, A.; Ritson, C.; De Boer, M.; Thompson, N. Who is at risk and what do they know? Segmenting a population on their food safety knowledge. *Food Qual. Prefer.* **2007**, *18*, 205–217. [CrossRef]
31. Kendall, H.; Brennan, M.; Seal, C.; Ladha, C.; Kuznesof, S. Behind the kitchen door: A novel mixed method approach for exploring the food provisioning practices of the older consumer. *Food Qual. Prefer.* **2016**, *53*, 105–116. [CrossRef]
32. U.S. FDA. 4 Basic Steps to Food Safety at Home. Available online: https://www.fda.gov/downloads/forconsumers/byaudience/forwomen/freepublications/ucm554423.pdf (accessed on 22 July 2019).
33. USDA, FSIS. Safe Food Handling Fact Sheets. Available online: https://www.fsis.usda.gov/wps/portal/fsis/topics/food-safety-education/get-answers/food-safety-fact-sheets/safe-food-handling (accessed on 22 July 2019).
34. Kish, L. *Survey Sampling*, 1st ed.; John Wiley & Sons: Hoboken, NJ, USA, 1995; pp. 359–377.
35. Särndal, C.; Swensson, B.; Wretman, J. *Model Assisted Survey Sampling*, 1st ed.; Springer: New York, NY, USA, 1992; pp. 303–333.
36. U.S. FDA. Refrigerator & Freezer Storage Chart. Available online: https://www.fda.gov/downloads/Food/FoodborneIllnessContaminants/UCM109315.pdf (accessed on 22 July 2019).
37. Moreb, N.A.; Priyadarshini, A.; Jaiswal, A.K. Knowledge of food safety and food handling practices amongst food handlers in the Republic of Ireland. *Food Control* **2017**, *80*, 341–349. [CrossRef]
38. Ergönül, B. Consumer awareness and perception to food safety: A consumer analysis. *Food Control* **2013**, *32*, 461–471. [CrossRef]
39. Tomaszewska, M.; Trafialek, J.; Suebpongsang, P.; Kolanowski, W. Food hygiene knowledge and practice of consumers in Poland and in Thailand—A survey. *Food Control* **2018**, *85*, 76–84. [CrossRef]
40. Gong, S.; Wang, X.; Yang, Y.; Bai, L. Knowledge of food safety and handling in households: A survey of food handlers in Mainland China. *Food Control* **2016**, *64*, 45–53. [CrossRef]
41. Sampers, I.; Berkvens, D.; Jacxsens, L.; Ciocci, M.C.; Dumoulin, A.; Uyttendaele, M. Survey of Belgian consumption patterns and consumer Behavior of poultry meat to provide insight in risk factors for campylobacteriosis. *Food Control* **2012**, *26*, 293–299. [CrossRef]
42. Leal, A.; Ruth, T.K.; Rumble, J.N.; Simonne, A.H. Exploring Florida residents' food safety knowledge and behaviors: A generational comparison. *Food Control* **2017**, *73*, 1195–1202. [CrossRef]
43. Badrie, N.; Gobin, A.; Dookeran, S.; Duncan, R. Consumer awareness and perception to food safety hazards in Trinidad, West Indies. *Food Control* **2006**, *17*, 370–377. [CrossRef]
44. Choi, J.; Norwood, H.; Seo, S.; Sirsat, S.A.; Neal, J. Evaluation of food safety related behaviors of retail and food service employees while handling fresh and fresh-cut leafy greens. *Food Control* **2016**, *67*, 199–208. [CrossRef]
45. Barjaktarović-Labović, S.; Mugoša, B.; Andrejević, V.; Banjari, I.; Jovićević, L.; Djurović, D.; Martinović, A.; Radojlović, J. Food hygiene awareness and practices before and after intervention in food services in Montenegro. *Food Control* **2018**, *85*, 466–471. [CrossRef]
46. Tan, S.L.; Bakar, F.A.; Karim, M.S.A.; Lee, H.Y.; Mahyudin, N.A. Hand hygiene knowledge, attitudes and practices among food handlers at primary schools in Hulu Langat district, Selangor (Malaysia). *Food Control* **2013**, *34*, 428–435. [CrossRef]
47. Rebouças, L.T.; Santiago, L.B.; Martins, L.S.; Menezes, A.C.R.; Araújo, M.d.P.N.; De Castro Almeida, R.C. Food safety knowledge and practices of food handlers, head chefs and managers in hotels' restaurants of Salvador, Brazil. *Food Control* **2017**, *73*, 372–381.

48. Trafialek, J.; Drosinos, E.H.; Laskowski, W.; Jakubowska-Gawlik, K.; Tzamalis, P.; Leksawasdi, N.; Surawang, S.; Kolanowski, W. Street food vendors' hygienic practices in some Asian and EU countries—A survey. *Food Control* **2018**, *85*, 212–222. [CrossRef]
49. Zhang, H.; Lu, L.; Liang, J.; Huang, Q. Knowledge, attitude and practices of food safety amongst food handlers in the coastal resort of Guangdong, China. *Food Control* **2015**, *47*, 457–461.
50. Robertson, L.A.; Boyer, R.R.; Chapman, B.J.; Eifert, J.D.; Franz, N.K. Educational needs assessment and practices of grocery store food handlers through survey and observational data collection. *Food Control* **2013**, *34*, 707–713. [CrossRef]
51. Al-Shabib, N.A.; Mosilhey, S.H.; Husain, F.M. Cross-sectional study on food safety knowledge, attitude and practices of male food handlers employed in restaurants of King Saud University, Saudi Arabia. *Food Control* **2016**, *59*, 212–217. [CrossRef]
52. Nesbitt, A.; Thomas, M.K.; Marshall, B.; Snedeker, K.; Meleta, K.; Watson, B.; Bienefeld, M. Baseline for consumer food safety knowledge and behaviour in Canada. *Food Control* **2014**, *38*, 157–173. [CrossRef]
53. Zanin, L.M.; Da Cunha, D.T.; De Rosso, V.V.; Capriles, V.D.; Stedefeldt, E. Knowledge, attitudes and practices of food handlers in food safety: An integrative review. *Food Res. Int.* **2017**, *100*, 53–62. [CrossRef]
54. Soon, J.M.; Wahab, I.R.A.; Hamdan, R.H.; Jamaludin, M.H. Structural equation modelling of food safety knowledge, attitude and practices among consumers in Malaysia. *PLoS ONE* **2020**, *15*, e0235870. [CrossRef]
55. Catellani, P.; Scapin, R.M.; Alberghini, L.; Radu, I.; Giaccone, V. Levels of microbial contamination of domestic refrigerators in Italy. *Food Control* **2014**, *42*, 257–262. [CrossRef]
56. Jackson, V.; Blair, I.; McDowell, D.; Kennedy, J.; Bolton, D. The incidence of significant foodborne pathogens in domestic refrigerators. *Food Control* **2007**, *18*, 346–351. [CrossRef]
57. Maffei, D.F.; Alvarenga, V.O.; Sant'Ana, A.S.; Franco, B.D. Assessing the effect of washing practices employed in Brazilian processing plants on the quality of ready-to-eat vegetables. *LWT Food Sci. Technol.* **2016**, *69*, 474–481. [CrossRef]
58. Ssemanda, J.N.; Joosten, H.; Bagabe, M.C.; Zwietering, M.H.; Reij, M.W. Reduction of microbial counts during kitchen scale washing and sanitization of salad vegetables. *Food Control* **2018**, *85*, 495–503. [CrossRef]
59. Gombas, D.; Luo, Y.; Brennan, J.; Shergill, G.; Petran, R.; Walsh, R.; Hau, H.; Khurana, K.; Zomorodi, B.; Rosen, J. Guidelines to validate control of cross-contamination during washing of fresh-cut leafy vegetables. *J. Food Prot.* **2017**, *80*, 312–330. [CrossRef] [PubMed]
60. U.S. FDA. Everyday Food Safety for Young Adults; Food Safety Quick Tips. Available online: https://www.fda.gov/food/buy-store-serve-safe-food/everyday-food-safety-young-adults (accessed on 22 July 2019).
61. USDA, FSIS. Danger Zone. Available online: https://www.fsis.usda.gov/shared/PDF/Danger_Zone.pdf (accessed on 22 July 2019).
62. Anderson, J.B.; Shuster, T.A.; Hansen, K.E.; Levy, A.S.; Volk, A. A camera's view of consumer food-handling behaviors. *J. Am. Diet. Assoc.* **2004**, *104*, 186–191. [CrossRef] [PubMed]
63. Baptista, R.C.; Rodrigues, H.; Sant'Ana, A.S. Consumption, knowledge, and food safety practices of Brazilian seafood consumers. *Food Res. Int.* **2020**, *132*, 109084. [CrossRef] [PubMed]
64. Tomaszewska, M.; Bilska, B.; Kołożyn-Krajewska, D. Do Polish consumers take proper care of hygiene while shopping and preparing meals at home in the context of wasting food? *Int. J. Environ. Res. Public Health* **2020**, *17*, 2074. [CrossRef]
65. USDA, FSIS. Leftovers and Food Safety. Available online: http://www.fsis.usda.gov/wps/portal/fsis/topics/food-safety-education/get-answers/food-safety-fact-sheets/safe-food-handling/leftovers-and-food-safety (accessed on 22 July 2019).
66. Marklinder, I.; Ahlgren, R.; Blücher, A.; Börjesson, S.M.E.; Hellkvist, F.; Moazzami, M.; Schelin, J.; Zetterström, E.; Eskhult, G.; Danielsson-Tham, M.L. Food safety knowledge, sources thereof and self-reported behaviour among university students in Sweden. *Food Control* **2020**, *113*, 107130. [CrossRef]
67. De Jong, A.; Verhoeff-Bakkenes, L.; Nauta, M.; De Jonge, R. Cross-contamination in the kitchen: Effect of hygiene measures. *J. Appl. Microbiol.* **2008**, *105*, 615–624. [CrossRef]
68. USDA, FSIS. Kitchen Companion, Your Safe Food Handbook. Available online: https://www.fsis.usda.gov/wps/wcm/connect/6c55c954-20a8-46fd-b617-ecffb4449062/Kitchen_Companion_Single.pdf?MOD=AJPERES (accessed on 22 July 2019).
69. Yang, H.; Kendall, P.A.; Medeiros, L.C.; Sofos, J.N. Efficacy of sanitizing agents against *Listeria monocytogenes* biofilms on high-density polyethylene cutting board surfaces. *J. Food Prot.* **2009**, *72*, 990–998. [CrossRef]

70. Beitrame, C.A.; Martelo, E.B.; Mesquita, R.D.A.; Rottava, I.; Barbosa, J.; Steffens, C.; Toniazzo, G.; Valduga, E.; Cansian, R.L. Effectiveness of sanitizing agents in inactivating *Escherichia coli* (ATCC 25922) in food cutting board surfaces. Removal *E. coli* using different sanitizers. *Ital. J. Food Sci.* **2016**, *28*, 148–154.
71. Verhoeff-Bakkenes, L.; Beumer, R.; De Jonge, R.; Van Leusden, F.; De Jong, A. Quantification of *Campylobacter jejuni* cross-contamination via hands, cutlery, and cutting board during preparation of a chicken fruit salad. *J. Food Prot.* **2008**, *71*, 1018–1022. [CrossRef]
72. Welker, C.; Faiola, N.; Davis, S.; Maffatore, I.; Batt, C.A. Bacterial retention and cleanability of plastic and wood cutting boards with commercial food service maintenance practices. *J. Food Prot.* **1997**, *60*, 407–413. [CrossRef] [PubMed]
73. Cho, T.J.; Kim, N.H.; Hong, Y.J.; Park, B.; Kim, H.S.; Lee, H.G.; Song, M.K.; Rhee, M.S. Development of an effective tool for risk communication about food safety issues after the Fukushima nuclear accident: What should be considered? *Food Control* **2017**, *79*, 17–26. [CrossRef]
74. Medeiros, L.C.; Hillers, V.N.; Kendall, P.A.; Mason, A. Food safety education: What should we be teaching to consumers? *J. Nutr. Educ.* **2001**, *33*, 108–113. [CrossRef]
75. Sivaramalingam, B.; Young, I.; Pham, M.T.; Waddell, L.; Greig, J.; Mascarenhas, M.; Papadopoulos, A. Scoping review of research on the effectiveness of food-safety education interventions directed at consumers. *Foodborne Pathog. Dis.* **2015**, *12*, 561–570. [CrossRef] [PubMed]
76. Kornelis, M.; De Jonge, J.; Frewer, L.; Dagevos, H. Consumer selection of food-safety information sources. *Risk Anal.* **2007**, *27*, 327–335. [CrossRef]
77. Liu, R.; Pieniak, Z.; Verbeke, W. Food-related hazards in China: Consumers' perceptions of risk and trust in information sources. *Food Control* **2014**, *46*, 291–298. [CrossRef]
78. Stephen, A.T. The role of digital and social media marketing in consumer behavior. *Curr. Opin. Psychol.* **2016**, *10*, 17–21. [CrossRef]
79. Godey, B.; Manthiou, A.; Pederzoli, D.; Rokka, J.; Aiello, G.; Donvito, R.; Singh, R. Social media marketing efforts of luxury brands: Influence on brand equity and consumer behavior. *J. Bus. Res.* **2016**, *69*, 5833–5841. [CrossRef]
80. Xiang, Z.; Magnini, V.P.; Fesenmaier, D.R. Information technology and consumer behavior in travel and tourism: Insights from travel planning using the internet. *J. Retail. Consum. Serv.* **2015**, *22*, 244–249. [CrossRef]
81. Al-Sakkaf, A. Domestic food preparation practices: A review of the reasons for poor home hygiene practices. *Health Promot. Int.* **2013**, *30*, 427–437. [CrossRef]
82. Fulham, E.; Mullan, B. Hygienic food handling behaviors: Attempting to bridge the intention-behavior gap using aspects from temporal self-regulation theory. *J. Food Prot.* **2011**, *74*, 925–932. [CrossRef] [PubMed]
83. Kim, N.H.; Hwang, J.Y.; Lee, H.G.; Song, M.K.; Kang, Y.S.; Rhee, M.S. Strategic approaches to communicating with food consumers about genetically modified food. *Food Control* **2018**, *92*, 523–531. [CrossRef]
84. Wilson, S.; Jacob, C.J.; Powell, D. Behavior-change interventions to improve hand-hygiene practice: A review of alternatives to education. *Crit. Public Health* **2011**, *21*, 119–127. [CrossRef]
85. Kendall, H.; Kuznesof, S.; Seal, C.; Dobson, S.; Brennan, M. Domestic food safety and the older consumer: A segmentation analysis. *Food Qual. Prefer.* **2013**, *28*, 396–406. [CrossRef]

Article

Food Innovation Adoption and Organic Food Consumerism—A Cross National Study between Malaysia and Hungary

Robert Jeyakumar Nathan [1], Soekmawati [1], Vijay Victor [2], József Popp [3,4], Mária Fekete-Farkas [3,*] and Judit Oláh [4,5]

[1] Faculty of Business, Multimedia University, Jalan Ayer Keroh Lama, Melaka 75450, Malaysia; robert.jeyakumar@mmu.edu.my (R.J.N.); soekma_wt@yahoo.com (S.)
[2] Department of Economics, CHRIST (Deemed to be University), Hosur Road, Bengaluru 560029, Karnataka, India; vijay.victor@christuniversity.in
[3] Faculty of Economics and Social Sciences, Szent István University, 2100 Gödöllő, Hungary; popp.jozsef@szie.hu
[4] TRADE Research Entity, North-West University, Vanderbijlpark 1900, South Africa; olah.judit@econ.unideb.hu
[5] Institute of Applied Informatics and Logistics, Faculty of Economics and Business, University of Debrecen, 4032 Debrecen, Hungary
* Correspondence: Farkasne.Fekete.Maria@szie.hu; Tel.: +36-20-970-4987

Abstract: In order to meet the rising global demand for food and to ensure food security in line with the United Nation's Sustainable Development Goal 2, technological advances have been introduced in the food production industry. The organic food industry has benefitted from advances in food technology and innovation. However, there remains skepticism regarding organic foods on the part of consumers, specifically on consumers' acceptance of food innovation technologies used in the production of organic foods. This study measured factors that influence consumers' food innovation adoption and subsequently their intention to purchase organic foods. We compared the organic foods purchase behavior of Malaysian and Hungarian consumers to examine differences between Asian and European consumers. The findings show food innovation adoption as the most crucial predictor for the intention to purchase organic foods in Hungary, while social lifestyle factor was the most influential in Malaysia. Other factors such as environmental concerns and health consciousness were also examined in relation to food innovation adoption and organic food consumerism. This paper discusses differences between European and Asian organic foods consumers and provides recommendations for stakeholders.

Keywords: organic foods consumerism; food innovation adoption; food security; circular economy; health consciousness; environmental concern

Citation: Jeyakumar Nathan, R.; Soekmawati; Victor, V.; Popp, J.; Fekete-Farkas, M.; Oláh, J. Food Innovation Adoption and Organic Food Consumerism—A Cross National Study between Malaysia and Hungary. *Foods* **2021**, *10*, 363. https://doi.org/10.3390/foods10020363

Academic Editor: Derek V. Byrne
Received: 24 December 2020
Accepted: 4 February 2021
Published: 7 February 2021

1. Introduction

The human population is still growing fast. Today, the global population is around 7.8 billion. This number is expected to increase by 10% (8.5 billion) by 2030, 26% (9.7 billion) by 2050, and 42% (10.9 billion) by 2100, according to the U.N. Department of Economic and Social Affairs [1]. The growing population increases the demand for food, sometimes leading to the irresponsible use of natural resources which are becoming scarce [2]. This rising demand exerts pressure on the environment, resulting in massive deforestation and the deterioration in biocapacity and marine ecosystems [3]. Due to the emergence of biological hazards that affect quality of life, health concerns are more prevalent among consumers now than ever before. Meeting food supply challenges and feeding the growing global population with good quality food, has emerged as the new global food security agenda.

The increasing demand for organic food may reflect consumers' concerns regarding the devastating effects of conventional agriculture on people's health and the environment [2]. Rimal et al. [4] and Saba and Messina [5] found that consumers purchased organic food as they perceived that the risk of pesticide contamination is relatively low in organic food and the growing of organic food is perceived as harmless for the environment. Growing buyers' interest in quality food, wellbeing, better health, and sustainable living make organic food a viable choice [6]. Organic food that is regulated is produced with less negative impact on nature as compared to the conventional food production process. The demand for organic food initiates the establishment of various organic farming techniques around the world, utilizing a minimum of synthetic inputs or none at all [7].

Besides health reasons, moral thought and responsibility towards the environment motivate some consumers to purchase organic food [8,9]. Hence, organic food has gained popularity and is seen as a way of life for some consumers [10]. Although organic food is positioned as a better food choice for health and the environment, the issue of its relatively higher cost is a hindrance to purchase for some. Surveys conducted in the United Kingdom, Japan, India, and Indonesia in 2015 revealed that consumers were willing to pay up to 30% more for fruits and vegetables as an act of social responsibility [11]. However, Timmins [12] noted that the advantages related to organic food were not sufficient for some purchasers to make the final decision to purchase organic food. Besides pricing concerns, the technology of producing organic food also draws consumer skepticism [13].

Overall, the demand for organic food globally is shaped by a number of economic, sociological, and psychological factors, which can vary from country to country and from type of commodity to food group [14]. Cross-national studies could aid in a better understanding global consumers' similarities and differences and pave the way forward towards a more sustainable food and future for all. A recent cross-national study by Boobalan et al. [15] compared Indian and American organic food consumerism and found key differences between consumers in these two large countries regarding the psychological benefits they acquire when purchasing organic food. Against this backdrop, this study aims to investigate the factors contributing towards the intention of consumers from Asia (Malaysia) and Europe (Hungary) to purchase organic food, taking into consideration the role of the food innovation adoption behavior of consumers.

To this end, this paper is presented in sections, as follows. This introduction addresses the aim and focus of this research. Section 2 presents the literature review and subsequently the conceptual framework based on the critical secondary research review. This is followed by the description of the research methodology in Section 3. The research findings and discussion are highlighted in Section 4. Section 5 demonstrates the cross-national comparison of the findings between consumers in Malaysia and Hungary. Section 6 presents the conclusion of the study and Section 7 the limitations of the research and suggestions for future study.

2. Literature Review and Theoretical Framework

According to the Oxford Dictionary, 'organic' simply means something that is derived from living matter. In the food and agricultural industry, the word 'organic' is a labeling term that is given by the regulators indicating the approval of methods for the production, handling, and processing of organic foods sold. Organic food cultivation integrates cultural, biological and mechanical practices that lead to resource conservation and recycling of resources which promote ecological balance and biodiversity conservation [16].

In Malaysia, organic certification is regulated under the Malaysian Quality Standard 1529:2015 which ensures that the practice of organic farming is based on the four principles of Health, Ecology, Fairness and Care. The Malaysian organic standard emphasizes the health of soils, plants, animals, and humans, and the well-being of the ecological system, the environment, as well as balance and fairness to the ecological system [17]. In Hungary, the procedures of organic products' certification, production, labeling and marketing are governed by the [18]. The EU Regulation 2018/848 (Article 1) describes organic production as *"an overall system of farm management and food production that combines best environmental and climate action practices, a high level of biodiversity, the preservation of natural resources and the application of high animal welfare standards and high production standards in line with the demand of a growing number of consumers for products produced using natural substances and processes. Organic production thus plays a dual societal role, where, on the one hand, it provides for a specific market responding to consumer demand for organic products and, on the other hand, it delivers publicly available goods that contribute to the protection of the environment and animal welfare, as well as to rural development".*

'Organic labelled' foods are produced without the use of pesticides and artificial nitrogen composts, antibiotics, synthetic hormones, genetic engineering, or other detrimental practices prohibited in the regulation [19]. The entire organic food value chain is regulated to ensure that it is environmentally safe and free from irradiation, industrial solvents and synthetic food additives [20]. Based upon the stringent regulatory framework for producing organic food, the 'organic' label thus gives assurance to buyers that it the food is produced without harming the environment and without chemical residues in food. It serves as an assurance that the food is free from toxic and harmful substances.

To obtain an organic certification, farmers need to ensure that their fields are processed naturally, and free from prohibited materials for at least three years [19], as healthy soil has a profound impact on the quality of crops. Organic farmers are also expected to use ethical practices in farming such as hand weeding, mulching, intercropping, using mechanical control against pests, spread yields, crop revolution, and thick planting, instead of using conventional pesticides, herbicides, and engineered nitrogen manures, in order to enhance soil health [21].

There is an increased interest in the study of organic food production as it is also linked to food security and the sustainable supply of food to promote the circular economy. Previous studies have shown that consumers were motivated to purchase organic food due to health and environmental concerns [20,22,23]. Studies also found that consumers' health consciousness predicted their consumption of organic food [24–26]. Subjective norms including the influence of family and friends, compounded with lifestyle trends, also show a significant influence on the intention to purchase organic food [27,28].

In this cross-national study, five determinants of organic food consumerism were measured to assess their impacts on consumers' purchase intention towards organic food. These factors were found to have common interest in research into organic food consumerism for both European and Asian consumers in recent literatures. The first four factors are health consciousness, environmental concern, perceived quality of organic food, and social lifestyle factors. The fifth factor that this study introduces to the literature is the impact of consumers' adoption of food innovation technologies on their organic food purchase intention. Food innovation adoption is introduced as both an independent variable and a mediating variable in this study in order to examine its wider role in organic food consumerism.

2.1. Organic Food Consumption Trends in Hungary and Malaysia

Despite the excellent agricultural conditions in Hungary and Malaysia, the proportion of land used for organic production is relatively low compared to conventional farming (4.0% of total agricultural land in Hungary according to the Central Statistical Office [29] and 0.1% in Malaysia according to Willer et al. [30]). Consumer spending on organic food is still lower than conventional food products and it is believed that by increasing the demand for organic food, better food sustainability can be achieved via a transformation of the food value chain [30].

Within the Asia-Pacific region, people consume organic produce because of its health benefits and its advanced biological farming techniques. Demand for, and the consumption of, organic foods and beverages in the Asia-Pacific region are predicted to grow from 2020 to 2025 [31]. In the Asia-Pacific region, Malaysia is one the countries offering great opportunities for organic food to flourish. Recently, Malaysian shoppers have become more cognizant of well-being, and hence have increased their consumption of organic substitutes for conventional food. Nevertheless, the supply of organic produce in Malaysia is unable to meet the local market demand, causing a nationwide shortage of organic food. Malaysia still vigorously imports organic food from Europe and North America [32]. In Europe, Christos and Athanasios [33] predicted a lack of supply of organic food, not a lack of demand for it.

The United States recorded the largest sales of organic food (43%) in 2017, followed by the European Union member states (38%) [34]. Among Central and Eastern European countries, Hungary ranked as the third largest market volume of organic foods in 2010 [35]. Hungary is among the largest exporters of organic food in Central Europe. In 2018, there were 3929 producers who cultivated a total of 209,382 hectares of organic-farming land in Hungary [36].

As regards the domestic consumption of organic food, studies found that European consumers were driven by health consciousness, environmental concern, quality of life, and technological development [24,37]. Although environmental and health consequences can influence organic consumerism, the affordability of organic food was significant in determining consumers' food choices, particularly those in Italy and Hungary [25]. In Hungary, innovation in the food industry has been evaluated favorably by consumers [38]. This implies the significance of technological innovation in the food industry in satisfying consumers' needs [38]. Interestingly, food-technology was found to be related primarily to environmental concern among Hungarian consumers [39]. Similarly, in Malaysia, the creation of organic food has turned into an inventive methodology for the food sector to meet the rising consumer demands for healthier food choices.

2.2. Consumers' Purchase Intention towards Organic Food

Consumers' purchase intention is explained simply as the possibility that a consumer will acquire a product [22]. This variable is used in social science and business literature to indicate the actual consumption behavior of consumers towards a product or service [40]. It represents the likelihood that a purchase would take place as a result of "the interaction between customer needs, attitude and perception towards the product" [41]. Purchase intention acts as a measure of consumers' attentiveness in acquiring a product and the possibility of actually purchasing it [42]. According to Park and Kim [43] and Shin [44], purchase intention can be treated as a predictor of the actual purchasing decision due to its inclination to approximate to the actual conduct of a consumer. Although having an intention to purchase would more likely lead to actual purchase, it cannot be assumed that all predictors used would lead to actual purchasing action. Behavioral intention is formed based on an individual's motivation to perform that behavior, taking into account alternative options and his or her currently active goals [45]. With the limitation of observing the actual purchase behavior of consumers, purchase intention is used in this study to measure the potential of consumers' purchases.

Gifford and Bernard [46] employed a two-limit Tobit model and found that purchase intention towards organic foods among consumers may be influenced by the perceived benefits of organic agricultural methods, and the perceived risk of purchasing food grown using conventional procedures. In addition, Verhoef [47] posits that consumers are not only motivated by their rational economic motives, but also by emotional motives when purchasing organic food. The study found that consumers were willing to pay premium prices for organic food due to emotional motives, such as fear, guilt and empathy towards the environment.

Based on the relevant previous works, this study identified five variables to form an organic consumerism framework to compare Malaysian and Hungarian consumers as regards their organic food purchase behavior. They comprise food innovation adoption, health consciousness, environmental concern, perceived quality, and social lifestyle.

2.3. Health Consciousness

Health consciousness means that an individual's orientation toward his or her efforts to prevent illness and improve overall well-being [48]. Iversen and Kraft [49] defined health consciousness as "a tendency to focus attention on one's health" (p.603). An individual's level of health could be assessed through how one searches for health information and incorporates it into daily life. Homer and Kahle [50] posit that there is a relationship between consumers' intrinsic motivation, such as self-fulfillment, and a sense of accomplishment in purchasing nutritional food.

Health-conscious consumers are cognizant of their wellness and this health concern drives them to continuously improve their health and quality of life. To measure health consciousness, Ellison et al. [51] used behaviors such as food consumption, exercise, and substance use as indicators. Since the concept of health consciousness is linked more to personal attributes, measuring one's health consciousness on a psychological basis would better predict diverse health behaviors and result in greater construct validity.

Health consciousness has been relevant in predicting purchase intention and behavior regarding organic food production since buyers are aware that their food intakes impacts on their health. Previous research done by Shaharudin et al. [52] identified that consumers' attention to their health was a primary motive for the purchase of organic food. From another study, 87% of consumers believed that organic food was a healthier choice as compared to conventional food [10]. Similarly, Michaelidou and Hassan [53] highlighted health consciousness as the most important motive in explaining consumers' attitudes and behavior towards organic foods.

Shaharudin et al. [52] found the most popular motive to purchase organic food was consumers' perception of organic food as a healthier option for them. They also identified that consumers' interest in health was their primary motive to purchase organic food. Although the inherent evidence of the health benefits of consuming organic food have not been validated by Meemken and Qaim [2], a positive relationship between consumers' health consciousness and their purchase intention has been frequently identified in previous studies. Thus, this study hypothesized that health consciousness would positively influence consumers' intention to purchase organic food (Hypothesis 1 (H1)).

2.4. Environmental Concern

Consumers who are environmentally conscious prefer to use certain products because they believe they can reduce ecological impacts [54]. Similarly, consumers of sustainable wines were willing to change their consumption behavior to minimize the negative impact on the environment [55]. This type of consumer, also referred to as green consumers, often determine their purchase behavior for the benefit of the environment. The more consumers are concerned about the environment, the more positive are their attitudes toward organic food [56].

Seventy-five percent of respondents in the study by Petrescu and Petrescu-Mag [8] believed that organic food contributes to environmental protection. Congruently Basha et al. [57], found consumers' attitude towards purchasing organic food was strongly influenced by their concern for the environment. Sogari et al. [55] investigated consumers' environmental concerns and their intention to purchase sustainable wines and found it was important for consumers to believe that sustainable wines truly benefitted the environment in order to form a positive attitude towards purchasing sustainably.

In this study, a positive impact of consumers' environmental concern on their intention to purchase organic food is presented for testing in Hypothesis 2 (H2).

2.5. Perceived Quality

Perceived quality has gained popularity in marketing studies as a predictor of purchase intention and consumers' satisfaction. It is considered a crucial key for business sustainability, especially in competitive markets [58]. Perceived quality is defined as the personal judgment of the quality and benefit of a product or service that consumers establish in their minds [26]. The value of a product, also known as product utility, is often evaluated based on its ability to meet consumers' needs, resulting ultimately in their satisfaction. Consequently, the higher the value a product has is in consumers' minds, the higher the price they are willing to pay for it.

Consumers who purchase organic food often appear to be particularly concerned about the quality of the foods they consume. Half of the consumers who participated in a survey conducted by Timmins [12] agreed that organic food had better quality and taste. However, the major barrier to organic consumption was still the higher price. Some consumers perceived that the benefits of organic foods were not sufficient to justify its higher price [12]. Although value-for-money is found to be important for some consumers, one previous study finds this does not translate into anti-organic attitudes [12]. The affordability of organic foods played a major role in influencing consumers' food selection, particularly those in Hungary [25].

The locality of the organic food supply could potentially off-set the high price concern linked to organic foods. Consumers believe that locally produced greens produce a smaller carbon-footprint and are thus more environmental friendly and sustainable [59,60]. Timmins [12] found that 60% of his respondents were interested in locally sourced crops. Although affordability could influence consumers' food selection, the perceived quality of organic foods was found to be significant in predicting consumers' purchase intentions. This study predicts a positive relationship between perceived quality and consumers' intention to purchase organic foods (Hypothesis 3 (H3)).

2.6. Social Lifestyle

Studies in psychosocial theories and health behaviors explore how cognitive and social factors affect human health and disease [61]. Social and lifestyle factors relate to how peers and the people who surround a person affect his or her decision making. Additionally, messages through the media as well as reference groups and celebrities can also influence an individual's decision making [45]. Previous studies have shown the strong impacts of social factors on an individual's decision making in a wide variety of situations including business, social and health decisions [62–64].

Petrescu and Petrescu-Mag [10] explain the positioning of foods as fashionable items and their consumption as a social phenomenon that can generate consumers' interest and in turn become a part of their lifestyle. The trend and image factors may also influence consumers' decision to purchase organic foods despite the higher price. For instance, trendsetters in Vietnam who enjoy cooking pay greater attention to healthy food and prefer organic foods [27]. Specifically, a study involving youngsters by Vermeir and Verbeke [28] found a strong impact of social influence on sustainable food consumption behavior among young adults in Belgium.

The media often broadcasts programs showing the enjoyment of food and cooking in such a way that boosts the importance of food in representing power, pleasure, cleverness, and beauty. Often, people strongly believe that "who you are" to some extent is reflected in "what you buy". Social status was often found to be a determinant influencing people's decisions to consume green products rather than their more luxurious, non-green counterparts [65]. Similarly, Sahelices-Pinto et al. [66] showed that the consumption of organic foods was influenced by both social factors and self-esteem, revealing the impact of organic consumption on boosting one's social identity. Thus, hypothetically, a positive relationship may be established between social lifestyle and consumers' intentions to purchase organic foods (Hypothesis 4 (H4)).

2.7. Food Innovation Adoption

Food security has become a vital point of focus globally [67,68]. It is included as being of paramount importance in the United Nation's Sustainable Development Goals, as Goal 2 [69]. Goal 2 calls on all the nations of the world to work together to end hunger, achieve food security and improve nutrition, and promote sustainable agriculture. Altogether, the SDG Goal 2 proposes 8 targets to be achieved globally by the year 2030. The third target in the goal (Target 2.3) aims to double agricultural productivity, while the sixth target (Target 2.a) specifically mentions increased investments in technology development. In order to meet these two targets, the pivotal role of technology and innovation in food production is highlighted.

Fortunately, innovations in digital technologies such as advanced data analytics, predictive modeling, robotics, and the Internet of Things (IoT) have increased the efficiency of modern farming. Biotechnology advances in food technology also assist in increasing the food supply. By utilizing food innovation technologies that provide timely and accurate data, farmers can significantly improve their farming processes and eventually improve productivity. The new application of digitization and IoT in farming makes it possible to assess the soil moisture level, temperature, and many more agricultural matrices in real time to facilitate farmers' timely and accurate interventions.

The EU Regulation 2018/848 (Article 24) reads: "In order to support and facilitate compliance with this Regulation, operators should take preventive measures at every stage of production, preparation and distribution, where appropriate, to ensure the preservation of biodiversity and soil quality, to prevent and control pests and diseases and to avoid negative effects on the environment, animal health and plant health. They should also take, where appropriate, proportionate precautionary measures which are under their control to avoid contamination with products or substances that are not authorized for use in organic production in accordance with this Regulation and to avoid commingling organic, in-conversion and non-organic products". Based on this article, organic farmers can still use preventive measures to ensure their crops are safe from pests and diseases. However, if unauthorized substances are used in any of these activities, the products can no longer be considered organic. Hence, technology-based preventive measures would be ideal in order not to contravene this article and lose the organic product label.

As consumers' food preferences move towards fresh and whole foods, food-processing technology is also forced to meet the highest environmental standards with minimal alteration in the qualities and original flavors of the foods. As a result, organic farmers and their distributors spearhead the trend towards sustainable food production and a more transparent value chain [70]. This move towards a sustainable cycle of production is also referred to as the circular economy where the main goal is to reduce waste in the food production lifecycle [71].

The adoption of food innovation technology may influence consumers' purchase of organic foods, as food innovation technologies are rapidly being introduced into organic farming. Accordingly, this study proposes the fifth hypothesis to measure the impact of food innovation adoption of consumers on their intention to purchase organic foods (Hypothesis 5 (H5)).

2.8. Food Innovation Adoption Behaviour as Mediator in Organic Foods Purchase Intention

It is crucial for the food sector to identify the important drivers of consumers' preferences for foods in these modern times [72]. Consumers have become increasingly conscious of what they eat for various reasons, including skepticism as to whether food technology really produces better quality foods that warrant the higher price. As the biggest stakeholder in the food supply chain, consumers' preferences and decision making in foods purchase make them a formidable force for the food industry to reckon with. Mindful consumers are looking for the move towards sustainable food production. Health consciousness, environmental concern, perceived quality of organic foods, and social lifestyle would hypothetically impact their food innovation adoption behavior.

This study postulates that food innovation adoption would have both a direct impact on the intention to purchase organic foods (Hypothesis 6, 7, 8, and 9), and mediate the impact of health consciousness, environmental concern, perceived quality and social lifestyle on consumers' intention to purchase organic foods (Hypothesis 10, 11, 12, and 13).

The following list presents Hypotheses 6 to 13 which are put forward for testing in this study:

Hypothesis 6 (H6). *There is a positive impact of consumer health consciousness on food innovation adoption.*

Hypothesis 7 (H7). *There is a positive impact of consumer environmental concern on food innovation adoption.*

Hypothesis 8 (H8). *There is a positive impact of consumer perceived quality of organic food on food innovation adoption.*

Hypothesis 9 (H9). *There is a positive impact of consumer social lifestyle on food innovation adoption.*

Hypothesis 10 (H10). *The impact of health consciousness on consumers' purchase intention is mediated by food innovation adoption.*

Hypothesis 11 (H11). *The impact of environmental concern on consumers' purchase intention is mediated by food innovation adoption.*

Hypothesis 12 (H12). *The impact of perceived quality on consumers' purchase intention is mediated by food innovation adoption.*

Hypothesis 13 (H13). *The impact of social lifestyle on consumers' purchase intention is mediated by food innovation adoption.*

The research variables and corresponding hypothesis are shown in Figure 1.

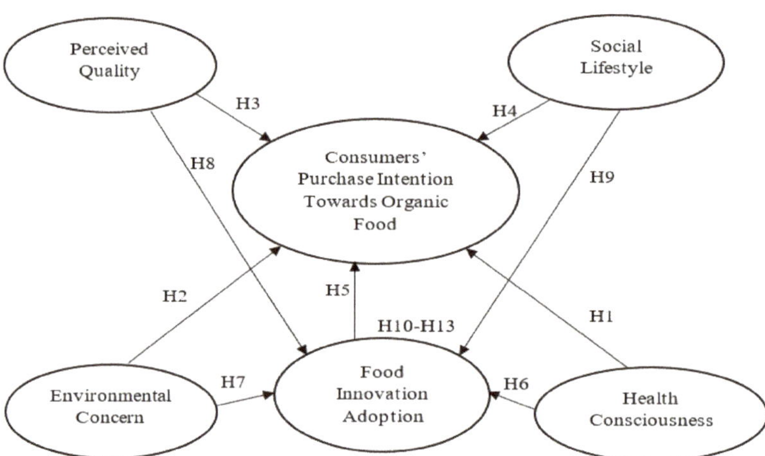

Figure 1. Research theoretical framework.

3. Methodology

According to Roitner-Schobesberger et al. [73], there have been numerous debates on buyers' views of organic foods in the United States and Europe; however, less has appeared in Asia despite the growing market for organic foods. For this reason, an analysis of organic foods consumerism in Malaysia—one of the leading contributors of agriculture in Asia—is selected for this study. Hungarian consumers in this study were chosen to represent organic consumerism in Europe. Although at this juncture, this comparison does not provide a holistic comparison between European and Asian consumers, in this pioneering cross-national study of organic foods consumerism, these two countries were chosen due to the proximity of the researchers to both countries to facilitate insightful data collection and to provide preliminary insights into this area of research. The findings of this research could warrant more comprehensive work in the future between Asia and Europe.

To conduct a cross-national comparison analysis of organic food consumerism in Malaysia and Hungary, this study utilized a research questionnaire as the data collection instrument for gathering primary research data. The participants in this study from both countries were approached randomly using the purposive sampling methodology and the classic mall-intercepted survey technique. The availability of organic food products in the areas where the respondents were approached was confirmed before administering the questionnaire to potential participants. Only participants who had prior experience of purchasing organic foods were selected as respondents. The survey was administered face to face among respondents in Malaysia and Hungary. Hardcopy questionnaire forms were used for data collection, which was carried out between June 2019 and March 2020 in both countries.

In total, 300 usable responses were obtained in Malaysia and 372 in Hungary. The filled questionnaires were carefully screened for missing data and mistakes in responses such as multiple responses for single response questions. Verified questionnaires were coded in the statistical software IBM Statistical Package for Social Sciences (SPSS), version 27, for descriptive analysis. Hypothesis testing and path modelling was done using Partial Least Square Structured Equation Modelling (PLS-SEM) using ADANCO PLS Software, Version 2.0. PLS-SEM was selected as the data analysis technique as the research model of this study is geared towards predictive modelling and testing the relationship between new constructs. Kline [23] pg. 286 recommends PLS-SEM as "well suited for where: (1) prediction is emphasized over theory testing and (2) it is difficult to meet the requirements for large samples or identification in SEM." Based on these criteria, the PLS-SEM technique was selected as the appropriate technique for hypotheses testing and path modelling for this study.

All measurement items of the research variables were measured using a five-point Likert scale based on the extent to which respondents agree or disagree with the particular indicator (item) statement in the questionnaire on a scale of 1 to 5; where (1) is Strongly Disagree, (2) is Disagree, (3) is Neutral, (4) is Agree, and (5) is Strongly Agree. This scale design is commonly used as measurement for social science studies. Churchill and Iacobucci [20] noted questionnaires using the Likert scale could provide appropriate measurements that would ease the process of tabulation and statistical analysis.

The indicators for Food Innovation Adoption (FIA), were self-developed for this study. These indicators were expert reviewed by two professors at the Multimedia University, Malacca Campus, Malaysia, who are specialists in technology adoption studies. Furthermore, the indicator statements were validated through a pilot study with data collected at the Multimedia University Malacca Campus among undergraduate students. The data from 200 samples showed the high reliability and internal consistency of the self-developed indicators, hence the indicator statements were incorporated in the final questionnaire for the productive phase of data collection in Malaysia and Hungary.

4. Results

Table 1 shows the respondents' demographic details.

Of the 300 respondents in Malaysia, 59% were females, and 71.3% were between 21 and 40 years old. Most of the respondents (83%) were single. In Hungary, 60.2% of the total respondents were males. As for ages, they had almost an equal number of respondents who were 21–40, 41–50 and 51–60 years old, with 36.8% between 21 and 40 years old. Lastly, more than half of the respondents in Hungary were married with children (53.2%), while in Malaysia this figure was about 15%.

Table 2 shows the research variables, indicator sources, aggregate means and standard deviations for both Hungarian and Malaysian data.

Table 1. Respondents' demographic information.

Demographic Factor	Options	Malaysia		Hungary	
		Freq.	Percentage (%)	Freq.	Percentage (%)
Gender	Male	123 (16.820 *)	41.0 (51.45)	224 (4.680 *)	60.2 (47.91)
	Female	177 (15.880 *)	59.0 (48.55)	148 (5.088 *)	39.8 (52.09)
Age	Below 20	65	21.7	4	1.1
	21–40	214	71.3	137	36.8
	41–50	13	4.3	123	33.1
	51–60	4	1.3	87	23.4
	Above 60	4	1.3	21	5.6
Marital Status	Single	249	83	108	29
	Married with children	46	15.3	198	53.2
	Married without children	3	1	28	7.5
	Single with children	2	0.7	20	5.4

Note: * Numbers in bracket represent national populations in millions. The Hungarian population statistics are obtained from [74]. The Malaysian population statistics are obtained from [75].

Table 2. Research variables and indicators with mean and standard deviation.

Research Variables	Indicators	Malaysia		Hungary	
		Mean	SD	Mean	SD
Health Consciousness (HC) Yang et al. [76]; Shaharudin et al. [52]	HC1—Healthy diet is an important factor when choosing what I eat	4.230	0.775	3.867	0.786
	HC2—I give a lot of attention to my health	3.826	0.837	4.11	0.745
	HC3—A healthy body is important to me	4.421	0.626	3.045	0.993
	HC4—Health concern is the reason for consuming organic food	3.919	0.874	3.140	1.00
	HC5—Proper nutrition is a key factor for purchasing organic food	3.909	0.848	4.196	0.909
Environmental Concern (EC) Yang et al. [76]	EC1—I am concerned about the state of our environment	3.993	0.798	3.457	1.065
	EC2—Environmental concerns affects my food choice	3.692	0.926	3.370	1.165
	EC3—Organic food is environmentally friendly	3.916	0.876	3.869	1.103
	EC4—Chemical fertilizers are harmful for the environment	4.143	0.910	4.382	0.8663
	EC5—Everyone should be concerned for our environment	4.568	0.707	3.471	1.067
Perceived Quality (PQ) Aulia et al. [58]	PQ1—Organic food is a healthier food option	4.220	0.788	3.353	1.078
	PQ2—Organic food has great nutritional benefits	4.153	0.880	3.251	1.059
	PQ3—Organic food has better quality due to its advanced cultivation methods	4.016	0.849	3.252	1.104
	PQ4—Though I may have to pay more, I get better quality organic food	3.879	0.926	3.225	0.9942
	PQ5—I am satisfied with organic food quality	3.923	0.939	3.733	0.9866

Table 2. *Cont.*

Research Variables	Indicators	Malaysia		Hungary	
		Mean	SD	Mean	SD
Social Lifestyle (SL) Basha et al. [57]; Falguera et al. [65]	SL1—Organic food is a trend in society	3.493	1.01	2.175	1.104
	SL2—My family influence me to consume organic food	3.177	1.14	2.046	1.036
	SL3—My peers influence me to consume organic food	2.959	1.04	2.754	1.22
	SL4—Celebrities often promote organic food consumption	3.214	1.09	3.807	0.935
	SL5—The lifestyle of consuming organic food is healthy	3.953	0.861	3.549	0.9742
Food Innovation Adoption (FIA) Self-Developed for this Study	FIA1—The way organic food is grown and processed influence me to consume organic food	3.721	0.908	2.843	1.132
	FIA2—The advantages of GM (genetically modified) foods outweighs potential disadvantages	3.476	0.931	3.495	1.036
	FIA3—Advances in food technologies have produced better quality food for the world	3.845	0.825	3.769	0.8930
	FIA4—Technologically superior organic food production improves food yields	3.815	0.860	3.939	0.9062
	FIA5—Innovation in food production is to be welcomed by all	3.922	0.796	3.877	0.9858
	FIA6—I support technology and innovation in food production	4.000	0.794	2.695	1.155
Consumer Purchase Intention Towards Organic Food (CP) Shaharudin et al. [52]	CP1—I purchase organic food frequently	2.966	1.12	2.587	1.143
	CP2—I will continue to purchase organic food	3.391	0.985	2.791	1.165
	CP3—I am willing to pay more for organic food than conventional food in the store	3.351	1.03	2.887	1.186
	CP4—I will recommend organic food to family and friends	3.738	0.918	2.195	1.194
	CP5—I consider myself a loyal organic food consumer	3.023	1.24	3.34	1.011

The measurement model is assessed via construct validity, convergent validity, and discriminant validity analyses. Before conducting hypotheses testing, it is essential to investigate the indicators' factor loadings. According Hair et al. [77], indicators with loadings below 0.50 should be removed from the path model due to the low predictability of the relevant variable. Thus, HC5, EC4, PQ5, SL1, FIA1, FIA3, and CP5 were removed from both the Hungarian and Malaysian path models in order to make identical comparison of path modelling for both countries (refer to Table 3).

For factor loadings that were above 0.50 but below 0.70, their variable's composite reliability (CR) and AVE are confirmed to exceed thresholds of 0.70 and 0.50 (Hair et al. [77] and Bagozzi and Yi [78]), assuring the path models' Reliability and Convergent Validity. As for the Cronbach Alphas, all values are above 0.70, fulfilling the satisfactory values, except for SL 0.673 (Malaysia) and 0.671 (Hungary), which were slightly below the 0.7 threshold; however, their CR and AVE are above threshold levels, hence fit for path modelling [79]. The statistics of all constructs and indicators are presented in Table 3.

Table 3. Internal consistency, composite reliability and convergent validity.

Variable	Indicator	Factor Loadings		Cronbach's Alpha		Composite Reliability		AVE	
		MD	HD	MD	HD	MD	HD	MD	HD
Health Consciousness (HC)	HC1	0.754	0.627	0.786	0.725	0.860	0.821	0.607	0.537
	HC2	0.768	0.655						
	HC3	0.650	0.809						
	HC4	0.801	0.820						
	HC5	0.720	-						
Environmental Concern (EC)	EC1	0.707	0.684	0.749	0.729	0.841	0.825	0.570	0.546
	EC2	0.720	0.815						
	EC3	0.766	0.607						
	EC4	0.635	-						
	EC5	0.767	0.826						
Perceived Quality (PQ)	PQ1	0.778	0.863	0.828	0.888	0.885	0.922	0.660	0.748
	PQ2	0.780	0.873						
	PQ3	0.817	0.890						
	PQ4	0.797	0.832						
	PQ5	0.843	-						
Social Lifestyle (SL)	SL1	-	0.673	0.673	0.671	0.787	0.752	0.515	0.513
	SL2	0.799	0.652						
	SL3	0.712	0.773						
	SL4	0.553	0.665						
Food Innovation Adoption (FIA)	FIA1	0.696	-	0.813	0.724	0.876	0.753	0.640	0.524
	FIA2	0.725	0.554						
	FIA3	0.739	-						
	FIA4	0.817	0.600						
	FIA5	0.798	0.603						
	FIA6	0.750	0.845						
Consumer Purchase Intention Towards Organic Food (CP)	CP1	0.847	0.919	0.896	0.914	0.923	0.919	0.706	0.796
	CP2	0.841	0.883						
	CP3	0.830	0.904						
	CP4	0.805	0.861						
	CP5	0.878	-						

Note: MD stands for 'Malaysian Data' (n = 300); HD stands for 'Hungarian Data' (n = 372).

A high inter-relationship and multi-collinearity between variables can lead to misleading findings, magnified standard errors, or weaker power of regression coefficients. According to Henseler et al. [80], when all the values of the Heterotrait-Monotrait Ratio of Correlations (HTMT) are lower than 0.85, this implies that the variables are conceptually distinct from each other. HTMT 0.85 is used in this study as the conservative criterion to assess discriminant validity [80]. From Table 4, it is observed that all HTMT values among the variables in this study are lower than the thresholds of 0.85, indicating the models are free from multi-collinearity.

Table 4. The Heterotrait-Monotrait ratio of correlations (HTMT).

	Malaysian Data						Hungarian Data				
	HC	EC	PQ	SL	FIA		HC	EC	PQ	SL	FIA
HC						HC					
EC	0.7260					EC	0.6272				
PQ	0.7212	0.7012				PQ	0.5968	0.8143			
SL	0.5757	0.3903	0.7851			SL	0.5162	0.7971	0.7720		
FIA	0.6335	0.6632	0.6631	0.6067		FIA	0.4229	0.5024	0.5512	0.5965	
CP	0.5099	0.3945	0.6405	0.6544	0.5104	CP	0.6738	0.7666	0.6357	0.8427	0.5773

Additionally, to assess the goodness of fit of the research model, the Standardized Root Mean Square Residual (SRMR) was calculated. The results show SRMR values of 0.0670 for Malaysia and 0.0541 for Hungary. The SRMR values for the Malaysian and the Hungarian models are within the threshold level of 0.08 (Hu and Bentler [81]), assuring the goodness of fit of the research models for both countries. The R square values for Malaysia are CP = 0.408 and FIA = 0.458; while R square values for Hungary are CP = 0.725 and FIA = 0.493. The research variables show a high variance explained in both models; especially with the Hungarian consumer purchase intention of organic foods, the model shows that the research variables account for approximately 73% of the variance.

Hypotheses Testing

For testing the hypotheses, bootstrapping with 5000 iterations was applied. The significance of the path coefficient is assessed to validate each hypothesis. The structural model for Hungary and Malaysia with the R square values, path coefficients, and factor loadings are presented in Table 5.

Table 5. Results of hypotheses testing for Malaysia and Hungary.

Hypothesis	Relationship	Malaysian Data		Hungarian Data	
		Path Coef.	*p*-Value	Path Coef.	*p*-Value
H1	HC → CP	0.600	0.107	0.185	<0.001 ***
H2	EC → CP	−0.011	0.860	0.117	0.014 **
H3	PQ → CP	0.271	<0.001 ***	0.187	0.002 ***
H4	SL → CP	0.306	<0.001 ***	0.113	<0.013 **
H5	FIA → CP	0.109	0.035 **	0.414	<0.001 ***
H6	HC → FIA	0.110	0.0402 **	0.188	<0.001 ***
H7	EC → FIA	0.341	<0.001 ***	0.014	0.852
H8	PQ → FIA	0.134	0.036 **	0.313	0.031 **
H9	SL → FIA	0.187	<0.001 ***	0.329	<0.001 ***

*** Significant at 1%; ** Significant at 5%

Based on the results of the Malaysian and Hungarian path analysis, it was found that the relationships between Health Consciousness (HC) and Environmental Concern (EC) regarding Consumer's Purchase Intention Towards Organic Foods (CP) are insignificant for Malaysia; however, these paths are significant for Hungary (Hypotheses 1 and 2 are partially supported—true only for Hungary). Perceived Quality (PQ), Social Lifestyle (SL) and Food Innovation Adoption (FIA) each show a significant impact on CP in both countries (Hypotheses 3, 4 and 5 are supported).

Among these five independent variables (H1 to H5), SL has the highest impact on CP (0.306) for Malaysia. However, for Hungary, FIA shows the highest impact on CP (0.414). It is interesting to note that the same factor (FIA), though significant, shows the lowest impact on CP in the Malaysian context (0.109). This indicates the difference in perception and role that FIA plays in these two countries.

On the other hand, PQ is found to be the second most important factor leading to CP in Malaysia and in Hungary. This finding highlights the consistent perception of users in both countries who tend to relate the perceived quality of organic foods with their purchasing intention.

As for the impact of the research variables on FIA as mediating variables, HC, PQ, and SL were found to be significant predictors of FIA in Malaysia and Hungary (Hypotheses 6, 8 and 9 are supported). Testing the impact of EC on FIA shows differing results for the two countries, where EC on FIA is significant in Malaysia but not in Hungary (H7 is partially supported). This shows that although EC leads to CP in Hungary, it does not significantly predict Hungarians' FIA behavior.

To further investigate the mediating effect of FIA in the relationships between each of the predictors HC, EC, PQ, and SL to CP, the significance of these indirect paths was tested. The results of the indirect effects are presented in Table 6.

Table 6. Indirect effects of factors towards CP through FIA.

Hypothesis	Relationship	MD		HD	
		Path Coef.	*p*-Value	Path Coef.	*p*-Value
H10	HC → FIA → CP	0.012	0.253	0.078	0.003 **
H11	EC → FIA → CP	0.037	0.082 *	−0.004	0.890
H12	PQ → FIA → CP	0.014	0.212	0.129	<0.001 **
H13	SL → FIA → CP	0.020	0.057 *	0.136	<0.001 **

** Significant at 1%; * Significant at 10%

The result of the indirect effects analysis reveals several significant paths. According to Hair et al. [77], it is necessary to evaluate indirect effects in order to determine whether a mediating effect is present. When both direct and indirect effects are significant, a partial mediation is observed; if the indirect effect is significant but the direct effect is insignificant, a full or indirect-only mediation is identified. However, when the indirect effect is insignificant, but the direct effect is significant, it indicates that a direct-only effect or no mediation effect is present [77].

From the mediation results above, it is observed that FIA is a significant mediator for EC and SL impacts on CP for Malaysia. However, it is not a significant mediator for PQ and HC. Reading this finding together with the earlier finding of the direct effect of HC on CP, it was also found not to be significant for Malaysia, while the direct effect of FIA on CP was significant. From these three findings, it can be deduced that for Malaysian consumers, health consciousness is an important reason that makes them consider accepting innovation in food production; however, health consciousness in itself is not the reason for purchasing organic foods.

As for the Hungarian data, the finding shows FIA as a significant mediator for HC, PQ and SL on CP. However, FIA is found not to be a mediator for EC. Compounding this finding with the direct impacts of EC on CP (significant) and FIA (not significant), it can be deduced for Hungarian consumers that environmental concern is an important factor of consideration for them when purchasing organic foods; however environmental concern in itself is not a reason for adopting innovation in food production. Based on this finding, Hypotheses 10, 11 and 12 are partially supported, while Hypothesis 13 is fully supported.

5. Discussion

The data obtained in both countries revealed that consumers in both countries have some commonalities and some key differences in their adoption of food innovation, as well as in the purchase of organic foods. This section presents a critical discussion of these findings for Malaysia and Hungary.

When it comes to the purchase of organic foods, both countries show different crucial determining factors that affect their decision making. To assist with visualizing the findings, Table 7 is based on the statistical results of path modelling coefficients in Table 5.

Table 7. Visual representation of Path Modelling Results showing the relative importance of the constructs.

Constructs	Malaysian Consumers		Hungarian Consumers	
	Food Innovation Adoption	Organic Food Purchase	Food Innovation Adoption	Organic Food Purchase
Health Consciousness	Important (4)		Important (3)	Important (3)
Environmental Concern	Important (1)			Important (4)
Perceived Quality	Important (3)	Important (2)	Important (2)	Important (2)
Social Lifestyle	Important (2)	Important (1)	Important (1)	Important (5)
Food Innovation Adoption		Important (3)		Important (1)

Note: Numbers in bracket show the ranking and relative importance of factors within the column.

For Malaysian consumers, SL is the most crucial factor, followed by PQ and FIA, in determining their organic foods purchase intention. Comparing this with Hungarian consumers, the result shows that FIA is the most crucial determinant for Hungarians, followed by PQ, HC, EC, and finally SL.

The social lifestyle factor is found to be the most important factor that contributes to the intention to purchase organic foods in Malaysia. The social lifestyle variable measures buyers' concerns regarding status and peer influences. Malaysian consumers demonstrate a greater tendency that social lifestyle will be a reason to purchase organic foods, which are more expensive than conventional foods. Social influence was also found in previous studies to impact on consumers' intention to purchase organic foods in Malaysia by Ayub [77,82] and in Pakistan [83]. In particular, peer pressure was found to be a significant determinant in persuading young Malaysian consumers to purchase green products in a previous study [84]. Malaysian consumers appeared to purchase organic products with the intention of fulfilling and expressing their social identity [85] which is found to be consistent with the findings of this study.

Other findings from the region, such as Nguyen et al. [86], reveal that organic foods label significantly contributed to buyers' favorable attitude to buying organic foods among urban Vietnamese consumers, while Fogarassy et al. [71] found that highly educated young people who are very conscious and live on good incomes may be the target group for circular innovation in Hungary. The study found that young consumers, the internet savvy, and software users living in cities buy organic foods and follow healthy lifestyle trends. Hence, having access to a more expensive food selection may be seen as a social symbol and a differentiator from the masses, as well as a sustainable lifestyle trend.

The perceived quality of organic foods is found to be the second most important factor that drives the intention to purchase organic foods in Malaysia. Malaysian consumers seem to compare conventional foods with organic foods based on this perceived main difference—its quality. In previous studies by Lee and Yun [87] and Lockie et al. [88], consumers were found to be committed to foods they perceived to be natural, nutritional and free of unnecessary processing as well as artificial additives. Organic plants contain lower levels of pesticide residues and minimum concentrations of nitrate and cadmium. Besides, organic animal products were also found to contain higher levels of omega-3 fatty acids. Overall, organic foods were associated less with allergies, eczema, and obesity. Although there was insufficient evidence to draw conclusions on the positive health outcomes of consuming organic foods in the study by Meemken and Qaim [2], the study found that consumers from Malaysia and Hungary do associate organic foods with higher quality.

Although health consciousness was expected to be an important reason for purchasing organic foods, this finding is contrary to the conventional wisdom. According to [84,89], health concerns were found to be more important than environmental issues for Indian consumers while they make purchasing decisions for organic foods. However, this study finds Malaysian consumers do not significantly associate health consciousness with their intention to purchase organic foods, but they do associate health consciousness with food innovation adoption, which is an important finding. FIA seems to fit the missing piece of the puzzle, in that it explains the inter-relationship between the health consciousness of consumers and their intention to purchase organic foods, as a mediator.

Food innovation adoption is the most crucial reason for the intention to purchase organic foods in Hungary. Hungarian consumers seem to show greater awareness of food innovations compared to Malaysian consumers. This is perhaps due to the greater usage of technology in the agricultural sector in Hungary and Europe in general, as compared to Asia where most countries still rely on human labor for agricultural output [90–92]. The labor intensity in Asian agricultural production could also be related to the type of crops they harvest. Rice cultivation is purportedly more labor-intense as compared to wheat production, contributing to the greater demand for human labor in Asian agriculture (Vollrath [93]).

A lesser emphasis on human labor in agriculture possibly allows European countries such as Hungary to focus more on food innovation technology. As a result, both capacity and performance in ecological innovations are found to be better in European countries, as compared to Asian countries [94]. The Hungarian data analysis shows the distinctly high impact of FIA (Beta Coefficient = 0.414) on consumers' intention to purchase organic foods. While FIA is a third important factor for Malaysian consumers, this finding shows a significant difference between European (Hungarian) and Asian (Malaysian) consumers. Food innovation adoption is an important determinant of intention to purchase organic foods among buyers in Europe, but not a strong determinant in Asia.

Although environmental concern was significant in determining Romanian consumers' eating habits (Oroian et al. [95]), this study finds environmental concerns do not have a substantial effect on Malaysian consumers' intention to purchase organic foods, and it is also the second least important factor that predicts the intention to purchase organic foods in Hungary. This inferior result of EC could be due to current consumers' motives in consuming organic foods, which is not primarily driven by their intention to protect the ecological environment. Rather, their motives are based on social lifestyle factors and perceived quality (for Malaysia) and food innovation, perceived quality, and health concerns (for Hungary). This finding is consistent with recent research that found health factor and maintaining social status in society take priority in consumers' minds over environmental safety [96].

The results for FIA reveal peculiar findings. EC is the most crucial determinant of FIA in Malaysia, while it is not significant in Hungary (but significant on CP in Hungary). Environmental concerns or ecological consciousness are seen as important determinants for FIA among Malaysians. There is a strong association between environmental protection and food innovation technology in Malaysian consumers' minds. These two dimensions are seen as highly connected. Conversely, for Hungarians, EC is seen as a 'distant factor' that has no direct impact on their food innovation adoption behavior. EC, although significant for Hungarians in their organic food purchases, is not something they associate with FIA. Perhaps Hungarian food consumers do not look at innovation in food technology as something that is truly protective and conserving the environment. This suggests a possible skepticism towards food innovation technology and production, which are perhaps not viewed as environmentally friendly albeit perceived to be producing good quality foods [97,98]. It is worth noting this major difference between Asia (Malaysian) and Europe (Hungarian) where Asian food consumers in this context associate environmental concerns with FIA, while European consumers seem not to associate the two.

Although Hungarian consumers do not associate EC with FIA, they strongly associate SL with FIA. Although social factors and status were not strong determinants of their intention to purchase organic foods, they are significantly more important in their FIA. This suggests Hungarian consumers consider social and lifestyle factors as trends that go together with innovation in the food sector. This possibly indicates that in their mind, innovation in food technology is just another social and lifestyle trend [99]. Social lifestyle trends are also found to be equally important elements in the Malaysian context which drive their adoption of food innovation. This could indicate a global trend of innovation in food technology being perceived by consumers as a social and lifestyle trend.

Hungarian consumers also show the high impact of PQ on CP, which indicates their high trust in food innovation technologies which are perceived to produce high quality foods, although they were skeptical about the environmental impact of FIA.

It was considered meaningful to include food innovation adoption as an important construct in the modelling of this study and to provide an understanding of the wider ecosystem of organic food consumerism. The indirect effect results show that food innovation adoption seems to significantly mediate the relationships between the independent and dependent variables for both countries in most relationship paths. For Malaysian consumers FIA effectively mediates the impact of EC and SL on CP, while for Hungarian consumers FIA effectively mediates the impact of HC, PQ and SL on CP. For an elusive construct such as the FIA which had previously been less understood, the findings of this research show its pivotal role in understanding the ecosystem of organic food consumerism.

6. Conclusions

Consumer consciousness towards a more natural lifestyle and consumption behavior has led to various attempts to incorporate technology and trends in food production innovations. Various studies have discovered that buyers were increasingly troubled about the kind of foods they consume daily [100,101]. The rising interest in nutritious foods is reflected in consumers' demand for organic food alternatives that promise better quality foods through the innovative use of technology and innovation in production. Food innovation serves the twin-role of providing high quality foods, as well as increasing the production of foods to meet rising global food demand.

Cross-national studies are gaining popularity as they are meaningful ways to provide new insights into consumer behavior by comparing consumer choices and actions in different cultures [87]. To add to this body of literature, this study measured important factors that influence the intention to purchase organic foods in Hungary and Malaysia, both countries that are strong in agricultural output in their regions. Additionally, this study identified food innovation adoption as an important variable to be included in the model as evidenced in recent food technology literatures, to better understand the organic food consumerism ecosystem impacted by food innovation technologies. We found food innovation adoption plays a critical role in explaining consumers' organic foods purchasing behavior in Hungary and Malaysia.

The marketing of organic foods could emphasize the quality of organic foods as this is found to be the biggest driver of the intention to purchase organic foods in both countries. Social and lifestyle factors are highly significant in driving purchasing intentions. Consumers associate organic foods with trends in society and see it as a lifestyle choice. This could be a persuasive narrative for governments, policy makers, organic food producers, and retailers in improving engagement with consumers to promote sustainable consumption behavior. This could also lead to greater involvement of organic food buyers in the organic foods value chain, which is desirable for consumers [102]. Organic food growers and retailers may provide more information and transparency regarding their cultivation process, which is often invisible to final consumers. This lack of transparency may be leading to skepticism towards food innovations that are utilized in the production of organic foods.

7. Limitations of the Study and Future Directions

Although the sample size obtained in this study was statistically significant, the demographics of the respondents from both countries were not similar. A more proportionate sampling of respondents based on national population statistics may provide more comparable data. Future studies could investigate demographic control variables as well as assess their moderating effects on food innovation adoption and the intention to purchase organic foods.

The purposive sampling methodology was used to select respondents in this study, due to the absence of a sampling frame. Future studies could collaborate with retailers to create a list of organic foods purchasers through customers' purchase records to target actual customers who have purchased organic foods to be included for data collection.

This study is also limited in measuring the consumer purchase behavior related to organic foods. We used purchase intention as a measure to estimate actual behavior. Future studies can address this limitation by measuring the actual purchase of organic foods.

There seems to be a higher level of skepticism, especially in Europe, regarding the relationship between environmental conservation and food innovation. More work is needed in this area to discover the reasons behind the skepticism and to further assess the impact of food technology and innovation on environmental protection and preservation in the context of organic foods.

Author Contributions: R.J.N., M.F.-F., J.O. and J.P. research idea conceptualization, R.J.N. and J.O. made the interviews and collected the research data, V.V., R.J.N. and S. designed the research methodology and did the formal analysis and initial drafting of the results. All authors have read and agreed to the published version of the manuscript.

Funding: The data collection for this study in Malaysia is funded by the Fundamental Research Grant Scheme, Ministry of Higher Education Malaysia (MOHE) FRGS/1/2016/SS01/MMU/02/6 (MMUE/160033).

Informed Consent Statement: Informed consent was obtained from all subjects involved in the study.

Conflicts of Interest: The authors declare no conflict of interest.

References

1. U.N. Department of Economic and Social Affairs. World Population Prospects 2019: Highlights. Available online: https://www.un.org/development/desa/publications/world-population-prospects-2019-highlights.html (accessed on 20 September 2020).
2. Meemken, E.; Qaim, M. Organic agriculture, food security, and the environment. *Ann. Rev. Resour. Econ.* **2018**, *10*, 39–63. [CrossRef]
3. Uniyal, S.; Paliwal, R.; Kaphaliya, B.; Sharma, R. Human overpopulation: Impact on environment. In *Advances in Environmental Engineering and Green Technologies (AEEGT) Book Series*; IGI Global: Hershey, PA, USA, 2017; pp. 1–11. [CrossRef]
4. Rimal, A.P.; Moon, W.; Balasubramanian, S. Agro-biotechnology and organic food purchase in the United Kingdom. *Br. Food J.* **2005**, *107*, 84–97. [CrossRef]
5. Saba, A.; Messina, F. Attitudes towards organic foods and risk/benefit perception associated with pesticides. *Food Qual. Prefer.* **2003**, *14*, 637–645. [CrossRef]
6. Shaqiri, F.; Musliu, A.Y.; Ymeri, P.; Vasa, L. Evaluating consumer behavior for consumption of milk and cheese in Gjilan Region, Kosovo. *Ann. Agrar. Sci.* **2019**, *17*, 375–381.
7. First, I.; Brozina, S. Cultural influences on motives for organic food consumption. *EuroMed J. Bus.* **2009**, *4*, 185–199. [CrossRef]
8. Eide, B.; Toft, M. *Consumer Behavior Theories—Purchasing Organic Food*; Aarhus University, Department of Business Administration: Aarhus, Denmark, 2013; Available online: https://www.academia.edu/8020018/Consumer_Behavior_Theories_Purchasing_Organic_Food (accessed on 20 September 2020).
9. Harper, G.C.; Makatouni, A. Consumer perception of organic food production and farm animal welfare. *Br. Food J.* **2002**, *104*, 287–299. [CrossRef]
10. Petrescu, D.C.; Petrescu-Mag, R.M. Organic food perception: Fad, or healthy and environmentally friendly? A case on Romanian consumers. *Sustainability* **2015**, *7*, 12017–12031. [CrossRef]
11. Miller, S.; Tait, P.; Saunders, C.; Dalziel, P.; Rutherford, P.; Abell, W. Estimation of consumer willingness-to-pay for social responsibility in fruit and vegetable products: A cross-country comparison using a choice experiment. *J. Consum. Behav.* **2017**, *16*, e13–e25. [CrossRef]
12. Timmins, C. *Consumer Attitudes Towards Organic Food*; IBERS—Aberystwyth University: Cardiff, UK, 2010; pp. 1–66.
13. Siegrist, M.; Hartmann, C. Consumer acceptance of novel food technologies. *Nat. Food* **2020**, *1*, 343–350. [CrossRef]
14. Rödiger, M.; Hamm, U. How are organic food prices affecting consumer behaviour? A review. *Food Qual. Prefer.* **2015**, *43*, 10–20. [CrossRef]
15. Boobalan, K.; Nachimuthu, G.S.; Sivakumaran, B. Understanding the psychological benefits in organic consumerism: An empirical exploration. *Food Qual. Prefer.* **2021**, *87*, 1–9. [CrossRef]
16. U.S. Department of Agriculture. What is Organic? Available online: https://www.ams.usda.gov/publications/content/what-organic (accessed on 20 September 2020).
17. Official Portal of Department of Agriculture MoAaFI. Malaysian Organic Certification Scheme. 2020. Available online: http://www.doa.gov.my/index.php/pages/view/377 (accessed on 20 September 2020).
18. European Parliament and of the Council. Regulation (EU) 2018/848 of the European Parliament and of the Council as of 30th May 2018. 2018. Available online: https://eur-lex.europa.eu/eli/reg/2018/848/oj (accessed on 12 January 2021).
19. Canada Organic Trade Association. What is Organic? Available online: https://www.canada-organic.ca/en/what-we-do/organic-101/what-organic (accessed on 20 September 2020).
20. Churchill, G.A.; Iacobucci, D. *Marketing Research: Methodological Foundations*; Dryden Press: New York, NY, USA, 2006.
21. Saffeullah, P.; Nabi, N.; Liaqat, S.; Anjum, N.A.; Siddiqi, T.O.; Umar, S. Organic Agriculture: Principles, Current Status, and Significance. In *Microbiota and Biofertilizers*; Hakeem, K.R., Dar, G.H., Mehmood, M.A., Bhat., R.A., Eds.; Springer: Berlin/Heidelberg, Germany, 2021; pp. 17–37. [CrossRef]
22. Dodds, W.B.; Monroe, K.B.; Grewal, D. Effects of price, brand, and store information on buyers' product evaluations. *J. Mark. Res.* **1991**, *28*, 307–319.
23. Kline, R.B. *Principles and Practice of Structural Equation Modeling*; The Guilford Press: New York, NY, USA; London, UK, 2015.

24. Naz, F.; Oláh, J.; Vasile, D.; Magda, R. Green Purchase Behavior of University Students in Hungary: An Empirical Study. *Sustainability* **2020**, *12*, 10077. [CrossRef]
25. Yeh, C.-H.; Menozzi, D.; Török, Á. Eliciting egg consumer preferences for organic labels and omega 3 claims in Italy and Hungary. *Foods* **2020**, *9*, 1212. [CrossRef]
26. Eftekhari, M.; Shaabani, M.; Lotfizadeh, F. The Effect of Perceived Quality, Perceived Cost and Repurchase Intention in the Insurance Industry. *Int. J. Manag. Sci. Bus. Res.* **2015**, *4*, 9–18.
27. Chi, M.T.T.; Lobo, A.; Nguyen, N.; Long, P.H. Effective segmentation of organic food consumers in Vietnam using food-related lifestyles. *Sustainability* **2019**, *11*, 1237.
28. Vermeir, I.; Verbeke, W. Sustainable food consumption among young adults in Belgium: Theory of planned behaviour and the role of confidence and values. *Ecol. Econ.* **2008**, *64*, 542–553. [CrossRef]
29. Central Statistical Office. Biogazdálkodás (2005–). 2020. Available online: https://www.ksh.hu/docs/hun/xstadat/xstadat_eves/i_ua001b.html (accessed on 20 September 2020).
30. Willer, H.; Schlatter, B.; Trávníček, J.; Kemper, L.; Lernoud, J. *The World of Organic Agriculture. Statistics and Emerging Trends 2020*; Research Institute of Organic Agriculture (FiBL): Frik, Switzerland, 2020.
31. ResearchAndMarkets.com. Global Organic Food and Beverages Market 2020 to 2025-Growth, Trends, and Forecasts. 2020. Available online: https://www.businesswire.com/news/home/20200817005391/en/Global-Organic-Food-and-Beverages-Market-2020-to-2025---Growth-Trends-and-Forecasts---ResearchAndMarkets.com (accessed on 20 September 2020).
32. Somasundram, C.; Razali, Z.; Santhirasegaram, V. A Review on organic food production in Malaysia. *Horticulturae* **2016**, *2*, 12. [CrossRef]
33. Christos, F.; Athanasios, K. Purchasing motives and profile of the Greek organic consumer: A countrywide survey. *Br. Food J.* **2002**, *104*, 730–765.
34. Nechaev, V.; Mikhailushkin, P.; Alieva, A. *Trends in Demand on the Organic Food Market in the European Countries, MATEC Web of Conferences*; EDP Sciences: Les Ulis, France, 2018; pp. 1–10. [CrossRef]
35. Redaktion. Organic Food in Hungary. Available online: https://organic-market.info/news-in-brief-and-reports-article/12628-Hungary.html (accessed on 20 September 2020).
36. Nieuwsbericht. Fields of Green: The State of Organic Farming in Hungary. Available online: https://www.agroberichtenbuitenland.nl/actueel/nieuws/2020/06/05/organic-farming-in-hungary-today (accessed on 20 September 2020).
37. Canova, L.; Bobbio, A.; Manganelli, A.M. Buying Organic Food Products: The Role of Trust in the Theory of Planned Behavior. *Front. Psychol.* **2020**, *11*, 575820. [CrossRef] [PubMed]
38. Tóth, J.; Migliore, G.; Balogh, J.M.; Rizzo, G. Exploring Innovation Adoption Behavior for Sustainable Development: The Case of Hungarian Food Sector. *Agronomy* **2020**, *10*, 612. [CrossRef]
39. Tari, K.; Lehota, J.; Komáromi, N. Analysis of food consumption in Hungary. In Proceedings of the ENTRENOVA—ENTerprise REsearch InNOVAtion Conference, Dubrovnik, Croatia, 7–9 September 2017; IRENET—Society for Advancing Innovation and Research in Economy: Dubrovnik, Croatia, 2017; Volume 3, pp. 26–32.
40. Ajzen, I. The theory of planned behavior. *Organ. Behav. Hum. Decis. Process.* **1991**, *50*, 179–211. [CrossRef]
41. Beneke, J.; de Sousa, S.; Mbuyu, M.; Wickham, B. The effect of negative online customer reviews on brand equity and purchase intention of consumer electronics in South Africa. *Int. Rev. Retail. Distrib. Consum. Res.* **2016**, *26*, 171–201. [CrossRef]
42. Wu, J.-H.; Wu, C.-W.; Lee, C.-T.; Lee, H.-J. Green purchase intentions: An exploratory study of the Taiwanese electric motorcycle market. *J. Bus. Res.* **2015**, *68*, 829–833. [CrossRef]
43. Park, J.-H.; Kim, M.-K. Factors influencing the low usage of smart TV services by the terminal buyers in Korea. *Telemat. Inform.* **2016**, *33*, 1130–1140. [CrossRef]
44. Shin, D. Beyond user experience of cloud service: Implication for value sensitive approach. *Telemat. Inform.* **2015**, *32*, 33–44. [CrossRef]
45. Ajzen, I.; Kruglanski, A.W. Reasoned action in the service of goal pursuit. *Psychol. Rev.* **2019**, *126*, 774–786. [CrossRef]
46. Gifford, K.; Bernard, J.C. Influencing consumer purchase likelihood of organic food. *Int. J. Consum. Stud.* **2006**, *30*, 155–163. [CrossRef]
47. Verhoef, P.C. Explaining purchases of organic meat by Dutch consumers. *Eur. Rev. Agric. Econ.* **2005**, *32*, 245–267. [CrossRef]
48. Hu, C.S. A new measure for health consciousness: Development of a health consciousness conceptual model. In Proceedings of the National Communication Association Annual Conference, Washington, DC, USA, 21–24 November 2013.
49. Iversen, A.C.; Kraft, P. Does socio-economic status and health consciousness influence how women respond to health related messages in media? *Health Educ. Res.* **2006**, *21*, 601–610. [CrossRef]
50. Homer, P.M.; Kahle, L.R. A structural equation test of the value-attitude-behavior hierarchy. *JPSP* **1988**, *54*, 638–646. [CrossRef]
51. Ellison, B.; Lusk, J.L.; Davis, D. Looking at the label and beyond: The effects of calorie labels, health consciousness, and demographics on caloric intake in restaurants. *Int. J. Behav. Nutr. Phys. Act.* **2013**, *10*, 21. [CrossRef]
52. Shaharudin, M.R.; Pani, J.J.; Mansor, S.W.; Elias, S.J. Purchase intention of organic food; perceived value overview. *Can. Soc. Sci.* **2010**, *6*, 70–79.
53. Michaelidou, N.; Hassan, L.M. The role of health consciousness, food safety concern and ethical identity on attitudes and intentions towards organic food. *Int. J. Consum. Stud.* **2008**, *32*, 163–170. [CrossRef]

54. Kaynak, R.; EKSI, S. Ethnocentrism, religiosity, environmental and health consciousness: Motivators for anti-consumers. *Eur. J. Bus. Econ.* **2011**, *4*, 31–50.
55. Sogari, G.; Corbo, C.; Macconi, M.; Menozzi, D.; Mora, C. Consumer attitude towards sustainable-labelled wine: An exploratory approach. *Int. J. Wine Bus. Res.* **2015**, *27*, 312–328. [CrossRef]
56. Honkanen, P.; Verplanken, B.; Olsen, S.O. Ethical values and motives driving organic food choice. *J. Consum. Behav.* **2006**, *5*, 420–430. [CrossRef]
57. Basha, M.B.; Mason, C.; Shamsudin, M.F.; Hussain, H.I.; Salem, M.A. Consumers attitude towards organic food. *Procedia Econ. Finance* **2015**, *31*, 444–452. [CrossRef]
58. Aulia, S.A.; Sukati, I.; Sulaiman, Z. A review: Customer perceived value and its Dimension. *Asian J. Soc. Sci. Manag. Stud.* **2016**, *3*, 150–162. [CrossRef]
59. Avetisyan, M.; Hertel, T.; Sampson, G. Is local food more environmentally friendly? The GHG emissions impacts of consuming imported versus domestically produced food. *Environ. Resour. Econ.* **2014**, *58*, 415–462. [CrossRef]
60. Polenzani, B.; Riganelli, C.; Marchini, A. Sustainability Perception of Local Extra Virgin Olive Oil and Consumers' Attitude: A New Italian Perspective. *Sustainability* **2020**, *12*, 920. [CrossRef]
61. Bandura, A. Health promotion from the perspective of social cognitive theory. *Psychol. Health* **1998**, *13*, 623–649. [CrossRef]
62. Ismail, I.S.; Farveez, A.A.N.; Nathan, R.J.; Mahadi, M. Parents' Purchase Behaviour and Buying Decision on Luxury Branded Children's Clothing. *Aust. J. Basic Appl. Sci.* **2015**, *9*, 87–92.
63. Ram, D.; Nathan, R.J.; Balraj, A. Purchase Factors for Products with Inelastic Demand: A Study on Tobacco Sector. *JEAS* **2017**, *12*, 1754–1761.
64. Victor, V.; Nambiar, A.S.; Nathan, R.J. Millenials' Intention to Wear Face Masks in Public during COVID-19 Pandemic. *Vadyba J. Manag.* **2020**, *36*, 43–48. [CrossRef]
65. Falguera, V.; Aliguer, N.; Falguera, M. An integrated approach to current trends in food consumption: Moving toward functional and organic products? *Food Control* **2012**, *26*, 274–281. [CrossRef]
66. Sahelices-Pinto, C.; Lanero-Carrizo, A.; Vázquez-Burguete, J.L. Self-determination, clean conscience, or social pressure? Underlying motivations for organic food consumption among young millennials. *J. Consum. Behav.* **2020**, 1–11. [CrossRef]
67. Gómez-Luciano, C.A.; Vriesekoop, F.; Urbano, B. Towards food security of alternative dietary proteins: A comparison between Spain and the Dominican Republic. *Amfiteatru Econ.* **2019**, *21*, 393–407. [CrossRef]
68. Popp, J.; Kiss, A.; Oláh, J.; Máté, D.; Bai, A.; Lakner, Z. Network analysis for the improvement of food safety in the international honey trade. *Amfiteatru Econ.* **2018**, *20*, 84–98. [CrossRef]
69. Achieve Food Security and Improved Nutrition and Promote Sustainable Agriculture. Available online: https://sdgcompass.org/sdgs/sdg-2/ (accessed on 20 September 2020).
70. Beck, A.; Cuoco, E.; Häring, A.M.; Kahl, J.; Koopmans, C.; Micheloni, C.; Moeskops, B.; Niggli, U.; Padel, S.; Rasmussen, I.A. *Strategic Resarch and Innovation Agenda for Organic Food and Farming*; TP Organics: Brussels, Belgium, 2014; pp. 1–61.
71. Fogarassy, C.; Nagy-Pércsi, K.; Ajibade, S.; Gyuricza, C.; Ymeri, P. Relations between Circular Economic "Principles" and Organic Food Purchasing Behavior in Hungary. *Agronomy* **2020**, *10*, 616. [CrossRef]
72. Loizou, E.; Michailidis, A.; Tzimitra-Kalogianni, I. Drivers of consumer's adoption of innovative food. In Proceedings of the 113th EAAE Seminar "A Resilient European Food Industry and Food Chain in a Challenging World", Crete, Greece, 3–6 September 2009; IAAE: Crete, Greece, 2009; pp. 1–10.
73. Roitner-Schobesberger, B.; Darnhofer, I.; Somsook, S.; Vogl, C.R. Consumer perceptions of organic foods in Bangkok, Thailand. *Food Policy* **2008**, *33*, 112–121. [CrossRef]
74. Hungarian Central Statistics Office. Population, National Event 1941 to 2020. Available online: http://www.ksh.hu/docs/eng/xstadat/xstadat_annual/i_wnt001c.html (accessed on 12 January 2021).
75. Department of Statistics Malaysia. Population and Demography. 2020. Available online: https://www.dosm.gov.my/v1/index.php (accessed on 12 January 2021).
76. Yang, M.; Al-Shaaban, S.; Nguyen, T.B. *Consumer Attitude and Purchase Intention towards Organic Food: A Quantitative Study of China*; Linn us University, School of Business and Economics: Kalmar, Sweden, 2014; Available online: http://urn.kb.se/resolve?urn=urn:nbn:se:lnu:diva-34944 (accessed on 20 September 2020).
77. Hair, J.F.; Risher, J.J.; Sarstedt, M.; Ringle, C.M. When to use and how to report the results of PLS-SEM. *Eur. Bus. Rev.* **2019**. [CrossRef]
78. Bagozzi, R.P.; Yi, Y. Specification, evaluation, and interpretation of structural equation models. *J. Acad. Mark. Sci.* **2012**, *40*, 8–34. [CrossRef]
79. Peterson, R.A.; Kim, Y. On the relationship between coefficient alpha and composite reliability. *J. Appl. Psychol.* **2013**, *98*, 194–198. [CrossRef]
80. Henseler, J.; Ringle, C.M.; Sarstedt, M. A new criterion for assessing discriminant validity in variance-based structural equation modeling. *J. Acad. Mark. Sci.* **2015**, *43*, 115–135. [CrossRef]
81. Hu, L.-T.; Bentler, P.M. Fit indices in covariance structure modeling: Sensitivity to underparameterized model misspecification. *Psychol. Methods* **1998**, *3*, 424–453. [CrossRef]
82. Ayub, A.H.; Hayati, Y.; Samat, M.F. Factors influencing young consumers' purchase intention of organic food product. *Adv. Bus. Res. Int. J. (ABRIJ)* **2018**, *4*, 17–26. [CrossRef]

83. Akbar, A.; Ali, S.; Ahmad, M.A.; Akbar, M.; Danish, M. Understanding the Antecedents of Organic Food Consumption in Pakistan: Moderating Role of Food Neophobia. *Int. J. Environ. Res. Public Health* **2019**, *16*, 4043. [CrossRef]
84. Sharaf, M.A.; Isa, F.M. Factors influencing students' intention to purchase green products: A case study in Universiti Utara Malaysia. *Pertanika J. Soc. Sci. Huminities* **2017**, *25*, 239–249.
85. Saleki, R.; Quoquab, F.; Mohammad, J. What drives Malaysian consumers' organic food purchase intention? The role of moral norm, self-identity, environmental concern and price consciousness. *J. Agribus. Dev. Emerg. Econ.* **2019**, *9*, 584–603. [CrossRef]
86. Nguyen, T.T.M.; Phan, T.H.; Nguyen, H.L.; Dang, T.K.T.; Nguyen, N.D. Antecedents of purchase intention toward organic food in an asian emerging market: A study of urban vietnamese consumers. *Sustainability* **2019**, *11*, 4773. [CrossRef]
87. Lee, H.-J.; Yun, Z.-S. Consumers' perceptions of organic food attributes and cognitive and affective attitudes as determinants of their purchase intentions toward organic food. *Food Qual. Prefer.* **2015**, *39*, 259–267. [CrossRef]
88. Lockie, S.; Lyons, K.; Lawrence, G.; Grice, J. Choosing organics: A path analysis of factors underlying the selection of organic food among Australian consumers. *Appetite* **2004**, *43*, 135–146. [CrossRef]
89. Yadav, R.; Pathak, G.S. Young consumers' intention towards buying green products in a developing nation: Extending the theory of planned behavior. *J. Clean. Prod.* **2016**, *135*, 732–739. [CrossRef]
90. Stevens, C.J.; Murphy, C.; Roberts, R.; Lucas, L.; Silva, F.; Fuller, D.Q. Between China and South Asia: A Middle Asian corridor of crop dispersal and agricultural innovation in the Bronze Age. *Holocene* **2016**, *26*, 1541–1555. [CrossRef]
91. Cséfalvay, Z. Robotization in Central and Eastern Europe: Catching up or dependence? *Eur. Plan. Stud.* **2020**, *28*, 1534–1553. [CrossRef]
92. Afandi, A.S.; Yunus, S.M.; Suhaimi, M.Y.; Tapsir, S. Technology adoption decision among food manufacturers: What are the critical factors. *Econ. Technol. Manag. Rev.* **2016**, *11*, 75–85.
93. Vollrath, D. The agricultural basis of comparative development. *J. Econ. Growth* **2011**, *16*, 343–370. [CrossRef]
94. Jo, J.-H.; Roh, T.W.; Kim, S.; Youn, Y.-C.; Park, M.S.; Han, K.J.; Jang, E.K. Eco-innovation for sustainability: Evidence from 49 countries in Asia and Europe. *Sustainability* **2015**, *7*, 16820–16835. [CrossRef]
95. Oroian, C.F.; Safirescu, C.O.; Harun, R.; Chiciudean, G.O.; Arion, F.H.; Muresan, I.C.; Bordeanu, B.M. Consumers' attitudes towards organic products and sustainable development: A case study of Romania. *Sustainability* **2017**, *9*, 1559. [CrossRef]
96. Paul, J.; Rana, J. Consumer behavior and purchase intention for organic food. *J. Consum. Mark.* **2012**, *29*, 412–422. [CrossRef]
97. Bildtgård, T. Trust in food in modern and late-modern societies. *Soc. Sci. Inf.* **2008**, *47*, 99–128. [CrossRef]
98. Bernstein, H. Food sovereignty via the 'peasant way': A sceptical view. *J. Peasant. Stud.* **2014**, *41*, 1031–1063. [CrossRef]
99. Voon, J.P.; Ngui, K.S.; Agrawal, A. Determinants of willingness to purchase organic food: An exploratory study using structural equation modeling. *Int. Food Agribus. Manag. Rev.* **2011**, *14*, 103–120.
100. Eze, U.C.; Ndubisi, N.O. Green buyer behavior: Evidence from Asia consumers. *J. Asian Afr. Stud.* **2013**, *48*, 413–426. [CrossRef]
101. Li, J.; Zepeda, L.; Gould, B.W. The demand for organic food in the US: An empirical assessment. *J. Food Distrib. Res.* **2007**, *38*, 54–69.
102. Busse, M.; Siebert, R. The role of consumers in food innovation processes. *Eur. J. Innov. Manag.* **2018**, *21*, 20–43. [CrossRef]

 MDPI

Article

Knowledge, Utility, and Preferences for Beef Label Traceability Information: A Cross-Cultural Market Analysis Comparing Spain and Brazil

Danielle Rodrigues Magalhaes [1,*], María del Mar Campo [1] and María Teresa Maza [2]

1 Department Animal Husbandry and Food Science, Instituto Agroalimentario IA2, Universidad de Zaragoza-CITA, Miguel Servet 177, 50013 Zaragoza, Spain; marimar@unizar.es
2 Department Agricultural Science and Natural Environment, Instituto Agroalimentario IA2, Universidad de Zaragoza-CITA, Miguel Servet 177, 50013 Zaragoza, Spain; mazama@unizar.es
* Correspondence: d.magalhaes@yahoo.com.br; Tel.: +34-998407183

Abstract: The consumer environment determines consumers' buying behavior and product preferences, and understanding these factors allows businesses in the industry to identify market demands. In view of the different contexts, Spain and Brazil, there are differences in the consumption of beef, in the production and the regulatory process concerning beef, and in particular the traceability system. The traceability system is mandatory in Spain and voluntary in Brazil. From these prerogatives, this cross-cultural study carried out through a self-administered questionnaire with 2132 Spanish and Brazilian beef buyers/consumers, aimed at comparing and understanding the familiarity with the bovine traceability system and traceability information of the label as a food security indicator. It is concluded that traceability information is well received by consumers as an attribute of credibility, and consumers are interested in ensuring that the item they buy is of known and reliable origin. But more incentives may help clarify the advantages of purchasing food with certified traceability, making it more effective for consumers to use this knowledge.

Keywords: beef; traceability system; marketing; consumer; safety food; cross cultural study; questionnaire

Citation: Magalhaes, D.R.; Campo, M.d.M.; Maza, M.T. Knowledge, Utility, and Preferences for Beef Label Traceability Information: A Cross-Cultural Market Analysis Comparing Spain and Brazil. *Foods* **2021**, *10*, 232. https://doi.org/10.3390/foods10020232

Academic Editor: Derek V. Byrne
Received: 18 December 2020
Accepted: 21 January 2021
Published: 23 January 2021

1. Introduction

Consumer purchasing behavior with regard to food products has been extensively studied. The context in which food products are purchased, the use of the product and the level of knowledge handled, i.e., the environment of consumption (socio-cultural, economic, technological, and political), define consumer behavior and preferences for products and processes influencing their choice at the time of purchase. It is a challenge for companies in the sector to consider all these factors, as consumer demand ultimately drives the continued investment of companies in research and innovation [1–5].

In view of the different contexts which, as already described, define the preference of the consumer for a particular product, the present work has been carried out in two countries, Spain and Brazil, in which there are differences in the consumption of beef, and in the production and in the regulatory framework which involves it, particularly with regard to food safety issues and with reference to traceability. The key focus of this analysis is on the fact that traceability is mandatory in Spain following European Union (EU) regulations and voluntary in Brazil [6–10].

The implementation of product traceability is mainly due to the food crises of the 1990s, particularly those caused by the appearance of bovine spongiform encephalopathy (BSE). Consumer trust was affected at that time, during which traceability systems proved to be inefficient, and a series of regulatory reforms were implemented, first in European countries, and later spread to other continents [11–14].

There is a great economic loss in any crisis facing the production system due to the massive loss of the product, in addition to the bad reputation suffered by the companies involved, which also creates an additional cost to regain consumer confidence in the product again [15]. Because of this set of guiding factors, traceability in this situation helps to improve the efficiency of internal management in terms of knowledge available on products and processes, which provides effective control of the food supply chain [14,16].

Traceability can be used as a method to comply with regulations and comply with food safety and quality standards, providing information on the origin, processing, and final destination of food to be related between producers and consumers [12,13]. It is also expected that if farmers and agro-food companies comply with legal traceability criteria, the role of health control would be facilitated. The more accurate the monitoring system, the faster it is possible to identify a producer and solve food safety or quality issues [17,18].

Product traceability is also important in a wider range of ideas because of the globalization of food marketing, as a result of the long distances that food travels from its origin to the consumer. Food monitoring is extremely important at this stage to discover the origin and authenticity of the food [19].

According to the activity or direction in which information is passed through the food chain, traceability can be classified into three types [20–23]:

- Internal traceability or process traceability: It is the ability to internally track the origin of products coming into the supply chain or leaving the company to control it. When using this traceability system, the main objective of industries is to enhance institutional management.
- Return traceability or supplier traceability: It is the ability of this system to find the origin and characteristics of a product based on one or more criteria at each point in the supply chain, allowing for effective product identification and management of quality and safety standards.
- Direct traceability or consumer traceability: It helps the consumer to establish trust, that is, to increase the credibility of purchased products. Foods must have a transparent tracking accompanied by information about their origin on product labels. The key theme discussed in this paper was this last example of traceability listed, which refers to traceability for consumers.

Traceability is necessary for verification of credence attributes such as origin. In general, it is considered that color, price and freshness of meat are search attributes, due to the fact that they are known before purchase, whilst taste and tenderness are experience attributes because they are only known after consumption. However, the greatest problem arises in the case of credence attributes, that is, those attributes that cannot be known even after having consumed the product or, on occasions, those with a high cost due to the adverse effects that may cause on the consumer [24]. Amongst these, animal welfare and environmentally friendly production methods, food safety, or origin can be found [25]. Grunert et al. [26] indicated the growing interest in the role of credence attributes play in consumer choice. In a recent study [27], traceability is considered a sustainability attribute like animal welfare and effect of greenhouse gas emissions.

From the literature reviewed, it is clear that traceability is highly valued by consumers for various reasons, since it is essential to know the origin of the animal and the places where the transformation processes have taken place [28]. The region of production or origin are some of the aspects most valued by consumers [29–31]. Many consumers show more confidence in meat produced in certain places precisely because they consider it safer. In the study of Loureiro and Umberger [28], indication of origin may only become a signal of enhanced quality if the source of origin is associated with higher food safety or quality. In this sense consumers value of country of origin depending on the number of other credence attributes included in product descriptions and the location of the consumer [32].

Therefore, traceability or certification of origin can only be used as an indication of consumer quality if it is associated to greater security. Security differs from many other attributes of quality, since it is a hard attribute to observe. A product may appear to be of high quality (i.e., color, texture, and flavor) but may not be healthy because it may be contaminated with undetected pathogens, toxic chemicals, or other health risks [13,28]. It performs in such a way that a tracked product can be considered of superior quality to a non-tracked product as a factor of competition in the industry [33].

Two lines of thinking are observed in investigations carried out on consumer confidence in traceability information: One in which it is argued that one of the ways to guarantee food safety comes from traceability, and the other in which the presence of traceability information on the labels does not translate into stronger consumer confidence.

Traceability, on the one hand, is used as a food safety method that provides product recall to identify the source of a problem [34,35]. It also acts as proof of the authenticity of the food and that the labels present in the packaging of meat products are capable of enhancing the consumer's well-being by improving the understanding of the origin of the product, indicating that the food has been checked during the manufacturing process [28,36,37]. The certification of the product makes the consumer sure that the product is from where it actually mentions on the label [13].

On the other hand, despite the recognized need for clear information on the quality of the entire food chain, supported by modern methods of monitoring and tracking, other quality attributes may compete with the label's traceability information. Such information, that may have little or no priority at the time of purchase, must be taken into account in any assessment of consumer preferences with regard to the certification of origin [38].

Furthermore, food quality is related to a proactive policy of creating requirements for the maintenance of a healthy food supply. However, the traceability information available on the labels does not always contribute to greater confidence, as the safety of a product is the responsibility shared by all actors in the food chain, including governments, industry and consumers, and is susceptible to failure [39].

Traceability system proponents believe that labels serve the right of the consumer to know about the origins of food products. Opponents are of the opinion that the labeling law is a protectionist measure, supported by distorted expectations of the quality of products imported. They find the labeling process expensive and complicated [40].

Amidst so many reflections on the advantages or disadvantages of traceability, do consumers know what the system of traceability is? Is this information, as it is currently shown, useful and relevant when deciding to buy beef? Do consumers trust the traceability information on the labels? Whenever necessary, do consumers know how to use this information (in a time of crisis, for example)? When a system is obligatory or voluntary, are there differences?

Communication is the crucial point of understanding between consumers and companies [38]. Once we know the types of information that different types of consumers require, we can begin to investigate whether the provision of specific information that is available through traceability will ultimately influence consumer confidence or not in making the final purchase decision [34].

Appropriate measurement methods are required to answer these questions in order to assess the value that consumers attach to information on traceability, but also to inform industry about market demand [18]. This would have significant consequences for public policy formulation [40]

The fact that the traceability system is mandatory in one country (Spain) and not in another (Brazil) could be the source of some differences in consumer perception. The purpose of this study was to understand the familiarity of Spanish and Brazilian consumers with the bovine traceability system and the label traceability information as an indicator of product safety in countries with different beef consumption, production and mainly different scale of the traceability system's implementation.

2. Materials and Methods

2.1. Experimental Overview

In general, meat consumption differs between regions due to different dietary patterns, income levels, and product availability [41]. This statement can be extended to places like Spain and Brazil, the focus countries of this study, which show differences in terms of consumption, production, and the traceability system of beef.

With regard to beef cattle production, Spain, considered one of the largest exporters both within and outside the EU, is of great importance at the European level [6]. In Brazil, on the other hand, beef production is of global importance, with the nation ranked among the world's largest producers and exporters of beef [8]. As far as consumption is concerned, three times more beef is consumed in Brazil than in Spain. The apparent consumption in Spain is 12.7 kg/inhabitant/year [7], while in Brazil the consumption of beef is basic and high (35.8 kg/inhabitant/year) and is the fifth largest user of this commodity worldwide [42].

The traceability of beef is another very important differential point between the two countries and the main theme of this study. In Spain, the mandatory traceability legally obligated is protected by the European framework [9]. On the opposite, in Brazil, where there is a voluntary traceability of beef [10], with the exception of the production of animals and products exported to countries requiring traceability. There is also a difference in the traceability information on the beef label that reaches the consumer in Spain and Brazil, in Spain this information is mandatory and required by law [43], while in Brazil this information is optional [10].

2.2. Consumers

The study was performed in 2016, with Spanish and Brazilian consumers reaching a total of 2132 regular beef purchasers. In Spain, 436 questionnaires were applied in the province of Zaragoza, and in Brazil, the questionnaires were administered in four different States—Minas Gerais (n = 424), Sao Paulo (n = 456), Parana (n = 406) and Santa Catarina (n = 410). Zaragoza was chosen because it is considered a model region in market studies in Spain due to its size and consumer behavior. In Brazil, the four regions were chosen because differ in consumption and beef production, complementing themselves to become representative of the country. The consumption of beef in the State of Minas Gerais is one of the lowest in the country, consumption in São Paulo is below the national average, in the State of Paraná is within the average and in Santa Catarina is above the average consumption. In relation to beef production, the State of Minas Gerais is the second largest producer of beef in the country, São Paulo is in ninth place, Paraná in tenth and Santa Catarina in thirteenth. The States of São Paulo and Minas Gerais are two of the main States where the largest meat processing plants in the country are gathered.

2.3. Online Questionnaire

The data of this study were collected through a self-administered questionnaire for dissemination in network designed with the support of the *Google.forms* software (Supplementary Materials, Web Application—Google Platform). Questionnaires were sent in each country's native language (Spanish or Portuguese).

Two types of non-probabilistic sampling were used for data collection: Chain sampling, used to classify subjects with specific characteristics, beef consumers in our case; and conventional sampling, in which subjects are selected for their accessibility [44]. The analysis has been planned to achieve descriptive and empirical goals. Closed questions with two or more alternatives and/or scales of responses were used for this [45].

The questionnaire initially presented questions about the knowledge and correct definition of the traceability concept, the variables that make up the current traceability system, and the level of credibility and importance of the traceability system for beef.

Four additional questions/topics were then dealt with, two of which were addressed to the Spanish consumer and two to the Brazilian consumer relating to purchasing beef with traceability information or Certificate of Origin (CoO). Spanish consumers were questioned if they had taken the traceability information on meat labels into account at the time of purchase. Brazilian consumers were asked if they had already purchased beef with traceability information on the label, since traceability information is not present in all products. As the response, positive or negative, was asked to justify the answer.

The other two questions were related to the importance of traceability or CoO information on labels. Six label models (Appendix A) with a different layout and traceability information set have been suggested for Spanish consumers. Participants were asked to order the labels from the most preferred to the least preferred from the layout and set of traceability information on each of the labels.

Each label contained the same "basic" product information such as the name of the product, the category, the recommendations for cooking, the expiration date, the price, etc. But with regards to the traceability information itself, three types of data were combined between the tags:

- "Traceability" includes the mandatory information provided by law: Traceability number of the animal, animal birthplace, animal fattening location, slaughterhouse, and cutting and boning room.
- Extra information, also referred to as "plus" in this study, was added to the label with mandatory traceability information. The breed and sex of the animal and the date of birth were included
- A "QR Code" (Quick Response Code) used on the labels, in this case, for the electronic reading of the traceability information of the product for sale.

Two types of labels (Appendix B) were sent to Brazilian consumers, one of which provided traceability information commonly found on beef labels sold in Brazil, i.e., an animal traceability number and a QR code for online access to animal origin information—CoO. And another label included traceability information on the label, just like the one supplied on the labels of beef sold in Spain, i.e., animal traceability number, animal birthplace, animal fattening location, slaughterhouse, and cutting and boning room. Brazilian consumers were asked which of the two labels provided the necessary information to satisfy their requirements.

Finally, participants from both countries were asked if, in case of suspicion that the meat they purchased has a problem of food safety quality or other aspects related to health, what action would they take with respect to the use of that food.

2.4. Data Analysis

Statistical analyses were performed using IBM® SPSS statistics version 22. Univariate and bivariate analyses were used to evaluate characteristics related to consumer behavior. Descriptive and chi-square analyses were used to evaluate characteristics related to consumer behavior: Knowledge, credibility, and importance of traceability and utility of information traceability on the beef label.

A maximum significance level of 5% or less was tested for acceptance or rejection of the null hypothesis. The conjoint analysis, using the dependency method, was used to determine the preference of different levels of Spanish beef label traceability information. As variables, the presence of mandatory traceability data, the presence of additional information and the presence of a QR code were analyzed, with two levels, omission/presence, in each variable.

3. Results and Discussion

3.1. Knowledge of the Traceability System for Beef

The use of traceability information at the time of purchase results depending on knowledge of the concept [35]. This investigation, in line with these findings, starts with an interest in understanding if Spanish and Brazilian consumers are familiar with traceability.

According to the results of this study, just over half of the Spanish consumer sample (52.8%) and 58.4% of Brazilian consumers say to be perfectly familiar with the concept of traceability. While about a third of consumers in both countries indicate they do not know traceability, the majority of respondents chose the correct definition of "monitoring of beef origin and products" representing 81.2% of Spanish consumers and 77.5% of Brazilians (Table 1).

Table 1. Knowledge and definition of the traceability concept for Spanish and Brazilians consumers.

Traceability Concept *	Country [1]		*p*
	Spain	Brazil	
Knowledge of the concept			
Yes, perfectly	52.8	58.4	
No, totally unaware	36.0	33.3	0.045
Unclear	11.2	8.3	
Definition of the concept			
Monitoring of beef origin and products	81.2	77.5	
Beef inspected by the health service	3.4	11.4	
Nutritional information by labels	2.3	1.3	
No chemical residues contaminants in beef	0.7	0.4	≤0.001
Branded beef	0.5	0.8	
Don't know	11.9	8.6	

[1] Total n = 2132, Spain n = 436, Brazil n = 1696. * Only one answer per question. (%).

Consumers with no traceability information end up not being interested in this subject [46]. In two studies evaluating knowledge of the concept of traceability, one in 2006 [47] and another ten years later, in 2016 [33], it was shown that considerable difficulty in hearing about traced meat was and continues to be associated with the lack of product availability and accessibility of product information.

It is worth noting that 11.4% of Brazilian consumers chose the traceability concept in this study: "Beef inspected by the health service" (Table 1). This choice is likely to refer to the quality inspection service for the production of food products of Brazilian animal origin (MAPA—Ministry of Agriculture, Livestock, and Supply) whose certificate seal is present in all packages of products of animal origin, with three levels of inspection: Federal Inspection Service (S.I.F), State Inspection Service (S.I.E) and Municipal Inspection Service (S.I.M) [48]. Our findings were confirmed by other studies carried out in Brazil that indicate that more consumers are aware of certificates of inspection for animal products than of traceability information or CoO [12,33].

For many consumers, the term that indicates information such as "monitoring" and "inspection" offers an additional impression of product protection, as is the case with traceability [33,46]. The results of this study suggest that consumers hesitated about the definition of traceability at first but once introduced to the definition, most will know how to identify it (Table 1).

3.2. *Aspects That Make up the Current Traceability System for Beef*

The ability to organize the chain efficiently is the basic requirement of the traceability system [47]. The introduction of traceability systems for the beef supply chain in some countries is based on mandatory legislation (such as Spain), while in others, regulations have been adopted on a voluntary basis, where market-dependent incentives for their implementation (such as Brazil) [33,47].

The first step towards the efficient use of traceability by consumers seems to be primarily related to informing consumers about the origin of the product, i.e., the various steps in which they show that the quality and safety of the product has been regulated [49].

Despite the different covering, one mandatory and the other voluntary, the regulations governing the traceability system in Spain and Brazil are the same. Consumers in both countries were therefore asked which aspects are incorporated into the existing traceability system. Table 2 shows a total of nine aspects, all of which are present in Spain and Brazil as part of the traceability system.

Table 2. Aspects assumed by Spanish and Brazilian consumers are part of the current traceability system for beef.

Aspects of Traceability System *	Country [1]		p
	Spain	Brazil	
Animal feed	49.5	75.9	≤0.001
Animal breed	61.5	76.2	≤0.001
Date of birth	64.9	78.9	≤0.001
Animal birthplace	48.4	72.4	≤0.001
Animal fattening location	49.3	75.5	≤0.001
Transport	44.7	73.1	≤0.001
Slaughter date	71.3	86.9	≤0.001
Slaughterhouse	65.4	87.7	≤0.001
Cutting and boning room	37.6	45.5	0.003
Don't know	17.2	4.6	≤0.001

[1] Total n = 2132, Spain, n = 436, Brazil, n = 1696. * Multiple answers possible. (%).

The findings indicate that more than half of the Spanish consumers surveyed identified only four aspects: The breed of the animal (61.5%), the date of birth (64.9%), the date of slaughter (71.3%), and the slaughterhouse (65.4%) (Table 2). From these results it can be concluded that the information that tends to be more common to Spanish consumers that is part of the traceability system is precisely the mandatory reference information found on the beef labels in this region, which are: Slaughter place and animal category of sales denomination provided for in Royal Decree 75/2009 [43].

A study has shown that specific aspects linked to the traceability system are less relevant for the general public/consumers and much more of interest to other market segments, such as producers, processors and the retail chain [47]. The stages of the process, which is quality related to safety which is a potential attribute of credibility to be valued at the time of purchase, are not evaluated out of their own, but what results from all of this [17,18].

In Brazil, more than half of consumers (45.5%) did not select just one attribute, "cutting room" (Table 2) The results found that the majority of Brazilian consumers are informed of the aspects that are part of the traceability system, according to researchers [50], because the information is known. This can be an indication that investments in the meat chain may bring benefits to retailers who need protection against the loss of consumer confidence.

3.3. Credibility and Importance of Traceability of Beef

The level of credibility and importance of beef traceability was evaluated in this study and the results (Table 3) showed that the credibility of traceability information is partial for 47.7% of Spanish consumers and 62.0% of Brazilians, whereas a greater proportion of Spanish consumers believe traceability information totally (28.0%) than Brazilians (14.6%). Spanish consumers also rely more than Brazilians on information such as: Origin and production of animals, marketing information, information on the slaughterhouse, and information on the animal itself.

Our study suggests that the fact that traceability information is provided more confidence by Spanish consumers than by Brazilians come from the fact that the traceability and labeling system in Spain is mandatory, and mainly because it is already consolidated, unlike in Brazil, where both traceability and labeling systems are voluntary, often only for marketing purposes, as some researchers suggest [37].

Table 3. Credibility and importance in the traceability information that Spanish and Brazilian consumers believe is appropriate to their way of thinking.

Credibility and Importance of Traceability *	Country [1]		p
	Spain	Brazil	
Total credibility	28.0	14.6	≤0.001
Partial credibility	47.7	62.0	≤0.001
No credibility	7.3	2.9	≤0.001
Credibility in the origin and production of animal	33.7	16.6	≤0.001
Credibility in marketing information	24.8	7.3	≤0.001
Credibility in slaughterhouse information	19.0	8.8	≤0.001
Credibility in the information of the animal itself	13.5	6.3	≤0.001
Traceability is important for consumers	59.6	50.9	0.001
Traceability is important for companies	33.7	31.1	0.301

[1] Total n = 2132, Spain n = 436, Brazil n = 1696. * Multiple answers possible. (%).

According to previous research [18,49], the credibility of traceability information is determined by its accessibility and the consumers' own reasoning. In order to become an attribute of confidence for the user, traceability information needs to be driven by the advantages that this information brings [27].

As far as the importance of the traceability system is concerned, more Spanish consumers (59.6%) and to a minor extent, Brazilians (50.9%) agree that traceability is important to consumers, that is to say, to themselves. To a minor extent, about a third of Spanish consumers (33.7%) and Brazilians (31.1%) agree that traceability is important for companies (Table 3). According to some researchers [13,18], although consumers partially believe in traceability information, they still believe that traceability remains important for them, data that corroborate our study.

A study conducted with Brazilian consumers [27] confirms the importance of attributes such as traceability in the decision-making process for purchasing beef. Other research, also carried out in Brazil [33,51] found that the majority of beef consumers, more than 90% of them, agreed that meat with any kind of certification has greater benefits than meat without any certification, and consumers believe that certificate products are better, higher quality and more secure.

Knowing the perceptions and requirements of consumers about food traceability is often an obstacle, because credit attributes can be interpreted with different ways, such as marketing purposes for example. Misinterpretations can lead consumers not to know the real reason for traceability, which is to improve food security against hazards that may occur at different points in the food chain [37,52].

For this reason, industry and government must focus their efforts to ensure that messages are properly transmitted and understood by consumers from credible attributes, such as traceability information [17,37].

In general, measures to minimize consumer inaccurate vision are straightforward and efficient communication, that is to say, in a format that is easily accessible and without overloading consumers. This suggests more visible labeling with certification assurances during buying, giving customers the ability to more clearly evaluate meat security [53].

3.4. Usefulness of Traceability Information on Beef Labels

Food labels aim to enable consumers to make informed decisions about experience and credibility attributes such as product quality, technology for production, and processing [52,54]. Given that consumer perceptions of food safety risk are strongly linked to the credibility of product attribute information, food labeling regulation needs to ensure the integrity of that information [37].

The information on the meat label is considered to have higher or lower levels of importance at the time of purchase. Research performed with meat consumers indicate that readily interpretable information, such as meat type/cut and expiration data, has a higher rating of significance and use compared to credit information indications, such as origin [17,46,47,55].

Our study evaluated the utility for Spanish and Brazilian consumers of traceability and/or CoO information on beef labels. Since traceability information is obligatory on beef labels in Spain and on a voluntary basis in Brazil, concerns about the utility of this information have been handled in accordance with the reality of the consumers in question (Tables 4 and 5).

Table 4. Opinion of Spanish consumers on information related to the traceability in beef labels.

Traceability of the Beef Label	Spain [1]
Do you take into account the data related to traceability on beef labels?	
Yes	57.3
No	42.7
In case of AFFIRMATIVE answer [2,*]	
I don't buy beef from unknown or suspicious countries	32.7
I want to know the origin of the product	22.9
I only buy beef originating from my country	20.7
I only buy beef originating from my region	13.0
I only buy beef from the EU	10.6
In case of NEGATIVE answer [3,*]	
It has never worried me	31.3
I don't know what to do with that information	21.3
Knowing there is traceability is enough	19.6
Traceability code must be on the products, but it is not my responsibility	11.7
This information is difficult to interpret	8.7
I don't have time to read it	4.9
I don't see the use of it	2.4

[1] Total $n = 436$, [2] only for people who said "yes" $n = 250$, [3] only for people who said "no" $n = 186$. * It is possible to mark several points. (%).

Table 5. Opinion of Brazilian consumers on information related to the traceability in beef labels.

Traceability of the Beef Label	Brazil
Do you buy or have you already bought beef with traceability information on the label? [1*]	
Yes	32.1
No	22.6
I don't know if I bought/didn't realize it at the time of purchase	45.3
If YES, how did you hear about this product? [2*]	
Knowledge of the traceability of agricultural products	46.4
Observing the identification of traceability on the label	31.7
Marketing in the establishment where I buy or in other channels of communication	17.1
In conversations with others	4.8
If NO, why never bought it? [3*]	
I don't know where it sells/I think it's not available in my city	84.6
I don't know where it sells and I'm not interested in buying it	9.1
I don't buy it because I don't see a difference between meat with traceability or without it	6.3

[1] Total $n = 1696$; Minas Gerais, $n = 424$; Sao Paulo; $n = 456$, Parana, $n = 406$; Santa Catarina, $n = 410$. [2] Total $n = 545$ (only for people who said "yes"): Minas Gerais, $n = 125$; Sao Paulo, $n = 161$; Parana, $n = 141$; Santa Catarina, $n = 118$. [3] Total $n = 383$ (only for people who said "no"): Minas Gerais, $n = 127$; Sao Paulo, $n = 70$; Parana, $n = 90$; Santa Catarina, $n = 96$. * Mark only one answer. (%).

Spanish consumers were initially asked if they had taken the traceability information on the packaging labels into account while purchasing beef (Table 4). Most respondents said yes (57.3%) compared to 42.7% who said that they did not. The results also showed that most Spanish consumers use beef traceability information to avoid purchasing products from unknown or suspicious countries (32.7%) or just want to know the origin of meat (22.9%). Beef originating in the country itself (20.7%), their region (13.0%), or the European Union (EU) (10.6%) are preferred, respectively.

Consumers who do not take traceability into account at the time of purchase are mainly due to never having been concerned with such information (31.3%), or because they do not know what to do with this traceability information (21.3%), for knowing that it is tracked is sufficient (19.6%), for thinking that this information is not their responsibility (11.7%), for not having time to read (4.9%), or not seeing any use of the traceability information for beef (2.4%) (Table 4).

In the EU, due to concerns on BSE, mandatory traceability information on labels for beef was introduced in 1997 in order to be used by consumers to infer product quality [56]. It has been shown [40,49] that the mandatory CoO benefits the domestic market in general if the preference for national products over imported products is high enough. According to our results, this is what, is happening within the Spanish market.

Also, according to a research on red meat in Spain [57], the origin of the meat was the most important factor in determining the consumer's purchase decision and one of the most important for German and Polish consumers in the decision to purchase of pork [26]. However, as other researchers have found out [46,58], the information indicating the product's visual quality is often more important than the signs on the labels relating to traceability and origin.

Other studies on the importance attributable to CoO information on the label [59] show that consumers can be pessimistic about this information because they do not believe, do not see advantages in the information offered, and do not want to pay for an attribute that does not value sufficiently.

In this research we found that traceability information is useful (of usefulness) to consumers who do not feel responsible for the label's traceability information, or do not know what to do about it (commitment and lack of knowledge); when purchasing beef. Other researchers [13] also indicate that consumers are not interested in the information on traceability itself, but in knowing that traceability has been established and that someone is inspecting the background of the meat. According to a study by Belgian consumers [47], about 70% of respondents prefer retailers or butchers to store all the necessary traceability information and to make it available at the consumer's request.

In Brazil, labeling with CoO information on beef is voluntary, and because not all establishments supply beef with this information, Brazilian consumers were asked whether they had ever bought beef with CoO information on the label.

We found that 45.3% of consumers say they do not know whether they have bought beef with CoO or not. Another 32.1% said yes, having already bought beef with CoO against 22.6% that said no (Table 5). A 46.4% of the customers who said they bought beef with CoO had previous knowledge of traceability from other products, and 31.7% did so by observing the CoO information provided on the label for themselves, another 17.1% due to marketing at the establishment or other communication channels, and 4.8% due to interaction/conversation with other individuals having been informed about beef CoO (Table 5).

On the other hand, in the case of consumers who said they had never bought beef with CoO on the label, the large majority (84.6%) of consumers said they did not know where it was sold and/or believed it was not available at their place of purchase; 9.1% do not know where to sell, nor are they interested in purchasing in the same way and 6.3% say they do not buy because they do not see a difference between a product with or without traceability or certification of origin (Table 5).

In a study conducted in Brazil [12], it was found that most consumers would like to have access to information on meat traceability and would be willing to pay more for this product, supporting Brazil's mandatory meat traceability. CoO beef products are gaining place on the shelves of large beef distribution chains, but the supply of these items is still limited, in most cases, to the packaging of vacuum-packed whole cuts [33].

In relation to the credibility and origin of beef, a vast literature on consumer preferences and on willingness to pay more for labeling initiatives is available [28,33,52,59]. The question is to what degree consumers actually value this information, considering the increasing use of CoO information on food labels [60,61].

The challenge, then, is how to communicate the use of food security technologies as consumers are willing to pay for these technologies [62]; and the best way has been to effectively label food [63].

The best way to provide this information on food labels is to confront costs vs. consumer attitudes/perceptions through the demand for origin information of the product, taking into account the costs of mandatory implementation of traceability or CoO, which are high [47,52]. Currently, the findings most commonly seen in research studies [61] are that consumer expectations for origin knowledge for meat products are clearly growing.

The usefulness of the traceability information for consumers depends on how far the traceability situation is in the country and the contribution of consumers' knowledge on this matter. For example, it was found that consumer risk perceptions in the US and Japan [52] are significantly affected more by their level of confidence in credit attribute information such as organic, natural, traceable, and country of origin, than by credibility in visual attributes such as color and texture.

Studies conducted in countries that do not have a mandatory traceability system until then, as in China [64] and Poland [65], indicate that regular meat consumers would benefit from the introduction of this system because this information could serve as a cognitive shortcut in evaluating the quality of beef and mechanisms of food safety in the public health perspective.

These findings are relevant to industry and policy, as they indicate that more advanced animal information quality tips give a sense of control throughout the food chain and help consumers infer the credibility of products [18,38].

Consumers can look for and expect more information about the CoO of a product, but if consumers do not value it sufficiently, there is no reason to force the provision of this information [66]. This does not mean that they are irrelevant, although not of general relevance to consumers, but from an informative research perspective that should be used to target specific consumer segments rather than all consumers [38].

3.5. Preference for Traceability Information on the Label for Beef

In order to satisfy consumer needs, the traceability information must be in the format desired by the user and capable of understanding their needs [52]. Preferences for beef products may not be similar among consumers in different countries [67]. Therefore, the preferences for how traceability and/or CoO information is presented on the beef label by Spanish and Brazilian consumers were also evaluated in this study (Tables 6–8).

Table 6. Opinion of Spanish consumers according to changes in the traceability information on beef label.

Changes in Traceability Information on Beef Labels *	Spain [1]
No need changes	47.9
No need for codes or symbols	36.0
Better represented by barcode/QR code	13.1
Don't know	3.0

[1] $n = 436$. * Mark only one answer. (%).

Table 7. Spanish consumers' preference according to the amount of information on the beef label.

Attributes	Importance Values [1]	Levels	Estimated Preference	ERROR	p
QR Code [2]	24.6	Yes No	−0.317 0.317	0.142	≤0.001
Information [3]	75.4	Basic Traceability Plus	0.226 1.198 −1.424	0.201	

[1] n = 436. [2] "QR code": Module to store information in a dot matrix or in a two-dimensional barcode. [3] Information—type of information on the label: "Basic": Name of the product, the category, the recommendations for cooking, the expiration date, the price, etc. "Traceability": Traceability number of the animal, animal birthplace, animal fattening location, slaughterhouse and cutting and boning room. "Plus": The "Traceability" information mentioned above, with the addition of information on: Breed, sex and date of birth of the animal.

Table 8. Brazilian consumers' preference according to the amount of information on the beef label.

Display of Traceability Information on Beef Labels *	Brazil [1]
I prefer it to be a combination of the two labels, with information I can see at the time of purchase (Spanish label) or that I can access it over the internet if I'm interested in product origin (Brazilian Label)	56.7
I prefer the Spanish label, because it brings traceability information in the product label at the time of purchase	31.6
I prefer the Brazilian labels, if I'm interested in knowing the information of traceability I can access it over the internet through my cell phone or computer at moment or after purchase	10.4
None of the labels, I don't see the need for traceability information for beef	1.3

[1] Total n = 1696; Minas Gerais, n = 424; Sao Paulo, n = 456; Parana, n = 406; Santa Catarina, n = 410. * Mark only one answer. (%).

First, Spanish consumers were questioned about the need for changes in the current display of traceability information on beef labels (Table 6). The results show that the information provided on labels is sufficient for approximately 48% of Spanish consumers interviewed and do not see the need for adjustments. Another 36% of consumers agree that the beef label does not require codes or indications to prove that this product has been traced. About 13% of consumers said that the traceability information on beef labels would be better represented using a barcode/QR code.

Six labels with different levels of traceability information have been presented, going deeper into preferences in the way in which beef traceability information is presented on labels to Spanish consumers (see set of labels Appendix A).

Regarding the preference for labels which present "traceability" information, the preference is higher, followed by labels with only "basic" product information. For Spanish consumers, labels with additional information on 'plus' traceability are not necessary. The results (Table 7) show that information written in the label (75.4%) is more important than that transmitted by a QR code (24.6%), in other words, the omission of the QR code is preferred.

The results suggest that Spanish consumers already seem to be used to the amount and way in which the traceability information is passed to them through the labels, because a large proportion of consumers believe that there is no need to modify the label and that the current available information is satisfactory. It is noted that Spanish consumers are satisfied with how traceability information is made available, without needing to add codes/symbols or even extra information. Also, the lack of information on traceability is not well accepted (Table 7).

In a similar study [47] which confronted consumers with different levels of label traceability information, the most preferred label was the one with traceability information on it. Participants also largely rejected the meat label without concrete information on traceability. In general, the results are indicative that a direct reference to origin may be an effective option when it comes to informing consumers about the traceability of meat.

It can be assumed at the time of purchase that consumers prefer a simpler presentation of traceability system information, that is, easy to understand and that gives them direct access to the information they need or prefer [68].

Studies [49] have shown a limited consumer interest in providing traceability information by codes, such as Radio Frequency Identification (RFID) tags (which have now been replaced by QR code), and a greater preference for simple labels or information written on product labels. This information corroborates the findings in this study.

In Brazil, in order to determine the preference for the display of beef traceability information on the labels, two labels were filed: One of them was a reproduction similar to that of a beef label sold to Spanish consumers (same traceability information) and the other was with information such as that given by most products traced in Brazil (traceability number of animal) (see labels in Appendix B).

The findings (Table 8) show that traceability information on beef labels would be better represented for 56.7% of the consumers questioned if it were a combination of labels from both countries i.e., with information available at the time of purchase (Spanish label) and also with access at any time to the origin of the product by purposes of a QR code (Brazilian label). The second opinion with the highest number of responses was from consumers who prefer the Spanish label (31.6%) versus 10.4% who prefer the Brazilian label and 1.3% who do not prefer any of them, saying that the label does not require any traceability information.

In an experiment [61], in which the willingness of the consumer to pay for CoO information on meat and meat products was investigated, it was shown that 41, 7% of participants preferred labels with a text and a symbol together, which was moderately more than just indicating with a symbol (40.8 percent), and 17.5% preferred text only.

Preferences on how the label should be interpreted with traceability information are a somewhat contentious subject. As researchers [31,35] say, consumers need simple and concise information because of the limited time available to purchase, but at the same time they will be interested in obtaining more elaborate and accurate information, that could be, such as through the Internet for example.

Other studies [28] have described that the greater the quantity of label information, the higher the purchase of a product.

Researchers [34] clarify more specifically that the preferences for any label information depend on the type of product under consideration. For instance, when products are new and unfamiliar, as well as fresh products, more detailed information is required, while concise information would be appropriate for regular purchases or non-perishable products.

3.6. Suspected Food Security Questioned

Food safety differs from other attributes of quality because it is a hard aspect to observe. A product may appear to be of high quality, but it may not be safe because it may be contaminated and go unnoticed before the product is ingested and the effects are manifested [13].

For this reason, food producers and policy makers are of particular interest in the consumer response to food safety information, then we investigate the attitude of consumers when there is a suspicion that purchased beef has a safety risk, quality, and other aspects (Table 9).

The majority of surveyed Spanish consumers (72.5%) and 41.5% of Brazilians answered that they do not use the product and throw it away when faced with a potential food danger. More than half (53.2%) of Brazilians claim, that in addition to not using meat, they call the customer information phone number available on the label. A minority of 8.9% of Spaniards and 5.3% of Brazilians use meat because quality and safety are guaranteed due to the fact that it is monitored/traced (Table 9).

Table 9. Consumer behavior when beef is suspected of presenting a problem in terms of safety, quality and other aspects.

Attitude of the Consumer Towards Suspect Beef Quality *	Country [1]		p
	Spain	Brazil	
I don't use it and throw it in the trash	72.5	41.5	
I don't use it, but I keep it and call customer service phone number	18.6	53.2	≤0.001
I use it because safety and quality are guaranteed by the producer	8.9	5.3	

[1] Total n = 2132, Spain n = 436, Brazil n = 1696. * Mark only one answer. (%).

The results show that more Brazilian consumers report concerns to those responsible for the product than Spanish consumers (Table 9).

According to researchers, the most immediate reaction when there is suspicion regarding safety and quality is not to buy the product again. Of the product in hand, that is, already purchased, most consumers throw away or call an information telephone, such as government agencies and consumer organizations, as well as complain directly to the retailer, which is called "complaint behavior". A minority uses the justification that, because it is traced, the safety and quality of the meat are guaranteed and can be eaten [34,35].

Traceability can have a tremendous value when talking about prevention in potentially hazardous products, such as meat, that lead to serious health consequences when contaminated [37].

Recalls are the main means by which consumers are warned that certain products are contaminated with harmful substances or microorganisms. Robust evidence has been found that a change in the way people interpret and react to beef recalls was induced by BSE cases in 2003 [35].

When consumers begin to respond to recalls, this means that consumers reject contaminated meat, there is an increase in purchases from companies that do not involved in the recall, which provides incentives for the beef industry to work with more control guarantees for reduced recalls and companies suffer less damage [35]. When everybody, and not just the industry, has access to information about the origin and quality of the product, the relationship between the parties (consumers and the supply chain) works best. This improves the brand name and increases its market credibility [37].

Currently, the attribute that people most associate with their own words to traceability is recall. However, consumers will only benefit from this information when it is reliable, otherwise consumers will likely lose confidence in the traceability system and the information it provides [34].

4. Conclusions

This cross-cultural research study aimed to compare Spanish and Brazilian consumers with the purpose to understand the familiarity with the bovine traceability system and the label traceability information as an indicator of product safety in countries with different beef consumption, production, and traceability system implementation.

Consumers do not have a clear understanding of the term "traceability systems" but may be able to indicate the utility of such a system. In the role of guiding and to be able for consumers understanding traceability, more effective public policy may help to explain and clarify the general benefits associated with the purchase of food with traceability certified.

It seems that Brazilians have more technical knowledge of the traceability system and its component aspects than Spanish consumers. It may be because of Brazilians' familiarity with the production of animals, since the country is an important producer/consumer of beef. On the other hand, Spanish consumers are less acquainted with the system itself, but far more familiar with the knowledge that reaches them via the labeling.

It is also clear that since traceability and labeling are mandatory in Spain, consumers have more credibility and offer the traceability system more importance, this can be a factor that can be taken into account in the decision-making process of a company looking to implement traceability where traceability is not established.

Consumers are interested in making sure that the item they purchase is of known and trusted origin. Consumers who do not take traceability information into account at the time of purchase are not feel responsible for traceability because they lack knowledge and are not engaged in product information. The implementation of a mandatory traceability system would only be interesting if consumers value this information. Again, public incentives would be a way to increase consumer demand for security.

With regard to the presentation of label traceability information, Brazilian consumers prefer to use combined information in the form of text and symbols (QR code). Traceability information in text form is well accepted by Spanish consumers. There is no successful acceptance of information provided with a QR code or an excess of written information.

Brazilian consumers are more concerned with reporting a potential risk in food, this may be an indicator that Brazilians will be more involved with respect to "complaint behavior" with compulsory traceability and this would bring benefits to public health in the case of any food issue that might arise.

Supplementary Materials: The following are available online at https://www.mdpi.com/2304-8158/10/2/232/s1, Annex 1: Self-administered questionnaire.

Author Contributions: Conceptualization, D.R.M., M.d.M.C., and M.T.M.; methodology, D.R.M.; software, D.R.M.; validation, M.d.M.C. and M.T.M.; formal analysis, D.R.M.; investigation, D.R.M., M.d.M.C., and M.T.M.; resources D.R.M.; data curation, D.R.M. and M.d.M.C.; writing—original draft preparation, D.R.M.; writing—review and editing, D.R.M., M.d.M.C. and M.T.M.; visualization, D.R.M.; supervision, M.d.M.C. and M.T.M.; project administration, M.d.M.C.; funding acquisition, D.R.M. and M.d.M.C. All authors have read and agreed to the published version of the manuscript.

Funding: D.R.M. was funded by the National Council for Scientific and Technological Development (CNPq-Brazil) through a doctoral scholarship 249957/2013-2.

Institutional Review Board Statement: This study was performed following the Regulation (EU) 2016/679 of the European Parliament and of the council of 27 April 2016 on the protection of natural persons with regard to the processing of personal data and on the free movement of such data, and the local independent Ethics Committee considered no necessary to submit a report.

Acknowledgments: Authors would like to thank Daniel Emygdio-USP, Jackeline kirinus-UDESC, Ivanor do Prado-UEM and Giovani Fiorentini-UNESP for assistance in submitting the questionnaires for this study.

Conflicts of Interest: The authors declare no conflict of interest.

Appendix A

Table A1. Distribution of traceability information on each of the labels presented to Spanish consumers.

Label	Traceability Label Information for Spanish Consumers		
	Traceability	Extras (Plus)	QR Code
A			x
B			
C	x		x
D	x		
E	x	x	
F	x	x	x

x—presence of information.

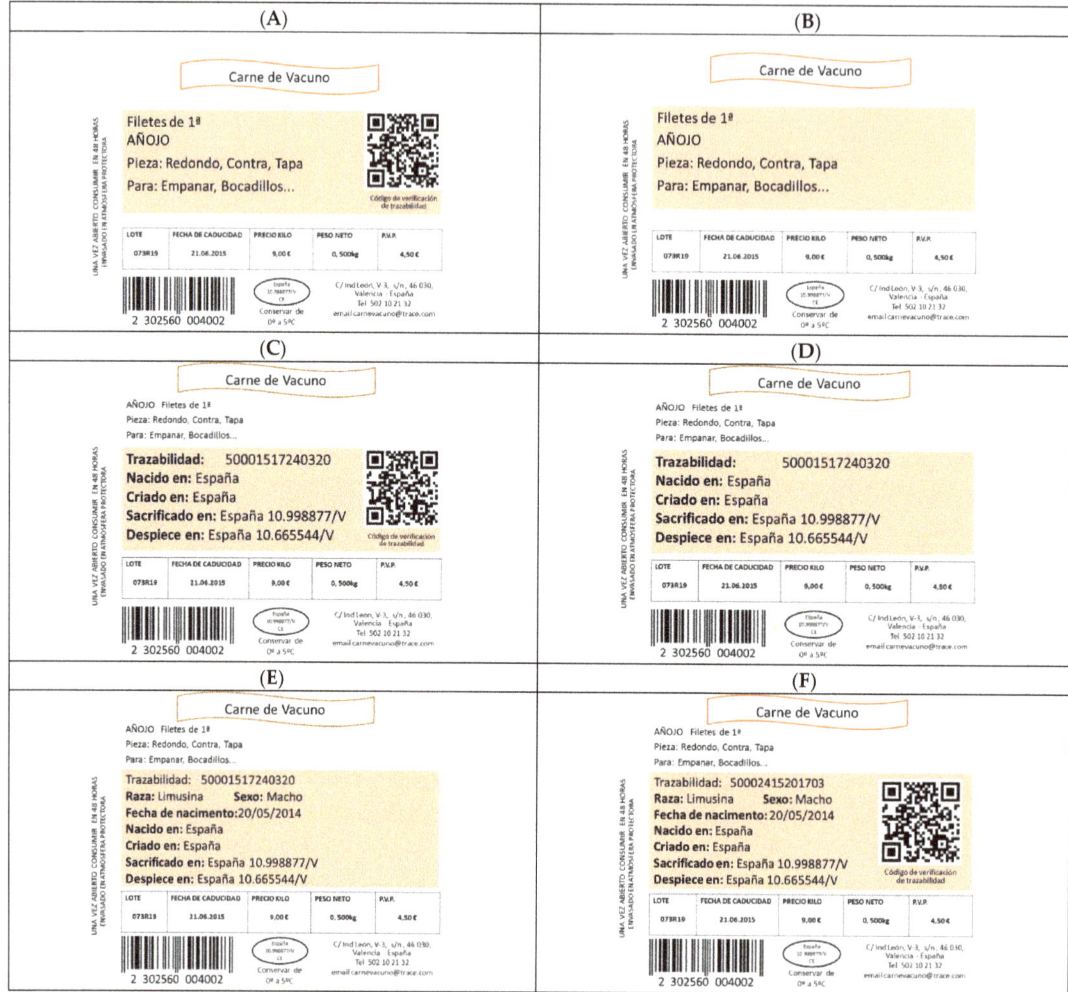

Figure A1. (**A**)—Traceability information available on the label: QR Code. (**B**)—Label without traceability information available. (**C**)—Traceability information available on the label: Written information (provided for by Royal Decree 75/2009) and QR Code. (**D**)—Traceability information available on the label: Written information (mandatory provided for by Royal Decree 75/2009). (**E**)—Traceability information available on the label: Written information (provided for by Royal Decree 75/2009) with additional information added by the authors. (**F**)—Traceability information available on the label: Written information (provided for by Royal Decree 75/2009) with additional information added by the authors and QR Code. Information contained within the colored card: Trazabilidad: (traceability number); Raza: (Animal breed); Sexo: (Gender); Fecha de nacimiento: (Date of animal birth); Nacido en: (Animal birthplace); Criado en: (Animal fattening location); Sacrificado en: (Slaughterhouse); Despiece en: (Cutting and boning room); Código de información de trazabilidad: (QR code).

Appendix B

Figure A2. Information contained within the colored card: Código de rastreabilidade: (Traceability number); QR code; Rastreabilidade: (traceability number); Nascido em: (Animal birthplace); Criado em: (Animal fattening location); Sacrificado em: (Slaughterhouse); Desossado em: (Cutting and boning room).

References

1. de Boer, J.; Schosler, H.; Aiking, H. "Meatless days" or "less but better"? Exploring strategies to adapt Western meat consumption to health and sustainability challenges. *Appetite* **2014**, *76*, 120–128. [CrossRef] [PubMed]
2. Henchion, M.; McCarthy, M.; Resconi, V.C.; Troy, D. Meat consumption: Trends and quality matters. *Meat Sci.* **2014**, *98*, 561–568. [CrossRef] [PubMed]
3. Kang, J.; Jun, J.; Arendt, S.W. Understanding customers' healthy food choices at casual dining restaurants: Using the Value-Attitude-Behavior model. *Int. J. Hosp. Manag.* **2015**, *48*, 12–21. [CrossRef]
4. Buitrago-Vera, J.; Escribá-Pérez, C.; Baviera-Puig, A.; Montero-Vicente, L. Consumer segmentation based on food-related lifestyles and analysis of rabbit meat consumption. *World Rabbit Sci.* **2016**, *24*, 169–182. [CrossRef]
5. Sans, P.; Combris, P. World meat consumption patterns: An overview of the last fifty years (1961–2011). *Meat Sci.* **2016**, *114*, 154. [CrossRef]
6. MAPAMA. Ministerio de Agricultura y Pesca, Alimentación y Medio Ambiente: El Sector de la carne de vacuno en cifras—Principales indicadores económicos. 2019. Available online: https://rica.chil.me/post/el-sector-de-la-carne-de-vacuno-en-cifras-principales-indicadores-economicos-265636 (accessed on 16 December 2020).
7. MAPA. Ministerio de Agricultura y Pesca, Alimentación y Medio Ambiente: Informe del consumo de alimentación en España. 2019. Available online: https://www.mapa.gob.es/eu/alimentacion/temas/consumo-tendencias/panel-de-consumoalimentario/ultimos-datos/default.aspx (accessed on 16 December 2020).
8. USDA. United States Department of Agriculture Economic Research Service: Annual and Cumulative Year-to-Date, U.S. Livestock and Meat Trade by Country. 2020. Available online: https://www.ers.usda.gov/data-products/livestock-and-meat-international-trade-data/livestock-and-meat-international-trade-data/#Annual%20and%20Cumulative%20Year-to-Date%20U.S.%20Livestock%20and%20Meat%20Trade%20by%20Country (accessed on 16 December 2020).
9. Reglamento (CE) n° 2629/97 de la Comisión de 29 de diciembre, por el que se establecen determinadas disposiciones de aplicación del Reglamento (CE) n° 820/97 y n°(CE) 1141/9 del Consejo en lo que respecta a las marcas auriculares, los registros de las explotaciones y los pasaportes en el marco del sistema de identificación y registro de los animales de la especie bovina—Agencia Estatal Boletín Oficial del Estado. Available online: https://www.boe.es/buscar/doc.php?id=DOUE-L-1997-82459 (accessed on 16 December 2020).
10. Instrução Normativa n° 51, de 1 de outubro de 2018—Ministro de Estado da Agricultura, Pecuária e Abastecimento, Diário Oficial da União. Available online: https://www.in.gov.br/web/guest/materia/-/asset_publisher/Kujrw0TZC2Mb/content/id/44306336/do1-2018-10-08-instrucao-normativa-n-51-de-1-de-outubro-de-2018-44306204#:~{}:text=%2D18%2C%20resolve%3A-,Art.,dos%20Anexos%20I%20a%20III.&text=2%C2%BA%20O%20SISBOV%20%C3%A9%20o,individual%20de%20bovinos%20e%20b%C3%BAfalos (accessed on 16 December 2020).
11. Pethick, D.W.; Ball, A.J.; Banks, R.G.; Hocquette, J.F. Current and future issues facing red meat quality in a competitive market and how to manage continuous improvement. *Anim. Prod. Sci.* **2011**, *51*, 13–18. [CrossRef]

12. Barcellos, J.O.J.; Abicht, A.D.; Brandao, F.S.; Canozzi, M.E.A.; Collares, F.C. Consumer perception of Brazilian traced beef. *R. Bras. Zootec.* **2012**, *41*, 771–774. [CrossRef]
13. Aung, M.M.; Chang, Y.S. Traceability in a food supply chain: Safety and quality perspectives. *Food Control* **2014**, *39*, 172–184. [CrossRef]
14. Charlebois, S.; Sterling, B.; Haratifar, S.; Naing, S.K. Comparison of Global Food Traceability Regulations and Requirements. *Compr. Rev. Food Sci. Food Saf.* **2014**, *13*, 1104–1123. [CrossRef]
15. Regattieri, A.; Gamberi, M.; Manzini, R. Traceability of food products: General framework and experimental evidence. *J. Food Eng.* **2007**, *81*, 347–356. [CrossRef]
16. Galvao, J.A.; Margeirsson, S.; Garate, C.; Vidarsson, J.R.; Oetterer, M. Traceability system in cod fishing. *Food Control* **2010**, *21*, 1360–1366. [CrossRef]
17. Britt, A.G.; Bell, C.M.; Evers, K.; Paskin, R. Linking live animals and products: Traceability. *Rev. Off. Int. Epizoot.* **2013**, *32*, 571–582. [CrossRef] [PubMed]
18. Macready, A.L.; Hiekeb, S.; Klimczuk-Kochańskac, M.; Szumiałc, S.; Vrankend, L.; Grunert, K.G. Consumer trust in the food value chain and its impact on consumer confidence: A model for assessing consumer trust and evidence from a 5-country study in Europe. *Food Policy* **2020**, *92*, 101880. [CrossRef]
19. Conchon, F.L.; Lopes, M.M. Rastreabilidade e segurança alimentar. *Boletim Tec. Ed. UFLA* **2012**, *91*, 1–25.
20. Opara, L.U. Traceability in agriculture and food supply chain: A review of basic concepts, technological implications, and future prospects. *J. Food Agric. Environ.* **2003**, *1*, 101–106.
21. Folinas, D.; Manikas, I.; Manos, B. Traceability data management for food chains. *Br. Food J.* **2006**, *108*, 622–633. [CrossRef]
22. Alfaro, J.A.; Rabade, L.A. Traceability as a strategic tool to improve inventory management: A case study in the food industry. *Int. J. Prod. Econ.* **2009**, *118*, 104–110. [CrossRef]
23. Hong, I.H.; Dang, J.F.; Tsai, Y.H.; Liu, C.S.; Lee, W.T.; Wang, M.L.; Chen, P.C. An RFID application in the food supply chain: A case study of convenience stores in Taiwan. *J. Food Eng.* **2011**, *106*, 119–126. [CrossRef]
24. Sepúlveda, W.; Maza, M.T.; Mantecón, A.R. Factors that affect and motivate the purchase of quality-labelled beef in Spain. *Meat Sci.* **2008**, *80*, 1282–1289. [CrossRef]
25. Becker, T. Consumer perception of fresh meat quality: A framework for analysis. *Brit. Food J.* **2000**, *102*, 158–176. [CrossRef]
26. Grunert, K.G.; Sonntag, W.I.; Glanz-Chanos, V.; Forum, S. Consumer interest in environmental impact, safety, health and animal welfare aspects of modern pig production: Results of a cross-national choice experiment. *Meat Sci.* **2018**, *137*, 123–129. [CrossRef] [PubMed]
27. Burnier, P.C.; Spers, E.E.; de Barcellos, M.D. Role of sustainability attributes and occasion matters in determining consumers' beef choice. *Food Qual. Prefer.* **2021**, *88*, 104075. [CrossRef]
28. Loureiro, M.L.; Umberger, W.J. A choice experiment model for beef: What US consumer responses tell us about relative preferences for food safety, country-of-origin labeling and traceability. *Food Policy* **2007**, *32*, 496–514. [CrossRef]
29. Bernués, A.; Olaizola, A.M.; Corcoran, K. Extrinsic attributes of red meat as indicators of quality in Europe: An application for market segmentation. *Food Qual. Prefer.* **2003**, *14*, 265–276. [CrossRef]
30. Maza, M.T.; Ramírez, V. Distintas consideraciones en torno a los atributos de calidad de la carne de vacuno por parte de industria y consumidores. *ITEA* **2006**, *102*, 360–372.
31. Stranieri, S.; Banterle, A. Consumer Interest in Meat Labelled Attributes: Who Cares? *Int. Food Agribus. Man.* **2015**, *18*, 21–38.
32. Ehmke, T.M.D. International Differences in Consumer Preferences for Food Country-of-Origin: A Meta-Analysis. In Proceedings of the Annual meeting American Agricultural Economics Association, Long Beach, CA, USA, 23–26 July 2006.
33. Magalhaes, D.R.; Lopes, M.A.; da Rocha, C.M.B.M.; Bruhn, F.R.P.; Borges, J.C.; da Cunha, C.F. Fatores socioeconômicos que influenciam na disposição de consumidores em adquirir carne bovina com certificação de origem em Belo Horizonte, Minas Gerais, Brasil. *Arq. Inst. Biol.* **2016**, *83*, 1–8. [CrossRef]
34. Van Rijswijk, W.; Frewer, L.J. Consumer needs and requirements for food and ingredient traceability information. *Int. J. Consum. Stud.* **2012**, *36*, 282–290. [CrossRef]
35. Taylor, M.; Klaiber, H.A.; Kuchler, F. Changes in U.S. consumer response to food safety recalls in the shadow of a BSE scare. *Food Policy* **2016**, *62*, 56–64. [CrossRef]
36. Moe, T. Perspectives on traceability in food manufacture. *Trends. Food Sci. Technol.* **1998**, *9*, 211–214. [CrossRef]
37. Glynn, T.T.; Schroeder, T.C.; Pennings, J.M.E. Factors Impacting Food Safety Risk Perceptions. *J. Food Agric. Environ.* **2009**, *60*, 625–644.
38. Lagerkvist, C.J.; Berthelsen, T.; Sundström, K.; Johansson, H. Country of origin or EU/non-EU labelling of beef? Comparing structural reliability and validity of discrete choice experiments for measurement of consumer preferences for origin and extrinsic quality cues. *Food Qual. Prefer.* **2014**, *34*, 50–61. [CrossRef]
39. Beulens, A.J.M.; Broens, D.F.; Folstar, P.; Hofstede, G.J. Food safety and transparency in food chains and networks—Relationships and challenges. *Food Control* **2005**, *16*, 481–486. [CrossRef]
40. Joseph, S.; Lavoie, N.; Caswell, J.A. Implementing COOL: Comparative welfare effects of different labeling Schemes. *Food Policy* **2014**, *44*, 14–25. [CrossRef]
41. OECD-FAO Agricultural Outlook 2018–2027—Food and Agriculture Organization of the United Nations. Available online: http://www.fao.org/publications/card/en/c/I9166EN (accessed on 16 December 2020).

42. IBGE. Instituto Brasileiro de Geografia e Estatística. Indicadores- Estatística da Produção Pecuária, Produção Animal: Bovinos. Available online: https://biblioteca.ibge.gov.br/visualizacao/periodicos/2380/epp_2018_4tri.pdf (accessed on 16 December 2020).
43. Real Decreto 75/2009, de 30 de enero, por el que se modifica el Real Decreto 1698/2003, de 12 de diciembre, por el que se establecen las disposiciones de aplicación de los Reglamentos comunitarios sobre el sistema de etiquetado de la carne de vacuno, y el Real Decreto 1799/2008, de 3 de noviembre, por el que se establecen las bases reguladoras para la concesión de ayudas destinadas a la reconversión de plantaciones de determinados cítricos.—Agencia Estatal Boletín Oficial del Estado. Available online: https://www.boe.es/boe/dias/2009/01/31/pdfs/BOE-A-2009-1602.pdf (accessed on 16 December 2020).
44. Grande Esteban, I.; Abascal Fernández, E. *Fundamentos y Técnicas de Investigación Comercial*, 12th ed.; ESIC: Madrid, Spain, 2014; ISBN 978-84-15986-02-7.
45. Coutinho, C.P. *Metodologia de Investigação em Ciências Sociais e Humanas: Teoria e prática*, 2rd ed.; Edições Almedina, S.A.: Coimbra, Portugal, 2014; ISBN 978-972-40-5610-4.
46. Verbeke, W.; Ward, R.W. Consumer interest in information cues denoting quality, traceability and origin: An application of ordered probit models to beef labels. *Food Qual. Prefer.* **2006**, *17*, 453–467. [CrossRef]
47. Gellynck, X.; Verbeke, W.; Vermeire, B. Pathways to increase consumer trust in meat as a safe and wholesome food. *Meat Sci.* **2006**, *74*, 161–171. [CrossRef]
48. Decreto N° 5.741, de 30 de março de 2006—Ministro de Estado da Agricultura, Pecuária e Abastecimento. Available online: http://www.planalto.gov.br/ccivil_03/_Ato2004-2006/2006/Decreto/D5741.htm (accessed on 16 December 2020).
49. Van Rijswijk, W.; Frewer, L.J.; Menozzi, D.; Faioli, G. Consumer perceptions of traceability: A cross-national comparison of the associated benefits. *Food Qual. Prefer.* **2008**, *19*, 452–464. [CrossRef]
50. Heyder, M.; Theuvsen, L.; Hollmann-Hespos, T. Investments in tracking and tracing systems in the food industry: A PLS analysis. *Food Policy* **2012**, *37*, 102–113. [CrossRef]
51. Cunha, C.F.; Spers, E.E.; Zylbersztajn, D. Percepção sobre atributos de sustentabilidade em um varejo supermercadista. *Rev. Adm. Empres.* **2011**, *6*, 542–552. [CrossRef]
52. Awada, L.; Yiannaka, A. Consumer perceptions and the effects of country of origin labeling on purchasing decisions and welfare. *Food Policy* **2012**, *37*, 21–30. [CrossRef]
53. Van Wezemael, L.; Verbeke, W.; Kugler, J.O.; Barcellos, M.D.; Grunert, K.G. European consumers and beef safety: Perceptions, expectations and uncertainty reduction strategies. *Food Control* **2010**, *21*, 835–844. [CrossRef]
54. Meyerding, S.G.H.; Gentz, M.; Altmann, B.; Meier-Dinkel, L. Beef quality labels: A combination of sensory acceptance test, stated willingness to pay, and choice-based conjoint analysis. *Appetite* **2018**, *127*, 324–333. [CrossRef] [PubMed]
55. Scozzafava, G.; Corsi, A.M.; Casini, L.; Contini, C.; Loose, S.M. Using the animal to the last bit: Consumer preferences for different beef cuts. *Appetite* **2016**, *96*, 70–79. [CrossRef] [PubMed]
56. Verbeke, W.; Roosen, J. Market differentiation potential of country-of-origin, quality and traceability labeling. *Estey Centre J. Int. Law Trade Policy* **2009**, *10*, 20–35.
57. Furnols, F.M.; Realini, C.; Montossi, F.; Sañudo, C.; Campo, M.M.; Oliver, M.A.; Nute, G.R.; Guerrero, L. Consumer's purchasing intention for lamb meat affected by country of origin, feeding system and meat price: A conjoint study in Spain, France and United Kingdom. *Food Qual. Prefer.* **2011**, *22*, 443–451. [CrossRef]
58. Sepúlveda, W.S.; Maza, M.T.; Mantecón, A.R. Factors associated with the purchase of designation of origin lamb meat. *Meat Sci.* **2010**, *85*, 167–173. [CrossRef]
59. Lusk, J.L.; Schroeder, T.C.; Tonsor, G.T. Distinguishing beliefs from preferences in food choice. *Eur. Rev. Agric. Econ.* **2014**, *41*, 627–655. [CrossRef]
60. Banterle, A.; Stranieri, S. The consequences of voluntary traceability system for supply chain relationships. An application of transaction cost economics. *Food Policy* **2008**, *33*, 560–569. [CrossRef]
61. Balcombe, K.; Bradley, D.; Fraser, I.; Hussein, M. Consumer preferences regarding country of origin for multiple meat products. *Food Policy* **2016**, *64*, 49–62. [CrossRef]
62. Klain, T.J.; Lusk, J.L.; Tonsor, G.T.; Schroeder, T.C. An experimental approach to valuing information. *J. Agric. Econ.* **2014**, *45*, 635–648. [CrossRef]
63. Britwuma, K.; Yiannaka, A. Consumer willingness to pay for food safety interventions: The role of message framing and issue involvement. *Food Policy* **2019**, *86*, 101726.
64. Ortega, D.L.; Wanga, H.H.; Wub, L.; Olynk, N.J. Modeling heterogeneity in consumer preferences for select food safety attributes in China. *Food Policy* **2011**, *36*, 318–324.
65. Żakowska-Biemans, S.; Pieniak, Z.; Gutkowska, K.; Wierzbicki, J.; Cieszyńska, K.; Sajdakowska, M.; Kosicka-Gębska, M. Beef consumer segment profiles based on information source usage in Poland. *Meat Sci.* **2017**, *124*, 105–113. [CrossRef] [PubMed]
66. Roe, B.E.; Teisl, M.F.; Deans, C.R. The economics of voluntary versus mandatory labels. *Ann. Rev. Resour. Econ.* **2014**, *6*, 407–427. [CrossRef]
67. Ardeshiri, A.; Rose, J.M. How Australian consumers value intrinsic and extrinsic attributes of beef products. *Food Qual. Prefer.* **2017**, *65*, 146–163.
68. Giraud, G.; Amblard, C. What does traceability mean for beef meat consumer? *Food Sci.* **2003**, *23*, 40–64. [CrossRef]

MDPI
St. Alban-Anlage 66
4052 Basel
Switzerland
Tel. +41 61 683 77 34
Fax +41 61 302 89 18
www.mdpi.com

Foods Editorial Office
E-mail: foods@mdpi.com
www.mdpi.com/journal/foods

www.ingramcontent.com/pod-product-compliance
Lightning Source LLC
LaVergne TN
LVHW070135170726
843515LV00007B/1802

* 9 7 8 3 0 3 6 5 1 3 4 0 9 *